Productivity in the Office and the Factory

Decision Making and Operations Management Series

Paul Gray, *Series Editor*

Gray/Jurison:	*Productivity in the Office and the Factory*
Couger:	*Creative Problem Solving and Opportunity Finding*
Segev:	*Corporate Strategy: Portfolio Models*
Segev:	*Navigating by COMPASS: Corporate Matrix Portfolio Analysis Support System*

Productivity in the Office and the Factory

Edited by
Paul Gray
Claremont Graduate School
and
Jaak Jurison
Fordham University

bf
boyd & fraser publishing company
I(T)P An International Thomson Publishing Company
Danvers Albany Bonn Boston Cincinnati
Detroit London Madrid Melbourne Mexico
City New York Paris SanFrancisco
Singapore Tokyo Toronto Washington

Executive Editor/Acquisitions Editor: DeVilla Williams
Production Editor: Jackie Bedoya
Production Services: Books By Design, Inc.
Composition: Coghill Composition Company
Interior Design: Barbara Werden
Cover Design: Rogondino & Associates
Cover Photo: Romilly Lockyer, The Image Bank
Manufacturing Coordinator: Gordon Woodside
Marketing Coordinator: Karen Grantham

Printed in the United States of America

For more information, contact boyd & fraser publishing company:

boyd & fraser publishing company
One Corporate Place • Ferncroft Village
Danvers, Massachusetts 01923, USA

International Thomson Publishing Europe
Berkshire House 168–173
High Holborn
London, WC1V 7AA, England

Thomas Nelson Australia
102 Dodds Street
South Melbourne 3205
Victoria, Australia

Nelson Canada
1120 Birchmont Road
Scarborough, Ontario
Canada M1K 5G4

International Thomson Editores
Campose Eliseos 385, Piso 7
Col. Polanco
11560 Mexico D.F. Mexico

International Thomson Publishing GmbH
Konigswinterer Strasse 418
53227 Bonn, Germany

International Thomson Publishing Asia
221 Henderson Road
#05-10 Henderson Building
Singapore 0315

International Thomson Publishing Japan
Hirakawacho Kyowa Building, 3F
2-2-1 Hirakawacho
Chiyoda-ku, Tokyo 102, Japan

1 2 3 4 5 6 7 8 9 10 K 8 7 6 5 4

Library of Congress Cataloging-in-Publication Data

Productivity in the office and the factory / editors, Paul Gray, Jaak Jurison.
p. cm.
Includes bibliographical references and index.
ISBN 0-87709-751-8
1. Industrial productivity. 2. Labor productivity.
3. Technological innovations. I. Gray, Paul, 1930–
II. Jurison, Jaak.
HC79.I52P75 1995
658.5—dc20 94-44917
CIP

Dedicated to our wives
Muriel and Siret

Contents

Foreword

THE title of this book sends a signal that what was divergent is now converging. Once upon a time, productivity in the office was an entirely different matter than productivity in the factory. Bankers and factory managers felt uncomfortable sitting together in the same classroom. Service and manufacturing productivity were divergent subjects because their underlying technologies and use of human resources were vastly different.

This state of affairs is no longer true. Blue collars and white collars—and what they stood for—are rapidly being replaced by assorted scholars in information systems. Because of information technology, the nature of jobs is changing for every employee in the world.

Training and learning pervade business organizations. Everyone has dual citizenship as student and teacher. Successful firms are managed by scholars who acknowledge that they are responsible for accelerated learning organizations. Information is the currency for managers, operators, programmers, and all others who are the doers and enablers for the new systems.

Bringing 41 readings together which support the fact that information is the new enabler for both factories and offices is a great contribution to the idea of unified smart systems functioning together. The editors have done this in a convincing way. They have started the book by defining, redefining, and explaining productivity so that it can be viewed as a term that applies to both the factory and the office.

The next stop along the way is critical. What has happened to change the world of the 1950s into the world of the 1990s? The new technologies—telecommunications, networks and the Internet, computers, image processing, and bar codes—share the common currency of information systems. This new relationship—which emerges early in the book and reappears throughout it—has brought about the marriage of office and factory productivity.

Throughout the book there are separate explorations of office productivity and factory productivity. Underlying similarities between these two categories of operating environments strike a chord. They are not only co-dependent. They are converging and merging—becoming each other.

The authors of these many articles have strikingly comparable concepts and coincidental language. The methods and procedures that stem from the new information technology are now in the mainstream. The fundamental forces that drive the effectiveness of operations and that account for productivity are commonly held and shared by all. Specialization, which is an attribute of mature technologies, is yet to come.

The systems for effective management in the twentieth century are going to change in the twenty-first century. Winners have the best quality and lowest costs. They are the fastest adapters, greatest benefit creators, and best value-added makers. Although the criteria are the same, the systems for winning have changed. Management with one foot firmly planted in the past century needs education in the new systems for success in the next century.

Managers need to know how the job is changing. Productivity improvement programs often fail so completely that they achieve productivity deterioration. Reengineering is badly handled and misunderstood by a broad base of advocates and causes harm despite its potential to make a major contribution to success. The culprit is not knowing what drives system success. The readings in this book point out this fact and help direct the focus of present and potential managers.

Knowing how to manage the implementation of new technologies is where the action is now and where it is going to be for the next decade. Knowing how to use the new technologies requires undoing the organizational knots that bureaucracies tied in the fabric of traditional management lore. This book provides a stimulating and valuable approach for getting one's act together to deal with this challenge of undoing before redoing.

Implicit in the structure of this book are the needs to redefine work, to relearn what jobs are, and to restructure how people and technology work together. No simple task and yet more than 50 people are contributors to this book, reflecting the power of the surge to bring about change.

Martin K. Starr
Columbia University

Preface

PRODUCTIVITY is a major concern of American business. Yet few schools offer a course in productivity per se. Furthermore, when a course in productivity is offered, it is usually taught in terms of manufacturing. As a result, business people and industrial engineers who are interested in functional areas other than manufacturing seldom examine—or even think about—productivity.

Is productivity increasing at all in the office and is it increasing fast enough in the factory? The answer to this question changed recently from no to yes. That is, the computer and communications *are* making a difference.

This book focuses on both white collar (office) and blue collar (factory) productivity. The basic theses of this book are:

- The same basic concepts can be applied to improving productivity in the office and in the factory.
- Computer and communications technologies are the key enablers to improving productivity in the years ahead.

Thus, this book is aimed at teaching about productivity and information systems, and how they intersect.

The intent is to show, using current literature, what productivity is, how information systems can assist productivity in the office and in the factory, how to make cost-benefit and value analyses for productivity improvement, and what management techniques are needed to implement productivity improvements.

The articles in this book are drawn from a variety of sources and discuss a range of topics: from productivity in general, to productivity in the office or in the factory. The important point is that the methods used to improve productivity rely on the same underlying technologies no matter where they are applied. The result is the convergence of the nature of office and factory work. In manufacturing, increased emphasis is placed on planning, coordinating, and managing flows of products and information. Touch labor is decreasing so that it is only a small percentage of total cost. Hence touch labor is becoming less and less relevant in determining factory productivity. In the office, through reengineering and similar efforts, increased emphasis is being placed on handling larger volumes of data and managing the flow of information. The revolution in productivity from information systems has spread from manufacturing to service industries. The ideas in this book apply to the production of services as much as to the production of goods.

The Nature of Productivity

Productivity measures how efficiently the organization uses its resources, be they human or capital. As the economy shifts to include an ever-larger proportion of service work, productivity becomes as important for the white collar worker as it has been for the factory worker.

Computer and communications technology can provide many opportunities to substitute capital for labor as a way to improve productivity. Yet its potential is broader than that. By performing many of the routine tasks, by providing memory of what was done, by organizing the work, the computer frees people to work at the next level of knowledge where better ideas pay off in competitive advantage.

The present is an important time. Gains from technology which only some users previously realized are now being achieved generally. The student and the practitioner need to know what these gains are and how to obtain them in their own work. However, simply throwing money at the problem is not enough. Managers must analyze the options available to select those options which make business sense. The manager must also look at how the options selected will be implemented. Implementation creates the challenge of matching people and technology. Implementation is an area that separates winners from losers.

Productivity, like quality management, is a way of thinking about the entire business operation. In fact, productivity and quality management complement one another. Improving quality results in improved productivity because producing a quality product at the outset is cheaper than fixing defects later. Conversely, using precision technology to increase productivity increases quality at the same time.

Classically, productivity is defined as

Productivity = Output/Input

where output is something countable (units produced, dollars) and input is the resources applied (people, energy, capital, etc.). This traditional view considers only the resources that produce countable output, and assigns the rest to overhead. The perspective of this book is broader in that it examines the dynamics of productivity and considers the use of technology to gain improvements.

The nature of enterprises changed during the last two decades. Communications now make it possible to do work at many locations rather than in a single place. Companies can have offices and factories everywhere. This globalization leads firms to look for the locations with the greatest advantage. Not only is globalization taking place in production, it also occurs in white collar work. Productivity increases are one way that industrial nations keep work within the country.

Organization of the Book

The book is divided into seven chapters.

CHAPTER	TOPIC
1	Overview of Productivity
2	The New Technologies
3	Office Productivity
4	Factory Productivity
5	Costs, Benefits, and Value
6	Implementation
7	Future Directions

The first chapter defines productivity and discusses the issues related to it. The second chapter, which describes new technologies that affect productivity, leads to chapters on increases in office and factory productivity made possible by these technologies. Chapters 5 and 6 consider the managerial aspects of productivity gains—including how to justify and implement change. The last chapter gives a glimpse of the managerial and technological future we can anticipate in the years ahead.

The editors of this volume considered the domain of productivity, from a manager's perspective. The topics covered in each chapter, shown

TABLE 1
The Domain of Productivity in the Office and the Factory

1 Overview of Productivity	*2 The New Technologies*	*3 Office Productivity*	*4 Factory Productivity*	*5 Cost, Benefits, and Value*	*6 Implementation*	*7 Future Directions*
1.1 Productivity Challenge	2.1 Computer Technology	3.1 Group Decision Support Systems	4.1 Just-In-Time Manufacturing	5.1 Economic Feasibility	6.1 Innovation Diffusion	7.1 The Corporation
1.2 Measuring Productivity	2.2 Telecommunications Technology	3.2 Groupware	4.2 Computer Aided Design	5.2 Cost/Benefit Analysis	6.2 Reengineering	7.2 The Technology
1.3 Productivity Payoff	2.3 Leading Edge Technologies	3.3 Telework	4.3 Flexible Manufacturing	5.3 The Case for Change	6.3 Factory Implementation	
1.4 Forecasting Productivity	2.4 Imaging	3.4 Executive Information Systems	4.4 Robotics	5.4 Non-Quantitative Justification	6.4 Project Management	
1.5 Competing in Time	2.5 Bar Coding	3.5 Electronic Data Interchange	4.5 Logistics	5.5 Decision Making	6.5 Integration	
1.6 Critical Success Factors	2.6 Virtual Reality	3.6 End User Computing	4.6 Interaction of Manufacturing and IS	5.6 Investment Criteria	6.6 Virtual Corporation	
	2.7 Internet	3.7 Sales Force Computing	4.7 Case Study			

in Table 1, were selected based on this analysis of the domain. Inevitably, some worthy topics were not included. In some cases, the reason for excluding them simply was that no appropriate reading was found.

The book contains 41 readings plus introductory material prepared by the editors for each reading. This introductory material, shown at the beginning of each section, indicates the relevance and significance of the issues involved. Questions for thinking about the issues presented and for homework are included at the end of each reading. Articles were selected for readability.

Audience for This Book

This book is addressed to business students and business people who need to understand the environment in which they work or will be working and to the industrial engineer who must understand the business environment of productivity. In a sense, this book spans the business and the technical culture and introduces each culture to the other.

The primary use of this book is for courses in

- Productivity
- Production
- Information systems

The book is the result of courses offered in California and New York. Most of the book has been used with Executive MBA students at the Peter F. Drucker Management Center of the Claremont Graduate School. Other portions were tested at Fordham University in the Graduate School of Business Administration. The result is a book that should appeal to:

- MBAs
- Executive MBAs
- BBAs
- Industrial Engineers
- Practicing Managers

By focusing on the cross-disciplinary nature of productivity, the book is designed for courses in quantitative methods and in information systems that seek to offer high relevance to practice.

Paul Gray
Claremont, California

Jaak Jurison
New York, New York

1 Overview of Productivity

PRODUCTIVITY measures how efficiently the organization uses its resources, be they human or capital. Productivity is important not only for the factory worker but also for the white collar worker. Managers should be concerned with raising productivity levels if their firms are to prosper. In Section 1.1, Peter Drucker explains that improving productivity is one of a manager's key responsibilities.

This initial chapter is concerned with the nature of productivity. It begins with the article by Drucker who challenges the professional community to extend the notions of productivity to knowledge workers in office and service work. Jaak Jurison then provides a conceptual foundation for defining and measuring productivity. Myron Magnet and Bill King, respectively, use recent data to show that the investments in information technologies pay off in increased productivity, particularly for service industries. The remaining two sections describe how innovative strategies and information systems techniques form the basis for productivity gains.

1.1

The New Productivity Challenge

PETER F. DRUCKER

Peter F. Drucker is a well-known management expert and professor whom many credit with defining the field of management. In this section, he contends that the greatest challenge facing managers is to raise the productivity of knowledge workers. He believes that the people qualified for knowledge work will always be a minority, outnumbered by those qualified only for low-skilled service jobs. Knowledge work differs from operator jobs in that capital cannot be substituted for labor. The implication is that knowledge workers have to "work smarter" to increase their productivity.

Drucker argues that we must be clever and redesign the jobs and processes performed by knowledge workers to increase their productivity. For example, major gains can be obtained by reducing paperwork. To find out how to improve productivity, quality, and performance, Drucker suggests that we ask the people who do the work. Drucker also advocates continuous learning and product improvement by doing better at those things we already do well.

Read this article carefully. Drucker introduces many of the ideas that you will see developed more fully in later sections. Drucker sees the big picture and puts it into context. He uses historical and social arguments to create important insights about productivity.

THE single greatest challenge facing managers in the developed countries of the world is to raise the productivity of knowledge and service workers. This challenge, which will dominate the management agenda for the next several decades, will ultimately determine the competitive performance of companies. Even more important, it will determine the very fabric of society and the quality of life in every industrialized nation.

For the last 120 years, productivity in making and moving things—in manufacturing, farming, mining, construction, and transportation—has risen in developed countries at an annual rate of 3% to 4%, a 45-fold expansion overall. On this explosive growth rest all the gains these nations and their citizens have enjoyed: vast increases in disposable income and purchasing power; ever-wider access to education and health care; and the availability of leisure time, something known only to aristocrats and the "idle rich" before 1914, when everyone else worked at least 3,000 hours a year. (Today even the Japanese work no more than

Source: Harvard Business Review (November–December 1991): 69–79.

about 2,000 hours each year, while Americans average 1,800 hours and West Germans 1,650.)

Now these gains are unraveling, but not because productivity in making and moving things has fallen. Contrary to popular belief, productivity in these activities is still going up at much the same rate. And it is rising fully as much in the United States as it is in Japan or West Germany. Indeed, the increase in U.S. manufacturing productivity during the 1980s—some 3.9% a year—was actually larger in absolute terms than the corresponding annual increases in Japan and Germany, while the 4% to 5% annual rise in U.S. agricultural productivity is far and away the largest recorded anywhere at any time.

The productivity revolution is over because there are too few people employed in making and moving things for their productivity to be decisive. All told, they account for no more than one-fifth of the work force in developed economies. Only 30 years ago, they were still a near-majority. Even Japan, which is still manufacturing intensive, can no longer expect increased productivity in that sector to sustain its economic growth. Indeed, the great majority of working people in Japan are knowledge and service workers with productivities as low as those in any other developed country. And when farmers make up only 3% of the employed population, as they do in the United States, Japan, and most of Western Europe, even record increases in their output add virtually nothing to their country's overall productivity and wealth.

The chief *economic* priority for developed countries, therefore, must be to raise the productivity of knowledge and service work. The country that does this first will dominate the twenty-first century economically. The most pressing *social* challenge developed countries face, however, will be to raise the productivity of service work. Unless this challenge is met, the developed world will face increasing social tensions, increasing polarization, increasing radicalization, possibly even class war.

In developed economies, opportunities for careers and promotion are more and more limited to people with advanced schooling, people qualified for knowledge work. But these men and women will always be a minority. They will always be outnumbered by people who lack the qualifications for anything but low-skilled service jobs—people who in their social position are comparable to the "proletarians" of 100 years ago, the poorly educated, unskilled masses who thronged the exploding industrial cities and streamed into their factories.

In the early 1880s, intelligent observers of every political persuasion were obsessed with the specter of class war between the industrial proletariat and the bourgeoisie. Karl Marx was hardly alone in predicting that the "immiserization" of the proletariat would lead inevitably to revolution. Benjamin Disraeli, perhaps the greatest of the nineteenth-century conservatives, was equally persuaded of the inevitability of class war. And Henry James, the chronicler of American wealth and European aristocracy, was so frightened by the prospect that he made it the central theme of *The Princess Casamassima*, one of his most haunting novels.

What defeated these prophecies, which seemed eminently reasonable, indeed almost self-evident to contemporaries, was the revolution in productivity set off by Frederick W. Taylor in 1881, when he began to study the way a common laborer shoveled sand. Taylor himself worked in an iron foundry and was deeply shocked by the bitter animosity between the workers and managers. Fearful that this hatred would ultimately lead to class war, he set out to improve the efficiency of industrial work. And his efforts, in turn, sparked the revolution that allowed industrial workers to earn middle-class wages and achieve middle-class status despite their lack of skill and education. By 1930, when according to Marx the revolution of the proletariat should have been a fait accompli, the proletariat had become the bourgeoisie.

Now it is time for another productivity revolution. This time, however, history is on our side. In the past century, we have learned a great deal about productivity and how to raise it—enough to know that we need a revolution, enough to know how to start one.

KNOWLEDGE and service workers range from research scientists and cardiac surgeons through draftswomen and store managers to 16-year-olds who flip hamburgers in fast-food restaurants on Saturday afternoons. Their ranks also include people whose work makes them "machine operators": dishwashers, janitors, data-entry operators. Yet for all their diversity in knowledge, skill, responsibility, social status, and pay, knowledge and service workers are remarkably alike in two crucial respects: what does not work in raising their productivity and what does.

The first thing we have learned—and it came as a rude shock—is what does not work. Capital cannot be substituted for labor. Nor will new technology by itself generate higher productivity. In making and moving things, capital and technology are *factors* of production, to use the economist's term. In knowledge and service work, they are *tools* of production. The difference is that a factor can replace labor, while a tool may or may not. Whether tools help productivity or harm it depends on what people do with them, on the purpose to which they are being put, for instance, or on the skill of the user. Thirty years ago, for example, we were sure the efficiency of the computer would lead to massive reductions in clerical and office staff. The promise of greater productivity led to massive investments in data-processing equipment that now rival those in materials-processing technology (that is, in conventional machinery). Yet office and clerical forces have grown at a much faster rate since the introduction of information technology than ever before. And there has been virtually no increase in the productivity of service work.

Hospitals are a telling example. In the late 1940s, they were entirely labor intensive, with little capital investment except in bricks, mortar, and beds. A good many perfectly respectable hospitals had not even invested in readily available, fairly old technologies: they provided neither x-ray departments nor clinical laboratories nor physical therapy. Today hospitals are hugely capital intensive, with enormous sums invested in ultrasound, body scanners, nuclear magnetic imagers, blood and tissue analyzers, clean rooms, and a dozen more new technologies. Each piece of equipment has brought with it the need for more highly paid people but has not reduced the existing staff by a single person. (In fact, the worldwide escalation of health-care costs is largely the result of the hospital's having become a labor-intensive and capital-intensive monstrosity.) But hospitals, at least, have significantly increased their performance capacity. In other areas of knowledge or service work there are only higher costs, more investment, and more people.

Massive increases in productivity are the only way out of this morass. And these increases can only come from what Taylor called "working smarter."[1] Simply, this means working more productively without working harder or longer.

The economist sees capital investment as the key to productivity; the technologist gives star billing to new machines. Nevertheless, the main force behind the productivity explosion has been working smarter. Capital investment and technology were as copious in the developed economies during the first 100 years of the Industrial Revolution as they have been in its second 100 years. It was only with the advent of working smarter that productivity in making and moving things took off on its meteoric rise.

And so it will be for knowledge and service work—with this difference: in manufacturing, working smarter is only one key to increased productivity. In knowledge and service work, working smarter is the only key. What is more, it is a more complex key, one that requires looking closely at work in ways that Taylor never dreamed of.

When Taylor studied the shoveling of sand, the only question that concerned him was, "How is it done?" Almost 50 years later, when Harvard's Elton Mayo set out to demolish Taylor's "scientific management" and replace it with what later came to be called "human relations," he focused on the same question. In his experiments at Western Electric's Hawthorne Works, Mayo asked, "How can wiring telephone equipment best be done?" The point is that in making and moving things, the task is always taken for granted.

In knowledge and service work, however, the first questions in increasing productivity—and working smarter—have to be, "What is the task? What are we trying to accomplish? Why do it at all?" The easiest, but perhaps also the greatest, productivity gains in such work will come from defining the task and especially from eliminating what does not need to be done.[2]

A very old example is still one of the best: mail-order processing at the early Sears, Roebuck. Between 1906 and 1908, Sears eliminated the time-consuming job of counting the money in incoming mail orders. Rather than open the money envelopes enclosed with the orders, Sears weighed them automatically. In those days, virtually all Sears customers paid with coins. If the weight of the envelope tallied with the amount of the order within fairly narrow limits, the envelope went unopened. Similarly, Sears eliminated the even more time-consuming task of recording each incoming order by scheduling order handling and shipping according to the weight of the incoming mail (assuming 40 orders for each pound of mail). Within two years, these steps accounted for a tenfold increase in the productivity of the entire mail-order operation.[3]

A major insurance company recently increased the productivity of its claims-settlement department nearly fivefold—from an average of 15 minutes per claim to 3 minutes—by eliminating detailed checking on all but very large claims. Instead of verifying 30 items as they had always done, the adjusters now check only 4: whether the policy is still in force; whether the face amount matches the amount of the claim; whether the name of the policyholder matches the name on the death certificate; and whether the name of the beneficiary matches the name of the claimant. What provoked the change was asking, "What is the task?" and then answering, "To pay death claims as fast and as cheaply as possible." All that the company now requires to control the process is to work through a 2% sample, that is, every fiftieth claim, the traditional way.

Similarly, a few hospitals have taken most of the labor and expense out of their admissions process by admitting all patients the way they used to admit emergency cases who were brought in unconscious or bleeding and unable to fill out lengthy forms. These hospitals asked, "What is the task?" and answered, "To identify the patient's name, sex, age, address and how to bill"—information found on the insurance identification cards practically all patients carry.

These are both examples of service work. In knowledge work, defining the task and getting rid of what does not need to be done is even more necessary and produces even greater results. Consider how one multinational company redefined its strategic planning.

For many years, a planning staff of 45 brilliant people carefully prepared strategic scenarios in minute detail. The documents were first-class works and made stimulating reading, everybody agreed. But they had a minimal impact on operations. Then a new CEO asked, "What is the task?" and answered, "To give our businesses direction and goals and the strategy to attain these goals." It took four years of hard work and several false starts. But now the planning people (still about the same number) work through only three questions for each of the company's businesses: What market standing does it need to maintain leadership? What innovative performance does it need to support that standing? And what is the minimum rate of return needed to earn the cost of capital? Then the planning people work with the operating executives in each business to map out broad strategic guidelines for achieving these goals under various economic conditions. The results are far simpler and much less elegant, but they have become the "flight plans" that guide the company's businesses and its senior executives.

WHEN people make or move things, they do one task at a time. Taylor's laborer shoveled sand; he did not also stoke the furnace. Mayo's wiring-room women soldered; they did not test finished telephones on the side. The Iowa farmer planting corn does not get off his tractor between rows to attend a meeting. Knowledge and service work, too, require concentration. The surgeon does not take telephone calls in the operating room, nor does the lawyer in consultation with a client.

But in organizations, where most knowledge and service work takes place, splintered attention is more and more the norm. The people at the very top can sometimes concentrate themselves (though far too few even try). But the great majority of engineers, teachers, salespeople, nurses, middle managers, and the like must carry a steadily growing load of busywork, activities that contribute little if any value and that have little if anything to do with what these professionals are qualified and paid for.

The worst case may be that of nurses in U.S. hospitals. We hear a great deal about the shortage of nurses. But how could it possibly be true? The number of graduates entering the profession has gone up steadily for a good many years. At the same time, the number of bed patients has been dropping sharply. The explanation of the paradox: nurses now spend only half their time doing what they have learned and are paid to do—nursing. The other half is eaten up by activities that do not require their skill and knowledge, add neither health-care nor economic value, and have little or nothing to do with patient care and patient well-being. Nurses are preoccupied, of course, with the avalanche of paperwork for Medicare, Medicaid, insurers, the billing office, and the prevention of malpractice suits.

The situation in higher education is not too different. Faculty in colleges and universities spend more and more hours in committee meetings instead of teaching in the classroom, advising students, or doing research. But few of these committees would ever be missed. And they would do a better job in less time if they had three instead of seven members.

Salespeople are just as splintered. In department stores, they now spend so much time serving computers that they have little time for serving customers—the main reason, perhaps, for the steady decline in their productivity as producers of sales and revenues. Field-sales representatives spend up to one-third of their time filling out reports rather than calling on customers. And engineers sit through meeting after meeting when they should be busy at their workstations.

This is not job enrichment; it is job impoverishment. It destroys productivity. It saps motivation and morale. Nurses, every attitude survey shows, bitterly resent not being able to spend more time caring for patients. They also believe, understandably, that they are grossly underpaid for what they are capable of doing, while the hospital administrator, equally understandably, believes that they are grossly overpaid for the unskilled clerical work they are actually doing.

The cure is fairly easy, as a rule. It is to concentrate the work—in this case, nursing—on the task—caring for patients. This is the second step toward working smarter. A few hospitals, for example, have taken the paperwork out of the nurse's job and given it to a floor clerk who also answers telephone calls from relatives and friends and arranges the flowers they send in. The level of patient care and the hours nurses devote to it have risen sharply. Yet the hospitals have also been able to reduce their nursing staff by one-quarter or one-third and so raise salaries without incurring a higher nursing payroll.

To make these kinds of improvements, we must ask a second set of questions about every knowledge and service job: "What do we pay for? What value is this job supposed to add?" The answer is not always obvious or noncontroversial. One department store looked at its sales force and answered "sales," while another in the same metropolitan area and with much the same clientele answered "customer service." Each answer led to a different restructuring of the jobs on the sales floor. But each store achieved, and fairly fast, substantial growth in the revenues each salesperson and each department generated, that is, gains in both productivity and profitability.

FOR all its tremendous impact, Taylor's scientific management has had bad press, especially in academia. Perhaps the main reason is the unrelenting campaign U.S. labor unions waged against it—and against Taylor himself—in the early years of this century. The unions did not oppose Taylor because they thought him antilabor or promanagement. He was neither. His unforgivable sin was his assertion that there is no such thing as "skill" in making and moving things. All such work was the same, Taylor asserted. And all could be analyzed step by step, as a series of unskilled operations that could then be combined into any kind of job. Anyone willing to learn these operations would be a "first-class man," deserving "first-class pay." He could do the most advanced work and do it to perfection.

To the skill-based unions of 1900, this assertion represented a direct attack. And this was especially true for the highly respected, extremely powerful unions that dominated what were then some of the country's most sophisticated manufacturing sites—the army arsenals and navy shipyards where nearly all peacetime production for the military took place until well after World War I. For these unions, each craft was a mystery whose secrets no member could divulge. Their power base was control of an apprenticeship that lasted five or seven years and admitted, as a rule, only relatives of members. And their workers were paid extremely well—more than most physicians of the day and triple what Taylor's first-class man could expect to get. No wonder that Taylor's assertions infuriated these aristocrats of labor.

Belief in the mystery of craft and skill persisted, as did the assumption that long years of apprenticeship were needed to acquire both. Indeed, Hitler went to war with the United States on the strength of that assumption. Convinced that it took five years or more to train optical craftsmen (whose skills are essential to modern warfare), he thought it would be at least that long before America could field an effective army and air force in Europe—and so declared war after the Japanese attack on Pearl Harbor.

We know now Taylor was right. The United States had almost no optical craftsmen in 1941. And modern warfare indeed requires precision optics in large quantities. But by applying Taylor's methods of scientific management, within a few months the United States trained semiskilled workers to turn out more highly advanced optics than even the Germans were producing, and on an assembly line to boot. And by that time, Taylor's first-class men with their increased productivity were also making a great deal more money than any craftsman of 1911 had ever dreamed of.

Eventually, knowledge work and service work may turn out to be like the work of making and moving things—that is, "just work," to use an old scientific management slogan. (At least this is what Taylor's true heirs, the more radical proponents of artificial intelligence, maintain.) But for the time being, we must not treat knowledge and service jobs as "just work." Nor can we assume they are homogeneous. Rather, these jobs can be divided into three distinct categories by looking at what productive performance in a given job actually represents. This process—defining performance—is the third step toward working smarter.

For some knowledge and service jobs, performance means quality. Take scientists in a research lab where quantity—the number of results—is quite secondary to their quality. One new drug that can generate annual sales of $500 million and dominate the market for a decade is infinitely more valuable than 20 "me too" drugs with annual sales of $20 million or $30 million each. The same principle applies to basic policy or strategic decisions, as well as to much less grandiose work, the physician's diagnosis, for example, or packaging design, or editing a magazine. In each of these instances, we do not yet know how to analyze the process that produces quality results. To raise productivity, therefore, we can only ask, "What works?"

The second category includes the majority of knowledge and service work: jobs in which quality and quantity together constitute performance. Department store sales are one example. Producing a "satisfied customer" is just as important as the dollar amount on the sales slip, but it is not so easy to define. Likewise, the quality of an architectural draftswoman's work is an integral part of her performance. But so is the number of drawings she can produce. The same holds true for engineers, sales reps in brokerage offices, medical technologists, branch bank managers, reporters, nurses, claims adjusters, and so on. Raising productivity in these jobs requires asking, "What works?" but also analyzing the process step by step and operation by operation.

Finally, there are a good many service jobs (filing, handling death claims, making hospital beds) in which performance is defined much as it is in making and moving things: that is, largely by quantity (for example, the number of minutes it takes to make up a hospital bed properly). In these "production" jobs, quality is primarily a matter of external criteria rather than an attribute of performance itself. Defining standards and building them into the work process is essential. But once this has been done, real productivity improvements will come through conventional industrial engineering, that is, through analyzing the task and combining the individual simple operations into a complete job.

DEFINING the task, concentrating work on the task, and defining performance: by themselves, these three steps will produce substantial growth in productivity—perhaps most of what can be attained at any one time. They will need to be worked through again and again, maybe as often as every three or five years and certainly whenever work or its organization changes. But then, according to all the experience we have, the resulting productivity increases will equal, if not exceed, whatever industrial engineering, scientific management, or human relations ever achieved in manufacturing. In other words, they should give us the productivity revolution we need in knowledge and service work.

But on one condition only: that we apply what we have learned since World War II about increasing productivity in making and moving things. The fourth step toward working smarter, then, is for management to form a partnership with the people who hold the jobs, the people who are to become more productive. The goal has to be to build responsibility for productivity and performance into every knowledge and service job regardless of level, difficulty, or skill.

Frederick Taylor has often been criticized for never once asking the workers he studied how they thought their jobs could be improved. He told them. Nor did Elton Mayo ever ask; he also told. But Taylor's (and Mayo's, 40 years later) methodology was simply a product of the times, when the wisdom of the expert prevailed. (Freud, after all, never asked his patients what they thought their problems might be. Nor do we have any record that either Marx or Lenin ever thought of asking the masses.) Taylor considered both workers and managers "dumb oxen." And while Mayo had great respect for managers, he thought workers were "immature" and "maladjusted," deeply in need of the psychologist's expert guidance.

When World War II came, however, we had to ask the workers. We had no choice. U.S. factories had no engineers, psychologists, or foremen. They were all in uniform. To our immense surprise, as I still recollect, we discovered that the workers were neither dumb oxen nor immature nor maladjusted. They knew a great deal about the work they were doing—about its logic and rhythm, its quality, and its tools. Asking them what they thought was the way to address both productivity and quality.[4]

At first, only a few businesses accepted this novel proposition. (IBM was a pioneer and for a long time one of the few large companies to act on this idea.) But in the late 1950s and early 1960s, it was picked up by Japanese industrialists whose earlier attempts to return to prewar autocracy had collapsed in bloody strikes and near-civil war. Now, while still far from being widely practiced, it is at least generally accepted in theory that the workers' knowledge of their job is the starting point for improving productivity, quality, and performance.

In making and moving things, however, partnership with the responsible worker is only the *best* way to increase productivity. After all, Taylor's telling worked too, and quite well. In knowledge and service work, partnership with the responsible worker is the *only* way.

The last component of working smarter is a two-part lesson that neither Taylor nor Mayo knew. First, continuous learning must accompany productivity gains. Redesigning a job and then teaching the worker the new way to do it, which is what Taylor did and taught, cannot by itself sustain ongoing learning. Training is only the beginning of learning. Indeed, as the Japanese can teach us (thanks to their ancient tradition of Zen), the greatest benefit of training comes not from learning something new but from doing better what we already do well.

Equally important is a related insight of the last few years: knowledge workers and service workers learn most when they teach. The best way to improve a star salesperson's productivity is to ask her to present "the secret of my success" at the company sales convention. The best way for the surgeon to improve his performance is to give a talk about it at the county medical society. We often hear it said that in the information age, every enterprise has to become a learning institution. It must become a teaching institution as well.

ONE hundred years ago, the signs of class conflict were unmistakable. What defused that conflict—and averted class war—was growth in the productivity of the industrial work force, something so unprecedented that even its prime mover, Frederick Taylor, had no term to describe it.

Today we know that productivity is the true source of competitive advantage. But what we must also realize is that it is the key to social stability as well. For that reason, achieving gains in service productivity comparable with those we have already achieved in manufacturing productivity must be a priority for managers throughout the developed world.

It is an economic truth that real incomes cannot be higher than productivity for any extended length of time. Unless the productivity of service workers rapidly improves, both the social and the economic position of that large group of people—whose numbers rival those of manufacturing workers at their peak—must steadily go down. At a minimum, this raises the prospect of economic stagnation; more ominously, it raises the prospect of social tensions unmatched since the early decades of the Industrial Revolution.

Conceivably, service workers could use their numerical strength to get higher wages than their economic contribution justifies. But this would only impoverish all of society, dragging everyone's real income down and sending unemployment up. Alternatively, the income of unskilled and semiskilled service workers could continue to fall in relation to the steadily rising wages of affluent knowledge workers. But this would lead to an even wider gulf between the two groups as well as to increasing polarization. In either case, service workers can only become increasingly bitter, alienated, and ready to see themselves as a class apart.

Fortunately, we are in a much better position than our ancestors were a century ago. We know what Marx and his contemporaries did not know: productivity can be raised. We also know how to raise it. And we know this best for the work where the social need is most urgent: unskilled and semiskilled service work—maintenance jobs in factories, schools, hospitals, and offices; counter jobs in restaurants and supermarkets; clerical jobs in insurance companies, banks, and businesses of all kinds. In essence, this is production work. And what we have learned during the past 100 years about increasing productivity applies to such work with a minimum of adaptation.

Further, a model of sorts exists in the steps some multinational maintenance companies have already taken to improve their employees' productivity. These U.S. and European employers have systematically applied the approach this article discusses to low-skilled service jobs. They have defined the task, concentrated work on it, defined performance, made the employee a partner in productivity improvement and the first source of ideas for it, and built continuous learning and continuing teaching into the job of every employee and work team. As a result, they have raised productivity substantially—in some cases even doubled it—which has allowed them to raise wages. As important, this process has also greatly raised the workers' self-respect and pride.

It is no coincidence that outside contractors achieved these improvements. Obtaining major productivity gains in production-type service work usually requires contracting it out to a company that has no other business, understands this work, respects it, and offers opportunities for low-skilled workers to advance (for example, to become local or regional managers). The organizations in which this work is being done, the hospitals that own the beds, for instance, or the colleges whose students need to be fed, neither understand it nor respect it enough to devote the time and hard work that are required to make it more productive.

The task is known and doable. But the urgency is great. To raise the productivity of service work, we cannot rely on government or on politics altogether. It is the task of managers and executives in businesses and nonprofit organizations. It is, in fact, the first social responsibility of management in the knowledge society.

NOTES

1. Among the few attempts to apply working smarter in health care are Roxanne Spitzer's *Nursing Productivity: The Hospital's Key to Survival and Profit* (Chicago: S-N Publications, 1986) and Regina Herzlinger's *Creating New Health Care Ventures* (Gaithersburg, MD: Aspen Publishers, 1991).
2. See Michael Hammer, "Reengineering Work: Don't Automate, Obliterate," *HBR* (July–August 1990), and Peter F. Drucker, "Permanent Cost Cutting," *Wall Street Journal* (January 11, 1991).
3. See Boris Emmet and John E. Jeucks, *Catalogues and Counters: A History of Sears, Roebuck & Company* (Chicago: University of Chicago Press, 1965).
4. In my 1942 book, *The Future of Industrial Man* (Westport, CT: Greenwood, 1978 reprint of original), and my 1950 book, *The New Society* (Westport, CT: Greenwood, 1982 reprint), I argued for the "responsible worker" as "part of management." Edwards W. Deming and Joseph M. Juran developed what we now call "quality circles" and "total quality management" as a result of their wartime experiences. Finally, the idea was forcefully presented by Douglas McGregor in his 1960 book, *The Human Side of Enterprise* (New York: McGraw-Hill, 1985, twenty-fifth anniversary printing), with its "Theory X" and "Theory Y."

QUESTIONS

1. Why is it important for developed economies to raise the productivity of service and knowledge work?
2. What does Peter Drucker mean by "working smarter"? Give some examples of "working smarter."
3. How can information technology contribute to productivity in service industries? What specific steps would you take to achieve these gains?

1.2

Defining and Measuring Productivity

JAAK JURISON

Once productivity is recognized as an important output characteristic, the next problem is how to define and measure it. The basic concept of taking the ratio of output to input is fine as long as both quantities involve one factor and have the same units. The problem becomes more complicated as soon as factors (such as capital and labor) can substitute for one another, technologies change, the product specifications change, the product is improved, or the price of products or factors changes over time.

Productivity is a relative measure. That is, a country's or a firm's productivity is measured relative to that of a competitor at a point in time, or against itself at some previous time. The most useful forms of productivity measures are time series that show how productivity changes over time.

Productivity is an indicator. Usually, improved productivity results in improved profitability, but not always. As Jurison shows in this section, the ultimate measure of productivity is the economic value added by the productivity gain.

INTRODUCTION

Productivity has been a major concern since people began working in an organized fashion. Productivity, in its broadest definition, is the relation between

- products and services produced, and
- the resources used to create them.

It is the most fundamental indicator of economic performance. Peter Drucker (in Section 1.1) considers productivity one of the key management responsibilities. What measures we choose and how we apply them determine how well managers are living up to this challenge.

The first estimates of productivity were made by the Bureau of Labor in the Interior Department during the mid–1880's. The focus of these studies was on labor productivity, motivated by the Bureau's concern for technological displacement of workers. Over the years, productivity calculations were refined and expanded to include a variety of measures. Extensive data are readily available today.

The Bureau of Labor Statistics is the primary source of productivity data in the United States, although several private organizations and research institutes also publish productivity statistics. While productivity was of interest to a handful of economists and researchers initially, it is now a major concern for all business managers.

The new heightened level of interest in productivity in the United States can be attributed to these factors:

1. While the United States still enjoys the highest labor productivity in the world, its productivity growth rate lags behind many leading industrialized countries.
2. Globalization and deregulation create increased pressures on operating margins. To respond, firms try to find new ways to improve productivity.
3. There is a growing concern about the seemingly low payoff from heavy investment in the service sector. Often the payoff from these investments requires considerable time before it occurs. While it is clear that investments need to be made to improve productivity, the expenditures are current and tangible whereas the benefits are in the future and may not be realized without aggressive management action.

The purpose of this paper is to define the concept of productivity, to describe how productivity is measured, and to explain how it is used to improve organizational performance.

WHY MEASURE PRODUCTIVITY?

Productivity measures the relation between products and services produced and the resources used to create them. Productivity tells you how effective you are in using resources for the production of various goods and services. The resources may be labor, capital, materials, energy, or information—or any combination of these. You can increase productivity by producing more with the same amount of resources or producing the same amount with fewer resources.

Productivity can be measured at many different levels, ranging from the individual worker to an entire nation. At the national level, productivity is an important indicator of a country's economic strength and is the key determinant of a nation's standard of living. It also measures the country's competitiveness in the global market. Failure by a nation to keep productivity abreast of that of other countries contributes to adverse balance of trade, unemployment, and other domestic problems.

At the industry level, a high degree of productivity is necessary to keep costs and prices competitive. Industries with a record of continuous productivity gains are able to survive and flourish (as long as they are not displaced technologically), while low-productivity industries experience business decline. With globalization, many industries in advanced industrialized nations are facing increased competition from low-wage developing countries. These industries can remain competitive only by increasing productivity at home to offset the low wage advantage of their competitors in the less industrialized world.

At the firm level, productivity is the main source for competitive advantage. High productivity is essential to profitability and survival. Firms with above average productivity tend to enjoy higher profit margins than their competitors. Failure to maintain adequate productivity growth is often the cause of business failures.

Finally, at the individual level, high productivity is necessary for maintaining one's living standard and is a source of personal satisfaction and self-fulfillment.

Productivity is equally important at every level, because in the final analysis, productivity and the living standard of a nation and its citizens depend on the productivity gains made at all levels within the nation.

CONCEPT OF PRODUCTIVITY

The concept of productivity is relatively simple and straightforward. But attempts to operationalize it, and to find and apply appropriate measures, are quite difficult.

The traditional approach to productivity is to

compare the quantity of output (desired results) with the quantity of one or more inputs (resources) used to produce the output. This relationship is expressed as a ratio:

$$\text{Productivity} = \frac{\text{Units of output}}{\text{Units of input}} = \frac{O}{I} \qquad (1)$$

The *output* usually represents some unit of production or desired results. Examples of output are: the quantity of products produced, number of reports prepared, number of contracts negotiated, number of customers visited, etc. The main requirement is that the output is expressed as a measurable quantity. At the firm level, output can be defined either as gross output or value-added output. Gross output is the total production of the organization, including intermediate goods and services that are purchased from the outside. Value-added output is defined as the output due to the efforts of the organization's own resources and excludes the cost of intermediate goods.

Inputs are the resources consumed during production of the outputs defined in the numerator. Typical inputs are labor hours, capital equipment, energy, raw materials, supplies, and electricity. Like outputs, inputs must be quantifiable and measurable.

Productivity measurements can be based on a *single input factor* or a combination of *multiple factors.* Single-factor or partial productivity measures indicate how well a firm is using that single factor in the production of its output. Multifactor measures take into account many or all input factors.

Before discussing single-factor and multifactor measures, it is necessary to consider the units in which these measures are described.

MEASUREMENT ISSUES

Measurement issues arise from two causes:

1. Multiple products
2. Different inputs

For example, Firm A produces 1000 small motors and 50 large motors and Firm B produces 800 small motors and 75 large motors with the same inputs. Simply adding the number of motors produced each month (1050 vs. 875) is not a good output measure of productivity to compare the firms. In the case of inputs, it makes no sense to add number of hours of manufacturing labor and number of BTU's of energy. To get around these problems, inputs and outputs are combined into a single set of units, usually dollars.

Example

In the motor example, if small motors sell for \$100 and large motors sell for \$5000, then Firm A produces $(100 \times 1000 + 50 \times 5000) =$ \$350,000 whereas Firm B produces \$455,000. Thus, although Firm A produces more units, it is less productive than Firm B.

Further, suppose Firm A uses 1000 labor hours at \$20/hour and its energy bill is \$10,000 and Firm B uses 800 labor hours at \$20/hour and its energy bill is \$14,000. Then Firm A's total inputs are $(1000 \times 20 + 10{,}000) =$ \$30,000 and Firm B's inputs are $(800 \times 20 + 14{,}000) =$ \$30,000. Thus, the two firms use different resources but have the same input.

In terms of productivity,

Productivity of A = \$350,000/\$30,000 = 11.7
Productivity of B = \$455,000/\$30,000 = 15.2

The important point is that converting all data to the same set of units makes it possible to compare firms with different inputs and outputs.

SINGLE-FACTOR MEASURES

The most common single-factor productivity measure is labor productivity. It can be expressed in terms of output per worker or per hour worked. Economists use this indicator to determine a nation's standard of living relative to other countries. At the firm level, particularly in labor intensive industries, labor productivity is used to determine the efficiency of the organization. Other single-factor productivity measures can be output per machine, output per ton of material or output per some other type of input.

While single-factor productivity measures are useful indicators for managers, they do have some serious shortcomings. The primary problem with defining productivity too narrowly is that it can lead to unsound management decisions and counterproductive behavior. It is easy to increase productivity of one factor by replacing it with another. For example, a firm may invest in new technology that reduces the number of labor hours needed to produce a particular product. Clearly, labor productivity has increased, but at the same time, capital input increased. Economists call this effect capital-labor substitution. The substitution effect accounts for most of the long-term labor productivity gains made in the U.S. manufacturing industry. As a result of capital investments in new technologies, the direct labor content in the U.S. manufacturing industry has declined to a point where it is only a small percentage of total production costs.

CAPITAL-LABOR SUBSTITUTION

To account for the capital-labor substitution effect, economists developed the capital-labor multifactor or total factor productivity factor (Grossman, 1993). This measure takes into account the contribution of both labor and capital in the production of goods and services. It is expressed as:

$$P_1 = \frac{O_v}{L + C} \qquad (2)$$

where P_1 = total factor productivity
L = labor
C = capital
O_v = value added output

The capital input C is the cost of all capital resources devoted to the production of goods and services, including working capital (cash, accounts receivable, and inventories) and fixed capital (structures and equipment). Note that this measure focuses only on labor and capital. It neglects the effects of intermediate goods and services. Therefore it is necessary to measure the value added output, O_v in the numerator.

MULTIFACTOR MEASURES

Total Productivity Measure

A broader view of productivity is to expand the notion of capital-labor substitution (Equation 2) to take into account all input factors (e.g., materials, energy, outside services). Such a total productivity index is still represented by a simple, single equation. Yet, it gives managers a handy scorecard to answer the question: "How are we doing?"

The concept of total productivity is particularly useful at the firm or business unit level of the organization. Partial productivity measures (such as given by Equations 1 and 2) are useful for day-to-day operational control at lower levels of the organization.

Total productivity is calculated from the formula:

$$P_t = \frac{O_t}{L + C + R + Q} \qquad (3)$$

where P_t = total productivity
O_t = total output
L = labor
C = capital
R = raw material and purchased parts
Q = other miscellaneous goods and services

Notice that in this case intermediate goods (such as purchased items and outsourced labor) are treated as input and therefore they must be also included in the output.

The advantage of a total productivity measure (over partial- or single-factor measures) is that it makes information available on trade-offs among factors and allows senior managers to make more informed business decisions.

PRODUCTIVITY OVER TIME

The relations discussed thus far are static. They tell you where your organization or industry is at a point in time. They are useful when comparing current performance against other organizations or industries. To understand whether productivity has improved or declined, measurements must be made over time.

Logically, productivity can be increased in several ways over time (see Figure 1):

a. Producing more output with the same level of input.
b. Producing the same level of output, but with less input.
c. Increasing both input and output, but increasing the amount of output more than the amount of input.
d. Decreasing both input and output, but decreasing input more than output.
e. Producing more output with less input.

As pointed out previously, productivity measurements have to be made in a single set of units, typically money. The value of inputs and outputs depend on price. Unfortunately, prices fluctuate over time due to such factors as inflation and competition. For example, a new entrant may force a firm to cut the price of its output. The actual productivity of the firm has not changed, but the formulas given thus far would imply productivity has decreased. Because input as well as output prices change over time, current prices must be deflated by their respective inflation factors. In effect, productivity changes over time are calculated by keeping prices constant. Various price indices can be used, such as consumer price index, producer price index, or an in-house developed product price or cost index. This method yields inflation-free answers. When coupled with price analysis (described next), it provides visibility into the firm's performance.

PROFITABILITY-PRODUCTIVITY RELATION

Although highly productive firms tend to be profitable, high productivity does not guarantee high profit performance. The following explains why. Profitability is defined, at the simplest level, as the ratio of sales to costs (productivity = sales/costs). But sales are equal to output units times output price while costs are equal to input units multiplied by their respective prices. Profitability can therefore be described as:

$$\text{Profitability} = \frac{\text{Output quantities} \times \text{Output prices}}{\text{Input quantities} \times \text{Input prices}} \tag{4}$$

By rewriting Equation 4 as a product of two ratios:

$$\text{Profitability} = \frac{\text{Output quantities}}{\text{Input quantities}} \times \frac{\text{Output prices}}{\text{Input prices}} \tag{5}$$

we see that the first ratio is productivity. The second ratio, (output prices to the price paid for inputs), is defined as *price recovery*. Equation 5 can now be rewritten to describe the fundamental relationship:

$$\text{Profitability} = \text{Productivity} \times \text{Price recovery} \tag{6}$$

This relation states that profitability comes from two sources:

1. Productivity and efficient use of resources.
2. Price recovery—the ability to maintain a favorable relation between prices paid for inputs and prices charged to customers.

Figure 2 describes the dynamic relations among the variables in Equation 5. In Figure 2, the vari-

Figure 1 Five Ways to Improve Productivity over Time

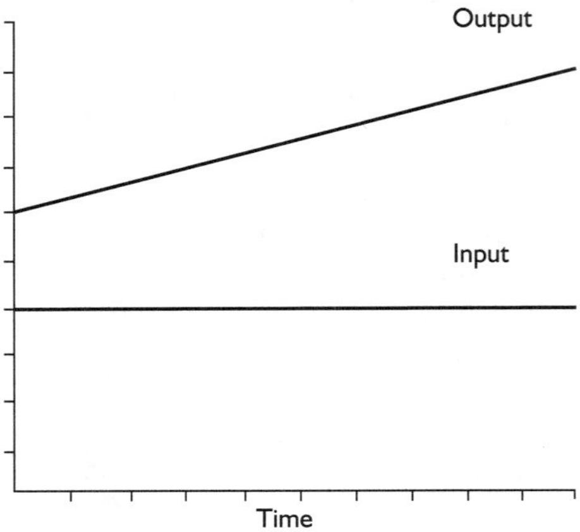

a. Increased output with constant input.

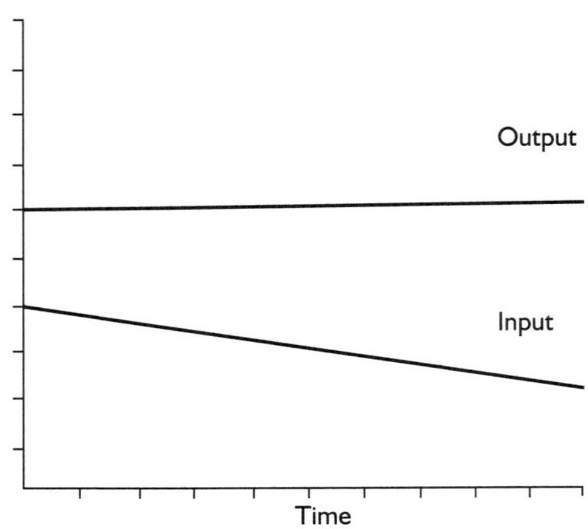

b. Constant output with reduced input.

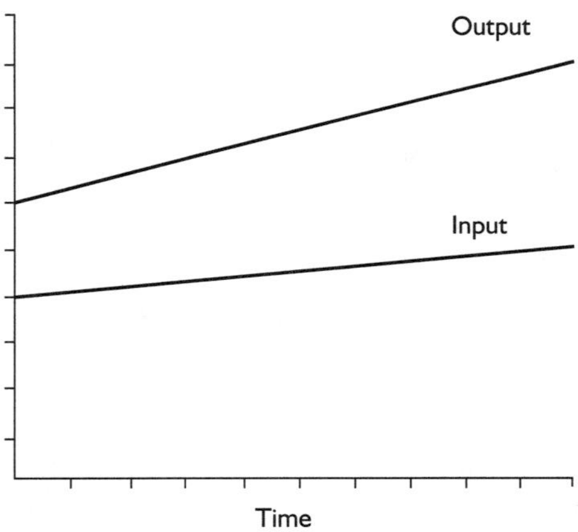

c. Faster output growth than input growth.

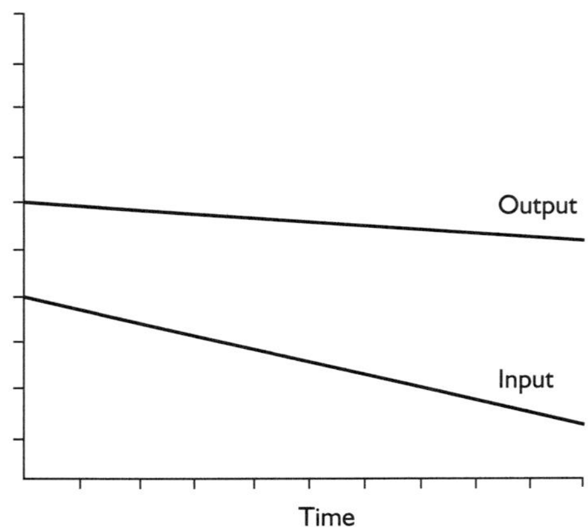

d. Slower output decline than input decline.

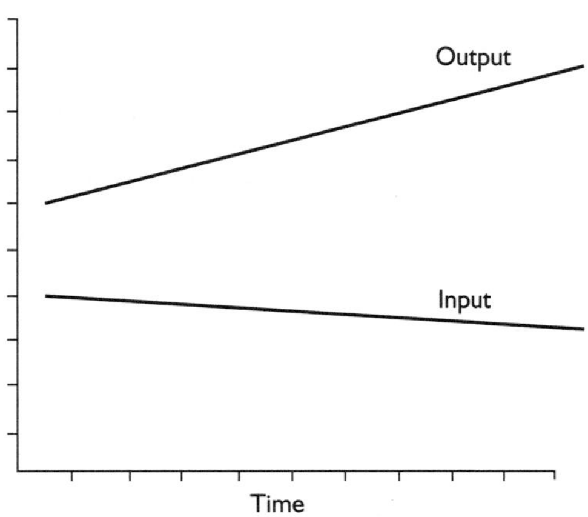

e. Output growth with input decline.

Figure 2 Relations among Productivity, Price Recovery, and Profitability

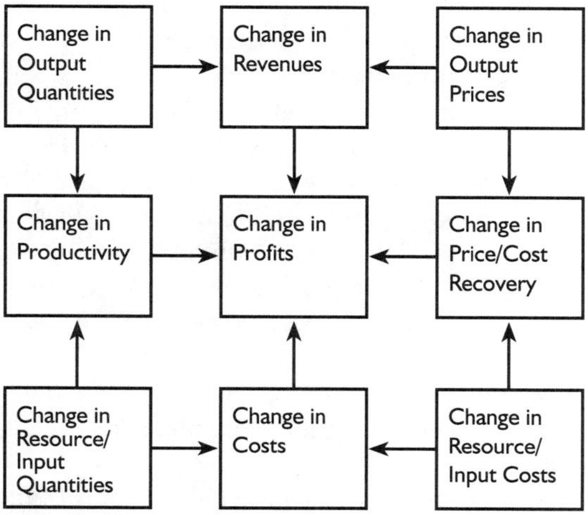

Source: Adapted from Sink, Tuttle, and Devries, 1984.

ables that can be controlled or are external are in the corners (output quantity, output price, resource/input cost, resource/input quantity). The business measures (revenues, productivity, profits, price/cost recovery, cost changes) are in the middle.

For example, suppose a raw material goes up in price (a change in resource/input costs). This results in an increase in cost and hence a reduction of profit. Figure 2 shows that to keep profits at the same level, a firm either has to increase revenues or increase productivity. When competition does not allow passing the cost on to the consumer, the only solution for the firm is to increase productivity. Thus, the model demonstrates the importance of productivity in a price-constrained environment.

USES OF PRODUCTIVITY MEASURES

Productivity measures can serve many purposes. Their five major uses and benefits are:

1. Productivity measures are indicators used by managers to assess progress in achieving goals. They provide managers early warning about potential problems.
2. Productivity measures are used in making decisions about capital investment, outsourcing, and acquisitions.
3. Productivity measures permit firms to compare themselves with others or with industry averages on efficiency and growth rates. The Bureau of Labor Statistics compiles and publishes industry productivity data that can be used as benchmarks.
4. Productivity measures provide perspective on a firm's financial data. For example, an organization may meet its targeted profit margins in the short run, yet experience declining productivity that could threaten its long-term survival.
5. Productivity measures heighten awareness. They define and communicate the purpose and meaning of productivity improvement (Kendrick, 1984).

PRODUCTIVITY MEASURES AND DECISION-MAKING

The primary motivation for measuring organizational productivity is to use the results in decision-making. For example, productivity data should be used in making capital investments, resource reallocations, and compensation decisions. Because these decisions have wide ranging consequences for the entire organization, productivity measures must be chosen carefully. First, the measures should focus on important goals and missions of the organization. While a single total productivity measure (such as Equation 3) provides a general aggregate measure of progress, it is usually an inadequate guide for corrective action. A family of single factor productivity measures (such as labor, capital, and materials productivity) help to identify the sources of aggregate productivity trends. Companies with successful productivity improvement programs use a combination of total- and single-factor measures.

In making decisions, managers face a tradeoff between accuracy and simplicity. Although accuracy is important, especially if productivity measures are used to determine compensation, simple productivity measures are usually more effective. It is more important for people to understand and apply a few simple measures than to have technically elegant and statistically accurate measures that nobody can comprehend. Complex measures are difficult to interpret and are frequently rejected by managers who must use them. A commonly heard complaint is "whoever designed these measures did not know anything about our business." The most effective productivity measures have low cost to obtain and are part of the regular work process (Brinkerhoff and Dressler, 1990).

EMERGING CHALLENGES: TIME, QUALITY, KNOWLEDGE WORK, AND VALUE ADDED

Conventional productivity measures are powerful tools when appropriately applied with good management judgment. They were developed for measuring and improving performance in manufacturing operations where outputs and inputs are easily quantifiable and the link between resource usage and outputs can be described and understood by everybody. But productivity measures face more serious challenges when dealing with:

1. Time
2. Quality
3. Knowledge work
4. Capital allocation

Managers are discovering that time is a source of competitive advantage. Because customers demand quality, quality is becoming equally important to productivity. In the rapidly growing service sector most workers are involved in knowledge work. They do not produce physical products—their output is information. Finally, the measures of productivity are linked to the value of capital and the ways of measuring the value of capital are changing.

Time

Many leading companies achieve competitive advantage by focusing on the management of time, whether it is in production, product development, distribution, or service. By measuring and managing time as a key resource in their business operations they outperform their competitors (Stalk and Hout, 1990).

Time differs from other resources. It cannot be purchased like labor or capital equipment. However, just because it cannot be bought, it is not free. If two companies are producing identical products with the same amount of resources, their traditional productivity measures are the same. But, if one firm ships products faster than its competitor, most managers would consider that firm more productive. That is, the less time it takes to get results, the more productive the organization.

Companies that recognize the key role of time in their operations usually have productivity measures for tracking performance based on time. These measures are based on the business objectives, and typically include such time-based measures as turnaround time, product development cycle, customer response time, and order fulfillment time. Time-focused companies also use other resources, especially capital, more productively.

Quality

The relation between quality and productivity is often misunderstood. Many managers used to think, and some still do, that quality and productivity are mutually exclusive, that a high level of quality can only be achieved at the expense of productivity. Current data show that improvements in quality actually lead to increased productivity (Deming, 1986). For example, defects are not free—it takes almost the same amount of resources to correct a defect as to produce another unit.

How should quality be factored into productivity considerations? The appropriate way to do this is to define output in terms of products or services that meet quality specifications. This approach focuses on producing only acceptable products, rather than the sheer number of units processed. But this raises an important question: who determines what is acceptable and what is not? The answer given by the total quality management movement is "the customer." Acceptability is determined by the customer of that product or service, whether the customer is another group within the firm or the consumer.

Knowledge Work

Traditional productivity measures which count the number of outputs are difficult to apply to organizations engaged in white collar and knowledge work. This work is done by office workers, who work with information, and professionals, who create new information and knowledge. Engineers, architects, lawyers, scientists, journalists, and managers all fall into this last category. Their output is mostly intangible. The process of measuring the output of individual people is more complex. Many analysts ignore the role of these professionals or treat them as overhead in companywide productivity measurements. This approach is acceptable for firms where the amount of white collar work is small or where it is possible to allocate the contribution of knowledge workers to output in a meaningful fashion.

The number and the size of firms totally engaged in knowledge work (e.g., law, architectural design, or research and development) is increasing in advanced industrial economies. Therefore it is becoming more important to measure and enhance the productivity of their work. Working harder or substituting capital for labor does not work in these industries.

As described by Drucker in Section 1.1, working smarter, working more effectively, is the key to raising the productivity of knowledge workers (Drucker, 1991).

Knowledge Worker Productivity

Industry experts argue over whether

- knowledge worker productivity is falling behind, or
- knowledge worker productivity is improving.

Currently available data and measures are simply not good enough to resolve the argument. In some areas, such as banking or financial services, customized productivity measures are being used (e.g. the number of deposits serviced per employee). These surrogate measures can be misleading unless they take minimum required quality into account.

Computers are a good example of the argument about knowledge worker productivity. The conventional wisdom of the mid-1980s was that computers did not pay back on their investment. However, by the early 1990s the conventional wisdom swung toward recognizing large gains from computer investments. Both sets of arguments are based on simplistic measurements. It was found that the payoff was not a simple capital substitution, but was dependent on organizational change and on training. The difficulties in judging gains from computer investment is typical of the difficulties in assessing knowledge worker productivity.

Economic Value Added (EVA)

Because profitability is related to productivity as defined in Equation 6, understanding productivity became an important management performance measure. Recent work focuses on an alternative way of measuring performance: Economic Value Added (EVA). EVA is calculated by the simple formula:

$$\text{EVA} = \text{Operating earnings} - \text{Taxes} - \text{Total capital costs} \quad (7)$$

If EVA is positive, the firm is creating value; if EVA is negative, the value of the firm is being destroyed.

EVA measures true productivity by taking into account the total cost of all operating capital. The total cost of capital includes the money provided by the shareholders (equity capital). The equity capital can be thought of as opportunity cost, what the shareholders would earn in dividends and stock appreciation by investing in an alternative portfolio of similar risk.

The cost of capital is expressed as a percentage rate (e.g., 8%). It is the weighted average return of both debt and equity. To convert this rate into the dollar amounts used in Equation 7, the cost of capital is multiplied by the dollars tied up in operations. Thus, in computing EVA, investments that are treated as operating expenses by traditional accounting measures like R&D, training, and software development must now be considered capital investments of some finite useful life.

EVA is a simple, but powerful measure of performance. It makes the cost of capital explicit at all levels in decision-making processes. It is particularly appropriate for capital intensive businesses. Companies that adopted EVA for their management decision making often see spectacular appreciation of their stock prices (Tully, 1993). EVA forces managers to think as stockholders of the company.

SUMMARY

Productivity is an important measure of economic performance. At the organizational level, it indicates how good the company is in taking inputs like labor and material and converting them into useful products. Despite its conceptual simplicity, productivity is difficult to operationalize.

There is no single, universal measure of productivity—organizations vary in terms of their nature and goals. What works for one firm may not be appropriate for another. Therefore different measures are needed in different organizations.

None of the productivity measures are perfect. They all have some shortcomings and limitations. By understanding the strengths and weaknesses of different measures, managers can choose the appropriate measures and use them to improve organizational performance.

Productivity should not be viewed simply as more efficient or intensive use of resources. The essence of productivity is not working harder, but working smarter. Working harder leads only to limited short-term gains because hard work cannot be sustained. Only by working smarter, reorganizing the work[1] and using our limited resources more effectively, are we likely to produce real sustainable productivity gains.

NOTES

1. See Section 6.2 where Davenport and Short discuss reengineering (i.e., the reorganization of work).

REFERENCES

Brinkerhoff, R. O. and Dressler, D. E. *Productivity Measurement.* Newbury Park, CA: Sage, 1990.

Deming, W. E. *Out of Crisis.* Cambridge, MA: MIT Center for Advanced Engineering Study, 1986.

Drucker, P. F. "The New Productivity Challenge." *Harvard Business Review* (November–December, 1991): 69–79.

Grossman, E. S. "How to Measure Company Productivity." In *Handbook for Productivity Measurement and Improvement,* edited by W. Christopher and C. G. Thor. Portland, OR: Productivity Press, 1993.

Kendrick, J. W. *Improving Productivity. A Handbook with Case Studies.* Baltimore and London: Johns Hopkins University Press, 1984.

Sink, D. C., Tuttle, T. C., and DeVries, S. J. "Productivity Measurement and Evaluation." *Productivity Review* (Summer 1984): 265–286.

Stalk, G., Jr. and Hout, T. M. *Competing Against Time.* New York: The Free Press, 1990.

Tully, S. "The Real Key to Creating Wealth." *Fortune* (September 20, 1993): 38–50.

QUESTIONS

1. Why is it necessary to differentiate between total output and value-added output?
2. Why do most companies use a family of productivity measures instead of a single total productivity measure?
3. Select an organization of knowledge workers (it could be your professors) and describe a set of productivity measures for the organization.

1.3

The Productivity Payoff Arrives

MYRON MAGNET

The assumption that the adoption of new technology will lead to immediate payoffs is not necessarily true. Despite massive investments in computers in the decade of the 1980's, the national productivity statistics hardly budged. Individual companies who mastered the computer were, indeed, reaping large rewards. However, the average firm with many fewer skills did not see these gains. People started talking about a "productivity paradox." A large part of the problem was in white collar service industries. The service sector spent over $750 billion on information technology hardware in the 1980's but had a lower productivity gain than in the previous 10 years and much lower than the gains in manufacturing productivity in the same period.

These statistics turned around in 1991 and the productivity paradox disappeared. Why did this happen? Magnet argues that technological inventions have a forty-year lead time in increasing productivity. He cites the example of the electric power industry which barely touched manufacturing for its first forty years. Similarly, the computer has been around for forty years and is only now fulfilling its promise.

Magnet gives us several insights:

- Technological revolutions do not happen automatically or instantaneously.
- To unlock technology's power, companies must restructure themselves before they can gain full benefit.
- Efforts to reorganize around information tools have begun to show up in the national productivity statistics.

Source: Fortune (June 27, 1994): 79–84.

LOOKING back, weren't we naive to think we had only to drop nearly $1 trillion of computer hardware into the slot, wait a little while for the machinery to clank and grind, and, presto, a big productivity boost would drop out like cold soda for parched lips? Technological revolutions like the one giving birth to America's new economy don't unfold like that, automatically or instantaneously. After more than a decade's anxious waiting—and many missteps along the way—we're only now beginning to see the payoff. Says MIT economist Erik Brynjolfsson, who recently published a computer productivity study: "Information technology, as of 1994, is one of the big contributors to growth."

To unlock an epochal technology's power, companies have learned that they must restructure themselves and how they work as they weave computers into their most basic processes. A technological revolution, in other words, is more than a merely technological matter: It entails an organizational transformation too. That's what U.S. business's recent frenzy of reengineering has been all about, as companies flatten and decentralize along a unifying nervous system of the new information technology (IT).

Examples: Before getting a big payback from its computers, Caterpillar had to radically redesign its work. One result: a factory that now produces twice as many truck engines with the same number of people. To achieve a big increase in sales, accounting giant Coopers & Lybrand needed to radically change the way it communicated internally. Now employees can use their PCs to swap crucial client information more efficiently. How these corporations and others use technology to transform their organizations and thus boost productivity is *the* story of the new economy, a tale that holds lessons for managers in any industry.

At long last efforts to reorganize around computers, faxes, networks, and other info-tech tools have begun to show up in the national productivity statistics. Nonfarm business productivity, which rose on average at a wan 1% a year during the Eighties, has grown at a more wholesome 1.6% a year so far in the Nineties. It hit a whopping 5.1% annual rate during the second half of 1993. Manufacturing productivity, rising at an average 2.4% a year in the Eighties, surged 3.8% annually from 1991 through 1993.

TABLE 1
Productivity Growth: 1970–1993

Year	*Manufacturing*	*Service*	*Total*
1970	98	79	89
1975	103	90	99
1980	101	92	99
1985	105	104	104
1990	122	108	110
1992	136	110	115

Index: 1982 = 100 all three cases.
Note: This table is based on data published with the original article.

Productivity in the service industries, which are notoriously hard to measure but an increasingly dominant and huge part of the economy, grew at a 1.5% annual rate in the Nineties, after declining in the late Seventies and remaining flat during the Eighties, estimates Morgan Stanley chief economist Stephen Roach. Some individual service industries have shown much stronger productivity growth, as measured by the Bureau of Labor Statistics: Productivity jumped 5% in the telecommunications industry in both 1991 and 1992, for example, and rose even faster in rail transportation and in certain retail sectors. Whatever the absolute growth rate of U.S. productivity in both manufacturing and services, it has been sufficient to keep America No. 1 on this front, well ahead of Japan and Europe.

TO understand vast economic transformations like the one we're living through, a little history comes in handy, because the way info tech may be driving productivity, at least in broad outline, has happened before. The saga of how American industry assimilated electricity after Thomas Edison introduced it to the world of business in the 1880s is especially instructive: The process took 40 years—and yielded almost no big productivity gains until it was complete. Similarly, there's little evidence that computers contributed much to U.S. productivity over the first 40 years of their history. But now the payback on information technology seems poised for takeoff (see chart).

When electricity was first introduced, captains of industry did nothing more than unhook their steam engines or water wheels and replace them with electric dynamos. "From the point of view of the internal organization of the factory, nothing changed," says Paul David, the Stanford economist who has traced this history. But ever so slowly, over the course of decades, manufacturers began to harness the benefits of electricity. The flexibility and the economies proffered by ever-improving electric motor technology gradually allowed yesteryear's managers to reorganize their factories more efficiently. Also, rural electrification, as well as providing low-cost power, enabled companies to build new, better-configured plants on cheaper land outside the cities.

All that didn't fall into place until the 1920s. And then manufacturing productivity, which had been puttering along at a sedate 0.3% to 0.5% annual growth rate since the 1890s, exploded. It roared along at over 5% a year throughout the Roaring Twenties, until the crash of 1929. Paul David ascribes three-quarters of that growth to electrification. "That was the fulfillment of the potential of the dynamo," he says, " but it took the reengineering and reconceptualization of manufacturing for it to happen."

With this history in mind, David says, it makes no sense today to expect "an immediate and tight linkage between the introduction of new technology and immediate productivity." First you have to clear away the existing structure, a tightly organized whole in which everything is so intricately adapted to everything else that demolition is tough. Then, says David, "there is a co-evolution of the technology and the organizational structure, which is likely in the early stages to require a lot of experimentation and learning."

That dual process, the laborious clearing away along with the slow, fitful reconstruction, is what was going on during the era of drooping productivity in the 1970s and 1980s. As new information systems went in and employees learned to use them—very inefficiently at first—the old systems kept on operating alongside until the new ones got up to speed. Meanwhile, the double costs smothered productivity. Mistakes occurred. New systems sometimes didn't work or didn't work well enough to make a difference. Even when everything went fine, companies were doing piecemeal rather than integrated automation, so that

TABLE 2

Electricity Boosted Productivity

For its first 40 years, electricity had little effect on manufacturing productivity, which barely increased at all. But by the 1920s, gains were up 5% a year.

Time Interval	*Average Manufacturing Productivity Growth Rate*	*Events During Time Interval*
1879–1889	2.5%–3.5%	1883 Electric Motors Used in Manufacturing
1889–1899	3.5%–0.5%	1891 A.C. Power Transmission for Industrial Use
1899–1909	0.5%	1900 Central Station Steam Turbine and A.C. Generator 1907 Advent of State-Regulated Utilities
1909–1919	0.5%–0.0%	1917 Generation of Utilities Exceeds that of Industrial Establishment
1919–1929	0.0%–5.0%	

the uncomputerized bottlenecks kept the productivity benefits from flowing through to the entire operation.

Don't forget the human factor either. In the face of the transformation to the new economy, managers have had a way of fighting hard to hold on to the information on which their power rested. Indeed, they long organized computer power to reinforce hierarchical, centrally controlled structures, restraining productivity rather than liberating it.

Managerial resistance might have held off computerization's full benefit even longer had not the external economic environment changed radically. During the Great Inflation of the Seventies, managers had no incentive to tighten their operations. On the contrary: Paid according to the size of their fiefdoms, they could add employees wildly and pass along the costs to buyers, concealed in the fog of inflation.

But competition sent that business culture the way of spats and bustles. Companies suddenly needed new efficiency to survive as the Federal Reserve Bank in 1980 started to curb inflation and stabilize the currency. The Fed's tightening lifted the inflationary smoke screen that had hidden gluttonous price increases, enabling tough, hungry foreign competitors to increase their presence in U.S. markets. And once the deregulation of the 1980s unshackled competition in the service industry, airlines, trucking, telecommunications, and financial service companies found they needed efficiency just as urgently as manufacturers did. "So managers began to say, 'Gee, we have all these computers out here; we should be able to do more with less,' " says Wayne Angell, Bear Stearn's chief economist. "Until then, the economic incentive had not been applied."

NOW, with competition the order of the day, company after company tells the same story of technological change combining with organizational change to yield higher productivity. A medical supply company, which chooses to remain nameless, provides an all too typical case study of how innovative technology remains inert until structural change breathes life into it. The company's plant introduced automation—but without any result. Disappointed top managers investigated and discovered that workers and line managers were using the new machinery as if it were the old equipment, taking no advantage of its flexibility. They were using it to do long production runs that ended up swelling inventories, as before.

Why? A closer look showed that the compensation scheme rewarded high output and minimal downtime. Even after management began to change that system, the culture stayed put. MIT's Brynjolfsson, who studied the factory, questioned

TABLE 3

How Soon Will Computers Do the Same?

Since Univac's introduction in 1952, info tech has had little discernible impact on productivity. But experts see the technology as set to drive growth dramatically.

Time Interval	*Average Productivity Growth Rate*	*Events During Time Interval*
1950–1960	3.0%–2.5%	1952 Univac I Commercial Electronic Computer
1960–1970	2.5%–1.5%	1969 Intel Memory Chip
1970–1980	1.5%–0.5%	1977 DEC VAX Minicomputer 1979 8088 Microcomputer
1980–1990	0.5%–1.0%	1985 Intel 386 Chip, Microsoft Windows 1989 Intel 486 Chip, Lotus Notes Groupware
1990–1993	1.0%–1.5%	1993 Intel Pentium Chip

Note: This table is based on data published with the original article. Values shown to the nearest 0.5%.

the workers during the transition. Says he: "They believed that being effective meant making sure the machine was always running, even if it meant building up inventories that weren't needed." What's more, they didn't like adjusting the machinery. "The workers did not feel they were being paid to make decisions," Brynjolfsson found to his surprise. "They were happy to daydream."

In response, the company gathered a team of skilled, highly motivated assembly workers and line managers and set up a pilot program with a new machine in a corner of the factory. Once the team succeeded in working out an efficient new system for using the machinery, management was able to propagate it throughout the plant, with gratifying results.

Caterpillar provides a much vaster and infinitely more complex companywide illustration of the same principle. Says Don Hamm, a division services manager: "Until three years ago we were investing very heavily in information technology and getting very little out of it. We couldn't harvest the benefits until we reorganized and refocused the company." Now Caterpillar invests less and gets results, measured in high customer satisfaction, rising share of market, and 27% greater sales with 29% fewer employees than a dozen years ago.

Before reorganization, computer spending had poured into building what Hamm calls "functional monuments"—hulking systems with gee-whiz technology that made Caterpillar's information technology staff proud but added little to company profits. Only when the company changed from a functional organization (marketing, manufacturing, and so on) to business units (tractors, engines) did all that change. Once business unit managers acquired profit and loss responsibility, they wanted to make sure they got their money's worth out of IT.

Now Caterpillar's computers supply business unit managers with information they need to make important decisions on issues like design alternatives for their tractors and engines, what sales strategies to undertake, and which suppliers are worth using again. Caterpillar also found that they could use IT to make their customers more productive, something that eventually leads to higher profits for everyone. For example, the company configures its best-selling 3406 engine differently for the various truck and boat manufacturers that buy it. Formerly, when ordering a custom engine, a truckmaker, say, had to specify Caterpillar's own part numbers for each component—a No. X turbocharger, say, or a No. Y flywheel housing—an error-prone and thus expensive and frustrating problem. Now, using a computer connected to a mainframe at Caterpillar's Peoria, Illinois, headquarters, customers can electronically order an engine with the desired specifications—horsepower, torque, rating—in plain English. Caterpillar's computer then translates that ordinary language into the company's internal language of part numbers.

Eventually, the trucker's electronic order ends up right on the factory floor. Workers read computer printouts to find out exactly what kind of engine to build. A computer-controlled monorail system and robots bring the assembler the engine, the necessary parts, and the computer-generated information about what to do, and he works at his own pace until he's satisfied that the job is right. Even some of his tools have computer controls. Each electric wrench automatically checks that it has been used to tighten—and to the correct torque—the bolt it is designed to tighten. If not, the monorail freezes and an alarm buzzes. Overall result: Caterpillar now needs only one shift to turn out a volume of engines that would have required two shifts a mere two years ago.

WHEN we think of information technology's productivity-enhancing power, our first thought is of how computers have let business cut legions of clerks, warehousemen, middle managers, and other workers in the past decade. But now Morgan Stanley's Roach argues that a new stage is opening in which productivity enhancements come from using technology not to cut costs but rather to provide new tools to the core of the work force in the new economy, the army of higher-paid technical, professional, and managerial

AMERICA MAY BE MORE PRODUCTIVE THAN YOU THINK

▪ ▪ ▪ ▪ ▪

If Hollywood gave an Oscar for economics, the person who figured out how to measure productivity more accurately would be a sure winner. It's a tough role—no one really knows what the new economy is doing to productivity. How, for instance, do we measure the value that increased quality gives to the goods we produce?

The productivity equation from the old industrial economy—output divided by input—doesn't take into account much of what has changed about output. Today, quality matters more than ever, with a more powerful car or computer representing more output than one that's less powerful. But measuring quality is a vexing problem because that measure should include especially hard to quantify factors that buyers value highly, factors like timeliness, variety, and customer service.

With that difficulty in mind, MIT economist Erik Brynjolfsson measures productivity by letting customers be the judge of how much output they are getting. Accordingly, on the theory that customers buy more of what they want, he investigated what effect information technology investments had on sales gains and on return on capital at 400 *Fortune 500* industrial and service firms from 1987 to 1991. "Returns on info tech," he reports, "were orders of magnitude larger than returns on other types of capital investments." Also, info tech accounted for half of all sales gains.

But so hard to measure is info tech's contribution to productivity that some experts, particularly the authors of a recent National Research Council report on the topic, believe that the long years of sagging productivity may represent only a failure of metrics, not a failure of technology to have boosted productivity. Says Dartmouth business professor James Brian Quinn, chairman of the report's panel: "Productivity figures understate significantly the benefits which have been achieved."

It's not only that the national accounts data the Bureau of Labor Statistics uses to calculate productivity cover only 42% of service-sector employment, leaving more than half the economy uncounted. Beyond that, pricing often skews the equation. If a product is greatly improved and its price drops—phone service being one obvious example—the productivity increase can be understated, because the price decrease may not fully reflect the improvements in the product.

Nor do the data capture the alternative costs of all this investment—the costs that would have been incurred had the investments not been made. What would be the cost in lost brokerage fees and higher capital costs for business, for instance, without the New York Stock Exchange's electronic trading system, which handles over 200 million shares a day, sometimes even 600 million, when in the pre-automation era a 12 million–share day could jam the system.

According to these arguments, the recent numbers that show rising productivity are just as faulty as the past decades' disappointing numbers and just as likely to understate real productivity growth. Says Quinn: "The cumulative effect of these investments seems to be picked up by the measures now, but we don't have the slightest idea whether performance improved by 5% last year or by 500%. We just don't know." The only thing we do know for sure, he says, is that "total U.S. productivity is a heck of a lot higher than any of us has been willing to say."

workers whose productivity so far has scarcely been touched by computerization.

Take for example the miniature computers that respiratory therapists at the University of California at San Diego Medical Center now carry in their lab coat pockets. Instead of waiting at the nurses' station for patients' records or rushing to their department for instructions on what treatments to give which patients, therapists call up the information on their hand-held computers, which pluck the data by radio frequency from a central computer. Says Rick Ford, the hospital's assistant director of respiratory care: "The access to information has maximized the time practitioners can spend at the bedside, where they're needed most."

As therapists treat patients, they send back the information via wireless communication to the main computer, so that the supervisor knows how much work remains to be done and how many part-time therapists he'll need to help out during the shift that afternoon or the next morning. Staffing now is precisely tailored to workload.

Another exciting new tool to enhance productivity is "groupware," the best-known example of which is Notes, the software product that Lotus introduced in 1989 but that has caught on only in the past year or so. Notes, which contains elements of an E-mail system, a document management system, and an electronic bulletin board, allows large numbers of people to communicate among themselves, both within and among companies. Coopers & Lybrand, for example, which just bought 28,000 copies of Notes, is embedding the software into the fabric of how it does business.

The Coopers & Lybrand unit that, among other services, puts businesses in touch with sources of capital and provides bankruptcy consulting, uses Notes to automate its sales efforts and to nudge its professionals to go out and hustle for business. Using a PC, employees enter on Notes the sales calls they've made, the results, and future appointments. All 750 employees of the Coopers & Lybrand unit can call up the information at any time, so managers can gauge how hard their subordinates are working—hello, Big Brother?—and how well they are succeeding. Employees can see—with behavior-shaping clarity—how their performance stacks up against their colleagues'. In the nine months the system has been running, sales are up over 20%.

In another Notes application, Coopers & Lybrand employees enter information about their own expertise and experience—what industries, what companies, what issues they've handled, and for how long—all constantly updated. So when an employee needs to find a colleague expert in insurance companies that serve, say, a Romanian airline, he or she need only fill in a few fields and click a mouse, and the appropriate names pop up.

The ultimate hoped-for result is a subtle but utterly fundamental organizational change. Explains Don Shay, a Coopers & Lybrand director: "Instead of being Lone Rangers, we want to bring in the cavalry and act as a group."

How much does all this matter? A new study of 65 Notes users by International Data Corp., a Framingham, Massachusetts, research firm, found that the average return on investment in the software—a rough but serviceable proxy for productivity increase—is 179%. Concluded the IDC researchers: "Notes may be the elusive Holy Grail of white-collar productivity."

Turning your organization topsy-turvy is never easy. In fact, it's a process that can be fraught with anxiety and uncertainty. The old ways of doing business disappear. Fewer people do more work. Those who survive must learn to master new information technologies like miniature computers and Lotus Notes, and then figure out how they fit into a new, more chaotic organizational structure.

In the emerging order of the new economy, those managers who understand how to use info tech as a productivity booster will do more than ensure the success of their own careers. For productivity growth directly affects the national standard of living, and the way we play the game may well determine the extent to which America can remain the dominant economic superpower in the years ahead.

QUESTIONS

1. Why has it been so difficult to prove that investments in information technology pay off?
2. Why have we seen smaller productivity gains from IT in the service industries as compared to manufacturing?
3. Can the payoff from IT investments be accelerated or increased? If so, describe three methods that could be used. If not, explain the barriers to gaining better payoff.

1.4 Forecasting Productivity: The Impact of IT

WILLIAM R. KING

William R. King reiterates the point made by Myron Magnet in the previous section: the projected productivity increases in manufacturing and in the office from applying information technology have materialized. The benefits come both from removing the need for people and the speeding of business processes. For example, King cites the significant reduction in the use of paper forms that resulted from sending large amounts of routine information electronically.

Forecasting is an essential business activity. Decisions are made based on what will be, not what has happened. To be useful, forecasts must be accurate both in what will happen and when it will happen. The changes that occurred were predicted accurately in terms of what they would be but not in terms of when they would occur. The forecasts failed to take into account the human dimension and time. Human beings are usually slower in adopting technology than technologists are in creating new technology. The lesson to be learned is that in forecasting, human factors must be considered and that technology may not move as fast as its enthusiasts predict.

Readers interested in this topic may refer to Section 6.1, which deals with modeling the impact of new innovations.

IN this article, I'd like to focus on the process of predicting the impact of information technology (IT) on individual and organizational behavior. I believe some useful lessons for information systems planners and managers can be drawn from such an assessment.

An interesting context in which to do this is to look at the predictions that were made in the 1950s and the 1960s about the future impact of computing on organizations. These early predictions remind me of Rodney Dangerfield—they didn't get much respect. And for many years they didn't seem to be coming true. Now, finally, many are beginning to look rather good.

In this, there are lessons for IS planners, who must make technological forecasts as a basis for their plans, and for managers, who should be aware that the conventional wisdom of the last

Source: Information Systems Management (Winter 1994): 68–70.

two decades seems to be becoming much less valid in the 1990s.

INFORMATION TECHNOLOGY'S IMPACT ON MIDDLE MANAGEMENT

First, let's consider the issue of IT's impact on middle management. The job of middle managers is often characterized as that of information processing and communication. Middle managers filter raw data on its way to higher-level management. They identify things that are out of control or need attention.

Certainly, said the early pundits, the number of middle managers required by businesses will be significantly affected by information technology; their information processing and communications functions could be handled automatically, or at least much more efficiently, by employing IT.

In their well-known 1958 *Harvard Business Review* paper, "Management in the 1980s," H. J. Leavitt and T. L. Whisler foresaw "the larger portion [of middle management] shrinking and sinking into a more highly programmed state and the smaller portion proliferating and rising to a level where more creative thinking is needed." For more than two decades after those predictions were made, this simply didn't happen. The number and proportion of middle managers did not decline, and in some areas of business, these measures showed increases rather than decreases.

THE IMPACT ON PRODUCTIVITY

The same is true of many forecasts of the impact of computers on productivity—both in manufacturing settings and in the office. The projected increases in productivity from production and office automation just didn't materialize.[1] Although capital investment per office employee went up significantly because of the acquisition of word processors, PCs, and the like, office productivity levels were uninspiring. Overall, industrial productivity remained flat during much of the 1970s and 1980s.

THE PAPERLESS OFFICE PREDICTION

The paperless office was another popular forecast of the 1960s. Computers were going to reduce all data to electronic pulses that could be displayed on screens and stored on electronic media, thus reducing the need for paper, file cabinets, and file clerks. Trees would be saved for posterity and the quality of the average worker's job would be enhanced. Paper shufflers would no longer be needed, and we would all be freed to do more useful and interesting things. Not only would productivity be enhanced, but effectiveness would increase—more profits would be the result of more thought.

This scenario showed no sign of happening as of only a few years ago. Just make a change in an airline ticket at a counter and watch the computer preside over the printing of volumes of paper that a clerk laboriously sorts into various piles, which are then carefully stapled together—one pile for accounting, one for audit, and so on. Businesses found that computers are magnificent devices to initiate the printing of great masses of paper, and until recently, few organizations have appeared to be able to meaningfully control that potential.

The result has been estimated to be more than four trillion paper documents in US offices with an astonishing growth rate of 22% per year in the late 1980s.[2]

FINALLY, THE PAYOFF

All these years, U.S. businesses have based their investment in IT more on faith than on demonstrated results. While productivity was stagnant, employee costs grew and paper proliferated, and the capital investment of U.S. business in IT grew to approach 50% of total capital investment. In effect, managers were betting the farm on IT and it did not seem to be paying off.

Now, at last, we may be beginning to see these

forecasts come to fruition. These IT investments may have finally begun to pay off. Consider the following facts:

- During 1991 and 1992, the productivity of U.S. firms (measured in terms of output per labor hour) has begun to move upward at a fast rate, with manufacturing productivity increasing fastest.
- Middle management ranks have been decimated across most sectors of U.S. business.
- The use of paper business forms is actually in decline.

Of course, IT didn't cause these things to occur; IT was the enabler that permitted them to happen.

AN ENABLER OF CHANGE

International competitiveness concerns have forced businesses to rightsize—a more goal-oriented term than *downsize*. Firms were able to cut middle management and worker ranks because of their investments in IT. If high levels of technology didn't exist or businesses hadn't invested in IT, they would have had a much more difficult time in resizing.

Expert systems, executive information system (EIS) technology, electronic mail, and other varieties of organizational decision-support and communications systems were the specific facilitators that permitted savings in middle management expenditures. Expert systems support the judgments that were once supported by staff studies and analyses. EISs provide the answers to questions that were once directed by phone or memo to subordinates. Executive users simply drill down from high-level summary screens to obtain specific detailed backup data on operating performance. The increasing ubiquity of E-mail lessens the telephone tag phenomenon and makes direct communications more efficient, even when there is no other sophisticated systems support.

The same competitiveness issues have pressured firms to reengineer and redesign their processes, thereby simplifying them and making more use of IT in performing them. IT also supports the reengineering process itself. This massive redesign of processes has led to more productivity and less paper.

This has led to some quantum changes in the way that some companies do business. For example, consider the New England clothing retailer that operates dozens of stores with little more than a cashier in each. The sole-source manufacturer's computer uses point-of-sale data to maintain stock records for each store, assesses national and local sales trends and consumer preferences, and initiates restocking, which is also physically handled by agents of the manufacturer.

The critical skill of retailing middle management—merchandising—is now handled on what Leavitt and Whisler called a "programmed" basis, with a consequent impact on middle managers, productivity, profits, and paper.

Redesigned processes that rely heavily on IT, coupled with software for designing business forms and networks over which electronic forms (e.g., invoices, expense accounts, and purchase orders) can be transmitted, have led to a 9% annual decline in sales of the ubiquitous multi-part paper business forms. Moreover, electronic forms totally eliminate the one-third wastage that is estimated to occur because printed forms become obsolete while in inventory.[3]

LESSONS FOR FORECASTING THE IMPACT OF TECHNOLOGY

What lessons can IS planners and managers learn from these experiences with forecasting the impact of technology? Here are a few.

Build It and They May Not Come

First, we need to question the "Field of Dreams" notion as it applies to IT: Build it and they will come. Just because a technology can have an identifiable impact on people and organizations doesn't mean that it will make that impact. Videotext has often been forecast to make newspapers obsolete. Clearly, this technology has that capability. But it hasn't happened just because it is possible.

Why? Because many significant behavioral and organizational changes require the confluence of several forces or technologies. A single technology is often a necessary, but not sufficient, condition for some individual or organizational change to occur. Reductions in middle management ranks and increases in productivity required technology but were not dictated by it. Many other forces—competitiveness concerns, desires to transform organizations, a focus on the customer, cycle time, and quality, to name a few—had to come to the fore before information technology could play the role that was simplistically assumed for it by forecasters of earlier decades.

Even in those cases where technology itself can produce change, several technologies may need to be integrated before significant change will occur. Our ability to digitize and process data did not lead to significant reductions in the use of paper in business. However, now that appropriate scanning and storage technologies have been developed (optical scanners and disk storage that can rapidly file and reference 40,000 pages using a PC), organizations are finally using integrated systems to realize these benefits.

These factors make the job of technology forecasting difficult because forecasters may not be well versed in all of the technologies or the areas that must complement technology to induce change. Indeed, these forces are often difficult to identify, much less to predict.

Forecast Functionalities and Capabilities, Not Broad Impacts

Technology forecasts are most reliably made in terms of functionalities and capabilities—the new things that technology enables us to do and the new activity rates that will be possible—rather than in terms of behavioral, organizational, or social impact. Because these broader impacts are much more complex, they are far more difficult and more risky to forecast.

Perhaps this is best illustrated by Leavitt and Whisler's "heroic" forecast that IT would cause large industrial organizations to recentralize. This has occurred in some fashion in some companies, but they were unable to foresee the vast possibilities of mass customization in products as well as in organizational forms and processes. For every clothing retailer that has centralized merchandising decisions in a far-off computer, there is a company that has empowered frontline salespeople and customer representatives by providing systems support. Some firms use the term "company in a box" to describe the fact that salespeople's notebook computers allow them to quote prices, delivery dates, and terms in real time, forever altering the old days of centralized decisions and delayed responses.

Beware of the Time Dimension in Forecasts

The time at which technological impact will be felt is especially difficult to forecast. I have no doubt that a large proportion of overall U.S. banking transactions will eventually be handled through automated teller machines. Indeed, the number of transactions made through ATMs is increasing rapidly. But most forecasts made early in the adoption phase of ATMs were that it would occur more broadly and would transform the banking business. The failure of the public to widely accept ATMs as anything other than cash machines, the significant proportion of customers who won't use them at all, the way banks reacted to these unforecasted events by instituting usage charges that inhibit (rather than motivate) broader usage, and the possibility that other changes may be required—all these factors make it extremely difficult to forecast when this transformation might occur. So while ATMs will inevitably radically transform banking, the timing of this transformation will be determined not just by this technology and its availability, but by management choices, customer responses, and possibly other technologies that are difficult to predict or foresee.

CONCLUSION

Past experience with some of the best-known forecasts should lead us to a better understanding of the possibilities of technology forecasts. If we remember that the "Field of Dreams" admonition usually does not apply to technology, that we should stick to forecasting functionalities and capabilities, and that the time dimension of a forecast is especially problematic, we can be better forecasters, even though our forecasts may be of less import. Assessing these forecasts in the light of present realities can also help us to grow more aware of the opportunities of the current business environment and to recognize that the tremendous commitment of U.S. business to IT investments may be well founded.

NOTES

1. K. Ramamurthy and W.R. King, "Do Organizations Achieve Their Objectives with Computer-Based Manufacturing Technologies?" *IEEE Transactions on Engineering Management* 39: 2 (April 1992): 475–91.
2. Estimates by Lawrence Livermore Labs and Coopers and Lybrand as quoted in "The Optical Disk: Big Files on a Small Budget," *Inc.* (May 1993).
3. "The Optical Disk," *Inc.* (May 1993).

QUESTIONS

1. Why did the predictions of Leavitt and Whisler made in 1958 fail to materialize?
2. Why is it so difficult to forecast productivity?
3. What do you think is the future for middle management? Give arguments to support your views.

1.5 Haste Makes . . . Money

ANNE STUART

Most companies use a standard method of competing, by having the lowest price or high quality, or by satisfying a niche market. However, with the availability of the computer and communications, some companies are competing on the basis of time. They use speed as a competitive weapon that requires being nimble as well as quick.

These companies not only race to get new products to market first, they compete by having quick turnaround on all orders. Their methods lead to additional benefits: greater variety, lower cost, and higher quality than the competition.

Quickness means that a company's computer-based design, manufacturing, distribution, and other systems, as well as their people, must work at top speed. In this article, Anne Stuart presents a wide variety of companies who have adopted speed as their way of competing.

IN 1985, when Henry F. Duignan took the reins at Ross Operating Valve Co., designing a new product to meet a client's specifications was a very big deal. "It would have taken us maybe nine months and $600,000 to create a new pneumatic valve for a customer," recalls Duignan, vice president and COO of the $100 million Troy, Michigan, manufacturer.

Today, it's a snap. "We can do it in about 24 hours and for about $2,000," Duignan says. And that's not for a rush order; it's simply business as usual. Using computer-aided design (CAD) workstations, the company's engineers can work online with customers to design prototypes—"virtual valves"—literally overnight.

The quick turnaround time has turned Ross around as well. When Duignan came on board, the company was slouching toward bankruptcy. Now, he says, "we're growing at about 500 percent a year." That's largely because Ross retooled its entire organization to operate under a simple, even simplistic, principle: Time is money.

Dozens of other industries and companies are also dusting off that ancient adage as they explore new ways to use time as a competitive weapon. But, like Ross, the top finishers in the race to market don't just run faster; they run better—while remaining poised to switch directions and take on new opportunities.

Or as Boston Consulting Group vice presidents Thomas M. Hout and George Stalk, Jr. put it, they run smarter. Hout and Stalk, co-authors of *Competing Against Time*, were most recently featured in a Harvard Business School video, *Time Based Competition*. In both, they contend that top time-based companies not only deliver

Source: CIO Magazine (May 1, 1994): 50–54.

faster but offer greater variety, lower cost and higher quality than slower rivals. Their rewards: greater growth and bigger profits.

For Information Technology (IT), the challenge is developing systems that help make contenders not only swift but flexible as well. At the same time, consultants warn, IT alone can't create a champion.

"Technology is inevitably a piece of the solution," says Laurence P. Chait, a director at consultancy Arthur D. Little Inc., in Cambridge, Massachusetts, and a vice president specializing in business-process redesign. But for significant change, he says, "there has to be something from the top that is pressing organizationwide." For instance, "companies will be disappointed if they get EDI [electronic data interchange] and don't put changes into organization and process."

Emphasizing speed isn't a new strategy; companies have always raced to get to market first. What's different is the scenery, a business backdrop that keeps shifting to reflect economic, social, and technological changes. Roger Nagel, deputy director of the Iacocca Institute and professor of manufacturing systems engineering at Lehigh University in Bethlehem, Pennsylvania, cites three major influences: worldwide communications, fast and flexible delivery systems, and the ability to treat each customer as an individual.

Such developments open new paths, but they also keep organizations scrambling to adapt while trying to maintain pace. Cutting-edge contenders don't just cross the finish line first; they're already warming up for the next race. "When you have fast development capability, your innovation possibilities go up," Stalk says.

As James Wetherbe sees it, companies that want to improve cycle time have two options. "One is being able to change directions quickly," says Wetherbe, director of the Federal Express Center for Cycle Time Research at the University of Memphis. "The other is being able to mass-customize." In other words, quickly create a prototype to customer specifications, then manufacture the product in any quantity.

That's what Ross does. "We can produce one of something or 500 of something and be profitably involved," Duignan says.

Every new wave of technology can help companies gain a short-term edge in the race to market. But experts warn that, from the marathon viewpoint, becoming fast and flexible involves more than beating the clock.

"It's about how fast you get onto the same wavelength as your customers," says David Jackson, management consultant for the worldwide organizational development practice at Digital Equipment Corp. in Hampshire, England. "It's about how fast you can translate the information of the customers about what they need, what they want, what they're trying to achieve, and bring that back into your organization, understand it and create something—product, service, support—that will address that need."

At the 73-year-old valve manufacturer, that happens at the starting line: the design stage. Customers participate in shaping the prototype, building their goals and knowledge into the product.

The relationship, which Duignan describes as partnering rather than serving a customer, also builds loyalty. "You don't have to re-earn that customer," Duignan says. "As long as you keep serving him, he's yours for life."

He acknowledges that the company invested heavily to overhaul and upgrade its operation, including Information Systems (IS). But he adds: "It's the cultural changes that rule, not the technological ones." At its heart, Ross's culture is horizontal. "We did away with hierarchy," Duignan says. "We're an extremely flat organization."

The product-development process works like this: A Ross engineer, known as an "integrator," works primarily with one representative from a customer company. The integrator directs engineering, manufacturing, and marketing from a single computer terminal. (Duignan built the staff of integrators by hiring and training young engineering wizards too inexperienced to know they wouldn't normally handle all three functions.) As the product evolves, the customer—monitoring the process online—can make changes with a telephone call.

Bill Davidow, co-author of *The Virtual Corporation*, says the Ross partnering approach will become more common as companies and suppliers become increasingly dependent on one another. "The wave of the future is for customers to have fewer suppliers and suppliers to have fewer customers," says Davidow.

Meanwhile, other companies have introduced a host of less radical efforts to make themselves nimble and quick.

Many simply streamline, cutting waste early on. Toyota Motor Corp. in Japan sped up assembly lines with "value engineering," a series of tiny timesaving design changes. For instance, the Japanese automaker now cuts sheet metal to precise measurements rather than leaving a narrow margin to trim and finish later.

Some strive for timesaving at the point of sale. Until 1989, Best Buy Co. Inc. of Minneapolis, the nation's second-largest consumer-electronics chain, ran its stores like many of its retail competitors. As Best Buy's CFO Allen Lenzmeier describes it, customers went through a 10-step purchasing process that began when they walked into the department and ended when they picked up merchandise. In between, salespeople made presentations, checked computers for available items, pitched extended warranties, and filled out paperwork; customers then moved through cashier lanes and, finally, retrieved their purchases at a warehouse window.

Best Buy redesigned stores to put all available items on display. Salespeople now serve as "product specialists," answering questions without making sales. Now customers complete transactions in four steps: enter the store, find the department, select the item, and carry it to the cash register. Simple and old-fashioned, but it works.

Primarily, company officials wanted to make shopping more pleasant by eliminating the hard-sell atmosphere. But they discovered that streamlining also cut time, particularly for customers who do their homework before buying. "In today's fast-track world, customers are much better informed," says Best Buy spokeswoman Laurie Bauer. "They come into a store knowing exactly what they want. Now you can walk in, grab it, and go."

Lenzmeier credits the approach with fueling the chain's growth from a $28 million company a decade ago to its current pinnacle as a $3 billion powerhouse with 151 superstores in 18 states.

Others, including the textile and apparel industries, are working feverishly to shorten the pipeline. Traditionally, it takes an average of 66 weeks to mass-produce and deliver a garment, says Joseph Off, managing director of Textile/Clothing Technology Corp. in Cary, North Carolina, a federally funded research group nicknamed TC^2. The long lead time frequently led to costly retail markdowns. But retailers and manufacturers have joined forces to shorten the cycle with point-of-sale information, more-frequent orders, and smaller inventories. Industry innovators are also experimenting with printing fabric after cutting, rather than before, and using flexible manufacturing lines staffed by operators cross-trained in multiple functions. "That way, you don't have one person sitting at the machine for hours, doing a repetitive motion," Off says of the latter. "And there is no idle time. Everybody works as a team."

Chrysler Corp. also used a team approach to cut product-development time from up to six years to 39 months for its LH line of large cars. The automaker replaced the traditional linear process with teams of several hundred employees and vendors working together, start to finish, in a single facility. Chrysler later streamlined that approach to develop its stylish subcompact Neon car in 31 months.

WHILE customers typically cheer new speed records, they also, understandably, raise questions about quality. Quality focuses on getting the job done right; speed focuses on getting it done first, or at least faster. But the two don't have to be mutually exclusive. Hout and Stalk cite the case of a backlogged Swedish hospital where medical personnel initially resisted timesaving measures proposed to alleviate budget pressures because they feared changes would hurt quality of care. But patients already suffered needlessly, waiting months or years for surgery. By overhauling scheduling, staffing, and preoperative procedures, the hospital cut the wait to weeks—and increased the number of operations per day by 30%.

Wetherbe cites studies purporting to show that, in many industries, only 3% to 5% of the time needed to complete a job actually involves productive work. The rest disappears into black holes of duplicated effort, misfiring communications, lost information, and linear processes where each step depends on the previous one.

In fact, experts say, quality complaints often stem from time-related problems, such as schedule setbacks and outdated information. And some companies that consistently set the pace for speed to market also regularly get high marks for quality; consultants cite in particular Boeing Co., Compaq Computer Corp., Honda Motor Co., and Microsoft Corp.

So why isn't every company fast and flexible? Because the typical organizational structure is just that: "Structured. Rigid. Difficult to break down," says Charles Savage, manager of Digital Equipment Corp.'s organizational development practice in Marlboro, Massachusetts. "It looks logical, but it's fragmented, with little communication or learning. It's set up to check and control, not to listen and facilitate. . . . We are discovering speed because we haven't had it. And we haven't had it because [the organization] hasn't been designed to be speedy; it's been designed to be fixed and solid and continuous."

It may be easier to climb the Hoover Dam than overcome resistance to tinkering with structure. "People get entrenched. They have a way of doing things, and they've convinced themselves it's the only way," says Chait, of Arthur D. Little. "Anytime you mention the word change, people wonder: 'Will I survive?' " says the Iacocca Institute's Nagel. "Not just, 'Will there be a job for me?' but 'Will I be able to do the job?' "

Speed depends on empowering employees at every level, Wetherbe says. For instance, front-line customer-service representatives at Federal Express Corp. are empowered to handle damage claims up to $100 without seeking higher approval, which means customers get their checks within a day.

In Nagel's view, companies seeking speed and agility must become more entrepreneurial and adopt the "virtual-company model," creating different structures to handle different opportunities, disbanding them when the project ends. In addition, he says, competitors must market workforce skills, knowledge, and experience. "People and information are the only two things you have to sell," Nagel says. "Invest in them."

Finally, he says, successful agile organizations anticipate their customers' future needs and try to meet them with customized solutions, rather than just products. The difference? If customers need to rebuild, retool, restructure, or retrain to use their new purchases, they've bought a product. If they deploy it easily and profitably, they've bought a solution.

Obviously, IT's role in boosting speed begins with helping information flow more quickly through fewer barriers. But experts say businesses don't need the most powerful architecture on the planet to become flexible and responsive. Instead, they call for installing modular systems or, as Jackson, the British Digital consultant, puts it, a "plug-and-play infrastructure," where components can be rearranged as needed.

"It's a shift from looking at IS as a jigsaw [puzzle], where you've got all the pieces, but they [each] only fit in one place, to a Lego model," he says, referring to the toy plastic building blocks that interlock in any number of combinations. Using a modular design, a company can modify one part of its system without having to redesign the whole thing.

And, another consultant says, IT needs to hit the gas itself. "The [current] IT takes too long to develop," says John Donovan, professor of business at MIT and CEO of the Cambridge Technology Group, a consulting firm in Cambridge, Massachusetts. "The entire infrastructure of your IT department has to change its mindset, from positioning as a service to the business side to partners with the business side."

Finally, Donovan says, CIOs must take a hard look at how they view themselves. "CIOs, in general, are managers, dispensers of caution," which slows the process, he says. "Every CIO in the world has to become a leader. Leaders are dispensers of hope."

QUESTIONS

1. How can time-based strategies provide competitive advantage?
2. How can information technology be used to implement time-based strategies? Give examples from several different industries.
3. Would you recommend time-based strategies for all types of businesses? Explain why or why not.

1.6 Critical Success Factors and Productivity

CHRISTINE V. BULLEN

Critical Success Factors (CSFs) describe the idea that every business has a small number of things that must go right if the business is to prosper. This section on CSFs was specially prepared by Christine Bullen who was part of the team at MIT that developed the concept.

Critical success factors exist at all levels—from the industry as a whole to the individual. They are a useful way to determine what steps are important for a business to improve productivity. CSFs are typically determined by interviewing people at all levels of the organization. The interview process itself adds value because it forces individuals to make their CSFs explicit. Once they are known, people tend to work toward achieving the goals implied.

INTRODUCTION

Enhancing productivity is the fundamental goal of management in today's global, streamlined, complex marketplace. Attempts to achieve improved productivity masquerade as a variety of corporate improvement programs "du jour": business process reengineering, total quality management, the development of the information technology infrastructure. Unfortunately, many of these programs have ultimately yielded little or nothing. For example, a study recently completed by CSC Index surveyed 497 large companies in the United States and 124 in Europe about reengineering. The report found that 85% of reengineering projects ended in failure (CSC Index, 1994).

The issue at the heart of these failures is not that the managers made poor decisions or that the consultants were wrong. The issue is that managers need help focusing on what is critical to manage in their environments. At the same time, resources and energy which are applied to activities that managers can control and improve will result in successes. Corporate life in today's business world is a complex mixing of scarce resources, unrealistic deadlines, instantaneous communication, demanding customers, and fierce competition. It is no wonder that managers may be confused about where to concentrate their attention.

The concept of Critical Success Factors (CSFs) is an approach to help managers work through the fog of complexity and identify those few areas that are critical to the survival of their roles and organizations (Bullen and Rockart, 1981).

HISTORY

The concept of critical success factors was introduced in 1979 (Rockart, 1979). Since that time numerous organizations have adopted the approach as a strategic planning framework for both information systems and corporate business planning. The CSF method directs managers to determine "those things which must go right in order to succeed in achieving [their] goals and objectives" (Rockart, 1979). The ultimate value that the CSF method brings is the ability to focus management attention on the tasks and activities that need to be done well to achieve success.

Examples of CSFs can be found on many levels. Individuals have critical success factors relating to their roles and styles. Departments and divisions of organizations identify CSFs resulting from their missions, products/services, customers, and the personal CSFs of their managers. Industry CSFs derive from their products/services, customers, and competing organizations. For example, the director of a major urban hospital identified CSFs covering a broad range of topics from "staying on top of the developing technology," to "the support of my spouse and family." In another interview, the vice president of R&D saw "partnership between R&D and Marketing" as a CSF for his division. A study of the grocery industry pointed out two CSFs for individual supermarkets: "the right mix of products for the ethnic groups of customers," and "no lost sales due to stock-outs." These examples demonstrate the range of activities and the complexity of ideas that the critical success factor method can reveal.

The method itself centers on an interview technique where managers move through the stages of 1. listing their goals and objectives; 2. identifying the critical success factors necessary to achieve the goals and objectives; and 3. suggesting ways in which the CSFs are to be measured. Skilled interviewers take managers through this process in about 1.5 hours using a toolkit of tested questions to elicit a full range of CSFs (Bullen and Rockart, 1981).

Prior to this concept of focusing management attention on the most critical activities, managers often tried to manage their organizations using traditional reports based on accounting information. While these were useful documents in reporting where a firm stood using generally accepted numerical indicators, they did not help managers gauge those factors they could actively manage and control. Unfortunately, this accounting-based information was the only form of management report available in organizations. Critical success factors changed that. The method helped to initiate a dialog between line management and technology management that viewed the information provided to managers in a new light: as information that facilitated the tasks of management. The use of CSFs to determine what should be reported and how it should be measured revolutionized computer-based information reporting systems. As a result the phrase has become widely used and the concept well understood and applied in every industry—organizations of any size, in locations around the globe (Forster and Rockart, 1989). To a certain extent, CSFs helped to usher in the age of executive information systems by demonstrating how managers, when provided with computer-based relevant information, can use it to improve their decision making (Rockart and Bullen, 1986).

HELPING INDIVIDUAL MANAGERS

The initial success that came with the use of critical success factors to design reporting systems led to their application in other ways. The CSF method helps individual managers to accomplish two vital tasks:

- CSFs help managers to *think through* the vast number of activities and issues they must perform, and to reduce that number to a manageable few.
- CSFs help managers *set priorities* on activities in order to focus on the most critical ones.

For example, a recent study at the MIT Center for Information Systems Research looked at the critical success factors for chief information officers (Rockart, 1994). This research was carried out through numerous interviews of CIOs in major corporations. The resulting CSFs present a clear picture of the challenges facing the information technology organization in today's firm. They also indicate where the CIO must concentrate effort to be productive, and therefore, succeed in carrying out the responsibilities of this role in today's organizational environment.

The following five critical success factors were found in this survey:

1. Build empowering relationships
2. Understand the business
3. Increase cost effectiveness
4. Challenge the status quo
5. Build centralized/decentralized organizations

The example of these five critical success factors for chief information officers demonstrates the power of the critical success factor method to help the individual manager focus on critical activities, and to set priorities for action. This example also demonstrates that with as few as five CSFs it is possible to capture the key strategic concerns of a manager. Moreover, with as few as five CSFs, the manager still has an enormous undertaking to manage these five areas well and realize success in each of them. These five factors will now be discussed in turn.

Five Critical Success Factors

BUILD EMPOWERING RELATIONSHIPS

The information systems function must act in partnership with the business function to create the strategic plan for the future of the firm. Business strategy and technology strategy must be aligned to succeed in today's global, highly competitive marketplace. Information technology is an integral part of the product and services of many organizations. Information technology allows organizations to change their strategies—even their basic businesses—and succeed. It is no longer a service that is linked after the fact to support a business decision.

For example, American Airlines found that their Sabre reservation system is more profitable than their airline business. Top management at American understands this fact and is beginning to view its primary business as that of a reservation system and not as that of an airline. The computers are more strategic to American Airlines than the airplanes!

In other industries, the capabilities of information systems surfaced as potential competitive weapons. MCI was able to win 10% of the long distance telephone service market with its "Friends and Family" program, based on the design of their billing system. Because AT&T traditionally left the billing for long distance service to the local telephone providers, they do not have the information systems capability to directly challenge this customer benefit. Instead, AT&T uses other incentives to meet this competition head on, such as across the board discounts on all long distance calls. This billing process turned out to be a strategic weapon and has helped MCI to maintain their competitive position in the market.

To leverage successfully the strategic role that information technology has today, the chief information officer must focus on building empowering relationships with line business areas of the organization.

UNDERSTAND THE BUSINESS

In the early days of information systems work in companies, computer systems were seen as a support service, not unlike heat, air conditioning, and electricity. This view was reinforced by the location of many computer rooms in the basements of office buildings. Most information systems professionals were also viewed as non-essential, highly specialized technologists who had no need to understand the business. However, as the strategic role of technology became more and more clear, the business role of information technology became more critical. Business acumen is now an essential characteristic that information technology professionals must have. In some organizations the chief information officer is chosen based on management skill, and technology experience is not required.

INCREASE COST EFFECTIVENESS

While this focus on technology as a strategic weapon was developing, organizations experienced a tight economy, resulting in a parallel focus on cost reduction and improved effectiveness. The introduction of the acceptability of "outsourcing" to deal with inefficient and ineffective technology areas fostered a healthy respect for well-honed information technology organizations. Today's chief information officer is expected to captain an organization that is of strategic importance to the survival of the firm, and is "lean and mean" at the same time.

CHALLENGE THE STATUS QUO

The charge of "business process reengineering" to "obliterate" old ways of doing things and create efficient, technology-driven, interorganizational processes (Hammer, 1990) has been co-led by the chief information officer in many organizations. S/he has been in a unique position to visualize how information technology can enhance and upgrade new approaches, create communication links between separate organizations, and revolutionize approaches to old problems. With empowered relationships in place, the CIO can challenge and renew the hidebound ways of the firm.

BUILD CENTRALIZED/DECENTRALIZED ORGANIZATIONS

The chief information officer must build global organizations that act locally, set universal standards so that local decision making can take place, and manage the communications infrastructure so that distributed processing can succeed. In today's corporation, information technology areas must act in both centralized and decentralized ways to support the strategic needs of the firm.

HELPING ORGANIZATIONS AND INDUSTRIES

In the same way that individual managers can think through their personal CSFs, groups of managers can think through their organization's CSFs, and groups of organizations can think through their industry's CSFs.

Once the top level managers in an organization have identified their personal sets of CSFs, these sets can be compared to see where there are overlapping concerns, related issues, and obvious conflicts. This process can be enlightening and often surprising to a group of managers who have worked together for years and assumed that they knew each other extremely well. For example, in one high technology engineering company, the CEO and his top management team met to review the critical success factor process that they had been undertaking. The CEO agreed to kick off the meeting by presenting his CSF list for discussion. The top level of managers were in an uproar when they saw the CEO's critical success factors. The CEO was confused by this reaction. What his management team had seen was that the CEO's concerns, as reflected in his CSFs, were very operational: his CSFs were almost identical to theirs! What became clear to everyone was that the CEO, who was also the entrepreneur who founded the firm, was reluctant to give up his hands-on control of the day-to-day activities. Apparently the top management team had been hampered by his controlling style without explicitly recognizing it. The CSF process, by focusing on what each person viewed as critical, revealed the conflict and made it possible for this management team to move ahead in a productive manner.

Determining industry critical success factors can be a useful process for helping organizations in the industry better understand their competition and the underlying assumptions in their products and in their consumer environments. This information can lead to strategic decisions about product development and marketing.

For example, the American automobile industry found itself in an awkward position in 1973 when it was caught with luxurious, gas-guzzling car models at the time of the Arab oil embargo. This accident of history launched the Japanese car success in the United States. This crisis illuminated an unstated, and probably unrecognized, assumption among U.S. car companies that the gasoline supply would never be threatened. Until the oil embargo, two of the CSFs for U.S. car companies were, no doubt, "style" and "engine power," reflecting their firm belief that these features sold their products in America.

Overnight, the oil embargo altered the desires of the car-buying public. Consumers did not want to be at the mercy of foreign powers controlling their ability to drive their cars. The automobile industry's critical success factors altered radically to "fuel-efficient cars" and "ugly is OK!" In the 1990's fuel efficiency persists as a CSF, but has been eclipsed by "safety" for most consumers. While the U.S. car makers could not have anticipated the oil embargo, if they had been going through the exercise of looking at what was critical in their industry for them to remain successful, they would have been in a better position to identify some of these important issues, and to build contingency plans.

INFORMATION SYSTEMS PRIORITIES

Another way in which the critical success factor method is used is in helping organizations determine their information systems priorities. This application has grown out of the original report-oriented use of CSFs. As managers identify their critical activities, the information systems required to support those activities can be determined. As the CSFs are put into priority order, these priorities can be applied in the information systems planning process to help determine which information systems (IS) projects should have precedence.

Too often the IS plan for systems development is determined by technology-oriented factors (such as sophistication of existing systems, desired skill acquisition, or "interesting" advanced technology) and politically oriented factors (such as the perceived power of the area requesting the project, who has funding available, or who "owns" the information technology [IT] organization). When information and communications technology are viewed as strategic tools for enhancing the organization's ability to compete and be productive, the focus on systems to support critical success factors can determine the priorities. This shift on factors determining what systems will be built is essential to improving productivity organizationwide.

Productivity Issues in Knowledge Work

Productivity is one of the traditional measures of an organization's effectiveness. It has also been management's foremost nemesis since organizational work has become more "knowledge work"-oriented, and therefore more difficult to see and measure its outputs. As early as 1982 researchers acknowledged that measuring output in knowledge work was the wrong approach (Packer, 1982). Packer proposed a technique using perceptual mapping to measure "outcomes." The creativity of this research was that it underscored the importance of defining productivity *within the context* of the organization trying to measure it. The notion that productivity is context-specific is one that most organizations are still struggling to understand.

Generalized notions of productivity make little sense in a knowledge work environment. For example, for many years researchers have been trying to assign measurable factors to secretarial work. Following the tradition of those who felt "lines of code" was an accurate measure of a programmer's skill (Brooks, 1975), "pages typed" and typing speed were thought to be appropriate measures of a secretary's value. However, in interviewing managers on the value of their secretaries, these measures are rarely mentioned (Bullen, Bennett, and Carlson, 1982). Rather, responses are more along the lines of how well s/he manages the office, handles telephone inquiries, manipulates the scheduling, and interfaces with the world outside the office door. While "pages typed" is easy to see and count as *output*, the relationship that a secretary creates between the manager and the outside world is an *outcome* that, while priceless, is difficult to measure.

It should be abundantly evident that if we have difficulty measuring the productivity of a secretary (one of the first levels of knowledge work), we are hopelessly lost in measuring the productivity of top management in an organization. However, the Packer approach proposed some useful guidelines. He suggested that if you interviewed knowledge workers about their activities, they could define what "productivity" meant for them. For example, if you interviewed product designers and discovered that they listed activities such as "ability to reuse parts of designs" and "quality of the final design" you could begin to construct a picture of productivity in their environment. And more importantly, you could begin to investigate ways in which to facilitate or improve that productivity. In this example, computer-aided design (CAD) software could be introduced to help improve these factors. An important addition to Packer's approach would be a way to determine from knowledge workers what their critical activities are.

Sherwood (1994) distinguishes among three important factors when measuring outputs in service industries (which are primarily the outputs of knowledge workers). He proposes that productivity should be measured using two approaches:

> The first approach is to measure the service transaction unit itself and make appropriate adjustments for changes in quality. . . . The second approach is to measure the outcomes resulting from the performance of the service in question.

He also proposes that changes in quality actually consist of two aspects: product innovations and process innovations. An example for illustrating how these three aspects work together can be found in the doctor-patient relationship. A patient's successful "outcome," when ill, is a cure. The ability of the doctor to bring about this outcome is based on both the process she uses to diagnose the patient (the process innovation) and the effectiveness of the drug (the product innovation) in curing what ails the patient. A logical conclusion from this work is that methods for understanding how to bring about process and product innovations will contribute to the ultimate goal of improving outcomes and, therefore, productivity in the service sector.

In Section 1.1 of this book, Peter Drucker defines the basic productivity issues:

> The single greatest challenge facing managers in the developed countries of the world is to raise the productivity of knowledge and service workers. This challenge, which will dominate the management agenda for the next several decades, will ultimately determine the competitive performance of companies.

Drucker argues that to improve productivity, organizations need to answer one important question: "What is the task?" Then the organizations must focus management attention on achieving the key task. The powerful example of the nursing shortage that he uses helps to make this clear. If nurses were not deluged with paperwork for various insurers, hospital administration, Medicare, Medicaid, and malpractice suit prevention, they would be able to carry out the tasks that they were trained for—caring for patients. Drucker's bottom line is that if organizations concentrated the work on the key task, then improved knowledge worker productivity would result. The key to Drucker's approach is understanding what the task or tasks are and focusing the organization on those tasks.

INTERSECTION OF CSFS AND PRODUCTIVITY

The critical success factor method has been proven over the course of fifteen years to achieve its goal of getting managers to identify those activities that must go well in order for them to succeed. That is, CSFs are being used to enumerate productivity factors in knowledge-work environments.

The context-specific nature of productivity requires that an understanding of what and how a knowledge worker does to achieve his/her goals and objectives is essential to any understanding of what it means to be productive in that environment. Once the nature of productivity is understood, measuring it becomes a much simpler task. Using the CSF interview approach helps to focus the knowledge worker's thinking on what is essential to improve productivity.

Sherwood's work directs us toward a method for understanding what product and process improvements are necessary. When a manager uses the critical factor approach, s/he can evaluate the processes being carried out with a goal of enhancing and streamlining them to focus on the critical factors. As demonstrated in the U.S. car maker example, when CSFs are established for an industry, they can contribute to a better understanding of product improvements.

Drucker's advice to determine what tasks an organization wants to achieve is another way of describing critical success factors. Presumably an organization that intends to remain in business wants to achieve those tasks that are and will be critical to its success in its industry.

EXAMPLES

The critical success factor method can be used to understand the nature of productivity. When the management of an organization pursues the notion of understanding and improving productivity, a clear view of what it takes to achieve goals and objectives can guide the process. For example, in the course of an interview, the director of a software development firm identified the "quality of the product" as a top CSF for his company. Immediately, a look of consternation came across his face as he said, "Well, that's not exactly true. It's not the quality itself, but rather what my customers think of the quality that really counts." Therefore, which is a "more productive" use of this top executive's time: pursuing the improvement of some abstract notion of product quality, or spending time with his key customers to build their confidence in his product?

The CEO of a start-up engineering firm identifies the morale of her professional engineers as a critical success factor. In her environment, she needs uniquely trained and experienced engineers. She spends a great deal of time searching for more professional engineers rather than managing the ones she has. Her assumption is that her employees are self-starters who don't need "hand holding." However, she also notices that the turnover rate is somewhat high. The CEO justifies this because of competition from other start-ups in the environment. Should she use her "productive time" to nurture the current employees or to seek out new ones?

The executive vice president of a consulting firm wants to introduce groupware technology to support the sharing of information across departments worldwide within the firm. He thinks the tasks he needs to focus on to get this done include choosing the right technology, getting it installed and fully operational as quickly as possible, and accomplishing all this for the least amount of money. Since his organization is generally computer-illiterate, he assigns his best technical people to review and choose the technology. When asked to consider the critical success factors necessary to make this technology roll out work, he includes the need for the employees of the consulting firm to "use and benefit from this technology." At this point he realizes that he has been putting all his time and considering himself quite productive in dealing with the technology aspects of the task. In fact, it is the organizational aspects that will determine whether his idea is a success. His productivity with respect to the time he has spent on this project will only be positive if the technology is accepted by the employees. He needs to reallocate his time and priorities to focus on understanding the information sharing needs and culture of his organization.

SUMMARY

In these examples, critical success factors help focus attention on the areas that need to be managed to increase knowledge-worker productivity. It is useless to the overall productivity of the organization to increase work on the wrong activities. Although this approach may create an *apparent* increase in productivity for the tasks being worked on, it will not truly have an impact on adding value to the bottom line of the organization. The critical success factor method identifies the *right* areas where knowledge-worker attention should be focused. It leads directly to what needs to be done and how to measure progress.

The recipe for increasing knowledge-worker productivity is straightforward. Follow these steps:

1. understand the context-specific nature of productivity;
2. understand the desired tasks;
3. focus on process improvement;
4. focus on product improvement;
5. understand the need to measure outcomes and not outputs;
6. use critical success factors to help managers think through tasks, processes, products, and outcomes,

and this goal will be achieved.

REFERENCES

Brooks, Frederick P., Jr. *The Mythical Man Month*. Reading, MA: Addison Wesley, 1975.

Bullen, C. V., Bennett, J. L., and Carlson, E. D. "A Case Study of Office Workstation Use." *IBM Systems Journal* 21, no. 3 (1982).

Bullen, C. V. and Rockart, J. F. "A Primer on Critical Success Factors," MIT Center for Information Systems Research Working Paper #69, June 1981.

CSC Index, "Reengineering Report 1994," *The Economist* (July 2, 1994).

Forster, N. and Rockart, J. F. "Critical Success Factors: An Annotated Bibliography," MIT Center for Information Systems Research Working Paper #191, June 1989.

Hammer, Michael. "Reengineering Work: Don't Automate, Obliterate." *Harvard Business Review* (July–August 1990).

Packer, Michael B. "Measuring Productivity in R&D Groups." *The MIT Report* X, no. 11 (November 1982).

Rockart, John F. Presentation at the MIT Sloan School Center for Information Systems Research Summer Session, June 14, 1994.

Rockart, John F. "Chief Executives Define Their Own Data Needs." *Harvard Business Review* (1979).

Rockart, J. F. and Bullen, C. V., eds. *The Rise of Managerial Computing*. Homewood, IL: Dow Jones-Irwin, 1986.

Sherwood, Mark K. "Difficulties in the Measurement of Service Outputs," *Monthly Labor Review* 117, no. 3 (March 1994): 11.

QUESTIONS

1. What are critical success factors? Give some examples of critical success factors for education in your degree program.
2. How can critical success factors improve productivity?
3. Are critical success factors more useful for improving productivity in offices than in factories? Take a position and defend your viewpoint. Give examples.

2 The New Technologies

THIS chapter introduces the new technologies that transform office and factory work. These technologies are used in two ways:

- To gain leverage by making existing processes more efficient.
- To create more effective ways of achieving organizational goals.

The computer and communications technologies that achieve these goals can provide many opportunities to substitute capital for labor to improve productivity. Yet their potential is greater than that. By processing many of the routine tasks, by providing memory of what was done, by organizing work, the computer frees people to work at the next level of knowledge where better ideas pay off in competitive advantage.

This chapter introduces the new technologies. It begins with three surveys that describe computer, telecommunications, and leading-edge technologies. It then presents four specific technologies: imaging, bar coding, virtual reality, and Internet. Each section is written for the non-technical user rather than the specialist. They stress how these technologies are applied and how they should be managed.

2.1 The New Computer Technologies

PAUL GRAY

This book is about the impact of the new office and factory technologies on productivity. Thus, you, the reader, need to know about these new technologies. What are they? What can they do? What are their limits? How do they impact the way people work? How do they affect the organization? This section describes the major technologies and what they do. As you read through the book, you will find references to these technologies. Refer back to this section whenever you find a term you don't understand.

TO understand the managerial implication of using technology to increase the productivity of white collar and factory workers, it is first necessary to know what the new technologies do. The initial applications were to improving the routine typing tasks associated with secretarial work. As the technology developed, it was applied to all levels of office work, from the simple creation and dissemination of routine letters and memoranda, to meetings among senior executives, to cooperative design.

The automation of office and factory work is the result of the merging of two technologies:

- The computer
- Communications

to create entirely new capabilities. In this article we discuss the following examples of the use of the new technologies:

- Word processing
- Desktop publishing
- Electronic mail
- Voice mail
- Facsimile, scanners
- Executive work stations
- Data bases
- Information retrieval systems
- Natural language and expert systems
- Telecommuting
- Computer conferences
- Video conferences
- Decision rooms
- Local area networks

We also briefly discuss the human factor considerations. This list is not exhaustive. A number of other computer technologies, particularly those used in manufacturing (such as robotics) are discussed in other chapters of this book.

The new technologies are in place to varying extents; word processing is nearing universality whereas telecommuting and video conferencing, for example, are still in their infancies.

In the discussion that follows, the focus will be on the functions performed rather than the specific implementations. The range of features available with these technologies will be described; the availability of particular features depends on the system realization. For many of the services, versions are available on computers ranging from mainframes to microcomputers. The telecommunications technology ranges from hard-wiring within a building, to the public telephone system, to satellite systems.

WORD PROCESSING

Word processing is a term introduced in the mid–1960s to refer to the use of on-line, computer-based text preparation systems. In such systems, a text file is entered into the computer and resides on a storage medium, typically disk. Conceptually, word processing works like audio or video tape in that revisions are made by retrieving the original text, making changes through deletions and insertions, and then storing the corrected version. In this way, the latest version is always available. Printed (often called "hard") copy is obtained by forwarding the text to a printer.

As the short description above implies, word processing is a software program for editing text. In addition to allowing text to be changed, most word processing software provides control over output format (e.g., margins, spacing, centering, page numbers, headers and footers), the ability to move text for "cut and paste," and global delete and replace (e.g., change 'colour' to 'color' every time it occurs in the text). Specialized capabilities include individualization of text (e.g., a form letter with the recipient's name referred to in the text), spellers to assist in locating and correcting typographical errors, automatic hyphenation, thesaurus, alphabetization of lists, creation of tables of contents and indexes, outlining, and merging of address lists with standard text for mass mailings.

High-quality printers, also computer-controlled and using a variety of technologies, including the laser, are available that allow word processing systems to create documents with a variety of type fonts and type sizes.

The principal economies achieved by word processing result from the ability to correct and change existing text and the ability to save text on electronic media rather than in file cabinets. Because word processing software is available on home computers and on microcomputers in the office, many professional and office workers create their own text and do not require the use of typists for routine work.

At the most sophisticated levels of word processing technology, text files created on one word processor can be sent to another word processor by telecommunications. This is one form of electronic mail. Software is also available that allows comments to be placed in the text with the name of the editor shown. These techniques lead to improved joint authoring of documents.

DESKTOP PUBLISHING

Desktop publishing is a natural extension of word processing. Desktop publishing refers to the creation and dissemination of documents using computers rather than printing presses. Most companies with word processing capabilities can also do desktop publishing.

Some additional equipment and software is required. The routine memos created on personal computers do not have the resolution or the quality associated with professional printing. To improve quality, companies invest in high resolution and color printers, and in specialized software languages that capture the additional resolution available. They also invest in the binding and other physical capabilities needed to create documents.

A particularly useful technology for desktop publishing is optical scanning and character recognition. These technologies allow the user to copy text and drawings directly into the computer from existing manuscripts and facsimiles without having to rekey or redraw. Scanning makes electronic copies. Optical character recognition creates text (with high accuracy) that can be manipulated on the user's word processor.

Photographs can be edited using a process called "morphing." In morphing, existing drawings can be modified in both color and content. While morphing makes it possible to improve both drawings and photographs, and to highlight specific features of a product, it carries with it the potential of fraud with graphics.

Many commercial publishers now use desktop publishing techniques to create documents. For example, most books and journals are now produced directly from text supplied on disks rather than being retyped.

ELECTRONIC MAIL

In electronic mail systems, the computer's memory is used as a message handler. As in standard mail systems, the originator creates a message, memo, letter, or report and addresses it. The text editing/word processing capabilities of the computer are used to produce the message. Mail can be sent to an individual or to a distribution list. The computer attaches the mail to each recipient's file and notifies them that they have new mail. In the best systems, notification is effectively on-line for people currently logged in. Others are notified when they next log in.

When called up on a computer screen, the message typically shows the name of the sender, the addressees, and the time it was sent. The recipient can create printed copy, keep a copy in his personal file, or delete the message. Most systems also provide reply and forwarding features. That is, the recipient can create an answer that is sent back to the originator and s/he can also forward a copy of the message to other people in the organization (with or without the reply).

Electronic mail provides extremely rapid communication both within a site and among remote sites in an organization. It is particularly useful for project coordination, especially when projects are changing rapidly and facing severe deadlines. When used in conjunction with software that keeps track of calendars, electronic mail has been found to be an effective vehicle for meshing calendars to establish meetings. In highly security-conscious organizations, only people tied to a particular computer or computer network can be reached. However, various commercial and non-profit electronic mail services are available that provide much greater connectivity. For example, the Internet connects people in a large number of networks world wide. Through "gateways," different networks communicate with one another. Thus, a person on one network can communicate with people on all the connected networks.

Electronic mail is effective when participants (or their assistants) either keep their terminal on-line or check regularly for messages. In the case of commercial services, communications and connection costs are incurred whenever an inquiry is made as to whether there is mail or not. The chief hazard of electronic mail is the ability to create and proliferate electronic junk mail within the organization. Administrative procedures are usually sufficient to control this problem.

Even if junk mail is controlled, some people receive too many messages. Systems are available that allow users to sort messages by their senders and their contents. Thus, messages from a supervisor can be moved to the top of the list to be seen first and routine announcements can be moved to the bottom.

VOICE MAILBOXES

Voice mailboxes are a form of electronic mail that transmits voice rather than text messages. They were developed by a number of communications firms as a solution to the "telephone tag" problem; that is, the problem that a number of telephone calls (stretching over extensive periods of time) have to be placed before two people can talk to one another.

From the point of view of the caller, the computer acts like a versatile telephone answering machine. A major advantage is that a message can be sent to many people as well as to a single person. The computer acts in a store and forward mode. It digitizes the voice message and forwards it to each addressee. When an individual dials the system, either to send or receive messages, they are told whether they have new messages and are able to listen to them if they so choose. As with electronic mail, replies can be sent, and messages can be deleted, kept for future reference, or forwarded to others. Hard copy is, of course, not available. Some systems also provide a "forced delivery" capability. That is, the recipient's phone rings at regular intervals until the message is received.

Experience with voice mailboxes indicates that efficiency gains result from reduced telephone tag and from shorter messages; people tend to eliminate small talk and extraneous information when speaking to a machine.

Typical applications of voice mailboxes have included notifying groups of people of changes (e.g., part-timers, extras, flight crews); quick reactions in decision making; communications across time zones; quelling rumors; and setting up meetings.

Voice mailbox systems provide a way of using the computer without requiring typing and are preferred by people who operate principally in a verbal rather than a written mode. Their effectiveness depends on inculcating the habit of regular use for sending and retrieving messages so that the speed advantages are realized.

Voice mailboxes are basically a large organization device since they require investment in a special-purpose computer and extensive associated communications equipment. However, some suppliers provide public voice mailbox service that can be used by small- and medium-sized organizations.

FACSIMILES, SCANNERS, AND OPTICAL CHARACTER READERS

A major cost in both time and money with word processing and electronic mail systems is keystroking to enter information. Voice mailboxes are one way to avoid keystroking. Facsimiles, graphical scanners, and optical character readers (OCRs) are others. In each case, images on a sheet of paper are read electronically and converted into digital form. The three units differ in purpose and use.

- Facsimiles digitize information for transmission over ordinary telephone lines and reconstitute it at the far end in its original form.
- Graphical scanners digitize the image and store it in the computer as a picture.
- OCRs scan text and convert it into standard code such as ASCII.

Although these technologies have been available for many years, only recently have the costs decreased and the technologies matured to make them routine office automation applications. Facsimiles offer the opportunity to send handwritten messages and signatures without intermediaries. Scanners and OCRs allow the direct entry of document information into the computer without rekeying. Combining facsimile with a scanner permits entering typed text directly into a remote computer. (See the previous discussion of desktop publishing for an application of scanners and OCRs.)

EXECUTIVE WORK STATIONS

Just as word processing represents the augmentation of secretarial capabilities with computers, executive work stations are designed to assist middle and upper managers in firms by providing them with computer capabilities for handling text and calculations. Executive work stations are essentially microcomputers with software and devices to aid in performing managerial functions.

Executive work stations are usually designed to require a minimum of typing for communicating with the computer. The rationale for this approach is that most executives do not know how to type well or prefer not to type. Two typical devices are the touch screen and the "mouse."

A touch screen, as its name implies, allows the user to touch the screen with his or her finger to select from among the alternatives shown. Typical alternatives include choice of information, selection of a value to be assigned, or an operating command such as NEXT SCREEN. Touch screens require careful preparation of alternatives, typically in menu form. Furthermore, the number of choices that can be made available are limited because there must be enough space between choices to accommodate people with thick fingers.

The mouse is an external device that is used to point to items on the screen. To the user, a mouse is a small box with one or more push buttons to be moved across a flat surface. The motion of the mouse is tracked (by converting its change of position electrically through following the motion of a ball bearing) by a pointer or indicator (i.e., cursor) on the screen. When the cursor is at the desired location on the screen, a button on the mouse is pressed to select the item at that location.

The mouse becomes particularly powerful when coupled with icons on the screen. The icon is a graphical representation of a file or function available on the computer (e.g., a document, a collection of files, a printer, a communications device, an electronic "in box" or "out box"). When an icon is selected by use of the mouse, it becomes available for manipulation. For example, if the icon represents a document, the document can be "opened" to make its text appear on the screen, copied, moved to another location, etc. If a document is moved on top of an icon representing a printer, a sequence of steps leading to obtaining a printed copy are initiated. If the document is moved on top of an item representing an out box, it can be mailed. A screen containing icons is, in effect, a desktop. Commands such as open, close, copy, and move are available either as function keys on a keyboard or as menu items on screen.

Development of the mouse and icon concept began in the 1960s in California at Stanford Research Institute by a team led by Doug Engelbart. Commercial development of executive work stations began in the late 1970s. Among the initial work stations on the market using the mouse and icon concept was the STAR by Xerox, followed several years later by Lisa and Macintosh from Apple Computer, and by Windows from Microsoft.

The initial implementations of work stations, while quite sophisticated, tended to have slow response times and to be relatively expensive. Product improvement increased speed and reduced cost. Executive work stations are available in both stand-alone and network configurations. The network configuration is designed to increase both effectiveness and efficiency within a firm by allowing the stations to communicate with one another and with other computer-based systems, to share computer storage, and to access a common printer. The network interconnection typically is through a local area network.

DATA BASES AND INFORMATION RETRIEVAL

Data bases are collections of information on a particular subject. You may have a data base of your friends' phone numbers. The conglomerate TRW maintains a database on the creditworthiness of most middle- and upper-income bracket people in the United States.

Data base technology started out with "flat files." In a flat file, data are stored one item after another according to some key. For example, your phone list may be organized alphabetically by last name. Because of the large amount of capital invested in the storage and retrieval of information from data bases, considerable work was done to get beyond the simple flat file. One such innovation is the relational data base. In a relational data base, information is organized as two-dimensional tables. Here the rows might represent the names and the columns represent multiple pieces of information about the individual such as address, sex, phone number, credit rating, etc. The idea is that it is easier to ask questions when data are grouped together. A specialized language, called Structured Query Language (SQL) is available to retrieve information from two-dimensional relational data bases. The next stage of data base development was the creation of multidimensional data bases which allow individuals to arrange data in a variety of ways. For example, you might want to store information about product, market, month, and units and be able to retrieve these four dimensions in any order without worrying, as you do in relational data bases, as to which pieces of information were paired in a table.

The ability of terminals in the office to communicate with remotely located computers led to the establishment of services that provide access to large public data bases. These data bases provide both current information (e.g., current stock prices, news reports, legal decisions) and library information. Some of these services also provide the capability for ordering goods and performing various transactions such as bank transfers. The growth of online transaction services has been relatively slow. Attempts to move these services into the home using television receivers (usually referred to as videotext) have met with low success in the United States (where they are privately sponsored) but have proven popular in France, where they are offered by the government's Post, Telegraph, and Telephone (PT&T) service.

Two emerging technologies in information retrieval are CD-ROM (compact disk, read-only memory) and high capacity storage devices. In the former, a compact disk is used to store large amounts of fixed data (e.g., an encyclopedia). The CD-ROM makes it possible to install very large software packages from a single compact disk rather than requiring ten or twenty floppy disks.

SOFTWARE INNOVATIONS

Innovative software packages allow office and factory workers to perform relatively complex model-based calculations without knowledge of computer programming. These packages range from those that create fairly simple "spreadsheets," to full-service, fourth-generation computer languages that permit expressing mathematical relations in nearly natural language form. The packages are available both on mainframes and on stand-alone microcomputers and are intended for on-line use. One of their main advantages is the ability to do "what if" calculations; that is, to alter one or more assumptions in the model and determine the effect of changes.

A spreadsheet is simply a two-dimensional matrix presentation of results, where each column typically represents one time period and each row represents the values of a variable. Intended primarily for financial calculations, the level of sophistication in spreadsheet packages has increased rapidly. Features available include various built-in mathematical, financial, and statistical functions; communication between the spreadsheet package and company data bases; graphical presentation of data in color; and the ability to treat text as output and as input, so that the spreadsheet can be integrated with word processing.

In the fourth generation of computer languages, such as IMPROV from Lotus, it is possible to express mathematical relations in direct, full-word format; for example:

$$\text{Sales} = \text{Market} \times \text{Market share}$$

These languages are typically non-procedural; that is, it is not necessary to define a quantity before it is used as long as it is defined at some point in the program. The advantage of non-procedural languages is that they are easier to use and can serve as a form of communications between the model builder and the user. The disadvantage is that they require more memory and have slower execution speeds than conventional procedural languages. Non-procedural languages typically produce spreadsheet outputs if no formatting instructions are given, but can be instructed to provide customized formatted reports. They contain extensive financial, statistical, and mathematical functions. Many are provided with built-in fixed time interval simulation options that allow risk analyses to be performed. Their "what if" capabilities include sensitivity analysis and goal seeking. These languages also permit consolidation of multiple reports into a single report.

NATURAL LANGUAGE SOFTWARE AND EXPERT SYSTEMS

As a spinoff from research in artificial intelligence, software is being developed that allows users to interact with the computer in nearly natural language, that is, in their ordinary way of speaking. The non-procedural languages discussed in the preceding passage were early harbingers of natural languages. Although implementations are still relatively crude and restrict the user to relatively narrow vocabularies that the computer can understand, the quality of products directed toward natural language input continues to improve.

Expert systems refer to software that

1. Contains an extensive set of information on a given subject (referred to as a knowledge base).
2. Can perform logical operations on the knowledge base interactively.

The knowledge base is obtained from experts in the subject. In a typical expert system the software asks the user questions designed to reduce the set of possible answers and, using built-in if-then-else rules, recommends a course of action. Expert systems have proven to be most useful for diagnosis situations such as determining infections.

TELECOMMUTING

For highly computer-intensive organizations, (e.g., insurance, banking, finance, administrative units of manufacturing organizations, computer support groups such as programmers) the availability of communications makes it no longer necessary to put the workers who interact with the computer and the computer itself in the same place. Thus, it is possible to place a central computer in one place and to put offices in multiple locations either within a metropolitan area or across the country. Telecommuting (also called "telework" and "flexiplace") is the concept of substituting communications for transportation by moving work to locations close to where workers live rather than having the workers commute each day to the computer location. Telecommuting can involve either work at home or at an office located close to the worker's home.

The advantages of telecommuting to a company include a reduced need for building large office buildings in the expensive real estate of downtown areas, access to new labor markets where transportation is a problem, the ability to time share jobs and equipment, greater employee morale, greater retention, reduced training, and reduced wages in outlying areas compared to downtown locations. From an employee point of view, the reduced need for commuting and the flexibility of hours are among the advantages. In a telecommuting environment, changing assignments involves changing the data set being used rather than changing office locations. The disadvantages of telecommuting include the physical separation of workers from their supervisors and the loss of an individual's visibility that could impede their vertical mobility in the organization. In a telecommuting environment, the supervision function can be divided between local supervision for punctuality, work habits, supplies, etc. and remote supervision for the technical content of the work. The computer/communications link, however, can be used for providing the technical supervision.

Surveys of telecommuters who work at home indicate that they have increased productivity, that remote supervision is adequate, and that there is little feeling of loss of socialization. However, to be successful in the work-at-home mode, the telecommuter must have regular, self-scheduled work habits.

ELECTRONIC MEETINGS

A typical conference is a meeting involving a group of people of near equal status in an organization for the purpose of exchanging information or reaching a decision. A conference is an intellectual activity, involving structured communication whose outputs lead to consensus and action. For conferences involving people whose normal work locations are geographically separate, conferencing methods include:

- Face-to-face
- Speakerphone
- Computer-based
- Video

In a face-to-face conference, people travel to the same location and all participants are in the same room. The major problems in face-to-face conferences are scheduling and cost. Tangible costs include transportation, hotel, food, and salary. Opportunity and intangible costs result from being away from the office and from the catch-up required upon return.

Research findings in Britain, Sweden, and the United States indicate that communications can be substituted for face-to-face meetings to a large extent. Initial contacts should be made face-to-face so that people can obtain visual images and develop mutual trust and respect. There also appears to be a periodic need (every 6 to 12 months) to refresh face-to-face contact. In the interim, electronic meetings serve as a viable substitute.

In speakerphone conferences, people at each location gather in a conference room. A telephone connection is used, with a loudspeaker attachment to each phone. The advantage of speakerphone meetings is that communications costs are minimized because only telephone bandwidths are required. The difficulties in such a meeting are identifying who is speaking and developing protocols for determining who is to be recognized next. Although relatively successful for two locations, speakerphone conferences are much more difficult for multiple locations. As a graphics supplement to speakerphone conferences, "electronic blackboards" have been developed that allow transmitting drawings as they are being made over a telephone circuit. The technique involves using pressure-sensitive digitized pads. Speakerphone conferences are sometimes augmented with a personal computer and/or facsimile at each location. The computers, also operating over phone lines, provide the ability to send and display text and pictures as well as voice.

In a computer conference, participants are separated in both space and time rather than being present in the same place at the same time. The participants use terminals and telecommunications to interact. Commercially available computer conference systems are usually implemented on large mainframes to store the central files and employ packet-switching networks for communications. Technically, computer conferences are an extension of electronic mail.

Although a computer conference can be on-line with all participants at terminals simultaneously, they are usually off-line and run over several days or weeks. The "conference" is really a file at the central computer containing the messages (also referred to as items) from participants. On a typical day, a participant reviews the items sent by other participants, responds to their items, and adds ideas by creating new items. These messages can be "public" so that they can be accessed by all participants or "private" in which case they are addressed to specific individuals like electronic mail. In addition to providing a record (including a printed record) of the meeting, computer conferences usually provide the ability to edit previous statements, to add responses directly at the end of the items, to vote on specific issues, and to keep records on participants.

Computer conferences have been used to exchange information, reach consensus, write joint reports and scientific papers, and for consulting activities. Experience indicates that if a computer conference involving many people is to be successful, the following factors must be present:

1. a strong moderator to ensure closure,
2. persistence and involvement by the participants, and
3. people willing to type at terminals.

Video conferences overcome the technical limitations of speakerphone conferences and the non-simultaneity and typing requirements of computer conferences. In video conferencing, people gather in instrumented conference rooms to obtain both sound and pictures. A variety of video conferencing systems have been developed.

These systems include:

- One-way video, one-way audio
- Compressed video
- Two-way, full bandwidth video

In one-way video, one-way audio systems, the activities of the participants at a central location are televised over closed circuits to all other locations; however, the people at the remote locations are only able to reply over speakerphones. Such one-way systems are particularly useful for meetings whose purpose is to transmit information to a large number of people. These systems have been used within metropolitan areas by a number of universities to televise courses to people working in industry. They have been implemented commercially by hotel chains over large geographic areas by using communications satellites for transmitting the video. One-way video systems have been used by companies for sales meetings and product introductions, and by politicians for fund-raising.

The cost of full video transmission can be overcome somewhat by using compressed video. These systems are based on the idea that visual images in a meeting change slowly and hence only the changes in the picture need to be sent. These systems give reduced quality images for rapid motion. However, they are usually adequate for meetings in which people are sitting around a table at each end talking. Both one-way video and compressed video systems are a compromise between speakerphone and full two-way video conferences.

In a video conference, participants gather at conference rooms with video transmission and receiving capabilities. A typical video conference room contains several cameras, two screens (one to show the picture from the remote location and the other to show what is being transmitted), a conference table for the participants and a spectator area. Some rooms contain voice-actuated microphones that allow the picture to follow the conversation, if desired. Controls are provided to the chairperson or his/her designee to select the picture to be transmitted. Visual aids (slides, flip charts, transparencies) are provided, as are facsimile and personal computers.

The foregoing image of two television studios communicating is the high-cost option for video conferencing. In the mid-1990s desktop video became available that can send images of both people and computer screens. Here a small camera is put on an individual's desk and transmits the picture of a single talking head. The system can also transmit what is on the person's computer screen. Each individual can display remote video on their computer. The net result is that people can collaborate directly. The costs of these "personal" videoconferencing systems is declining rapidly.

Experience with two-way video conferences indicates that they are successful when they involve fewer than 15 people, the people know one another, the meeting is highly structured, and they last less than three hours. These meetings also tend to be more efficient because the participants are aware of the costs involved. Although two-way video is generally not appropriate for initial contacts, delicate negotiations, or personnel matters, in certain circumstances it provides savings in such matters. A particularly useful example is the ability of people at small colleges to interview with large companies. The companies not only save the expense of sending a recruiter to a small school, they can also determine whether it is worth offering an interview trip to the applicant.

One-way video conferencing has proved useful is in personalizing senior managers to their employees, particularly for groups that are located a long way from corporate headquarters.

In summary, electronic meetings offer the following benefits.

- Improved communications. They add to existing communications. They provide faster response to problems. They not only serve as substitutes for face-to-face meetings but also allow pre-meeting communications and trip follow-up.
- Organizational effectiveness. It becomes

easier to reach an individual in an organization and to reach more people in the organization. As a result, more people know what is going on. Specialized skills can be brought into meetings more easily. Junior people can participate and can enhance their skills and knowledge.

- Economic benefits. In addition to net travel cost savings, there are productivity gains from having people in the office or the factory, rather than on trips. Losses from travel fatigue are also eliminated.

DECISION ROOMS

In both government and industry, decisions are usually taken only after building consensus within the organization. A significant portion of the consensus development process occurs in conferences. However, in most organizations, conference rooms contain little technology. Decision rooms are conference rooms containing the new technologies to support the decision making process. Specifically, executive work stations and video displays in a decision room make possible the retrieval of information from computers, the rapid evaluation of alternatives, the visual display of complex information in graphic form, and the participation of people located remotely through teleconferencing.

Decision rooms can be characterized by sponsorship and purpose. They can either be located within a firm or can be made available as an outside service. In the latter case, the equipment can either be brought into the firm's premises or members of the firm can travel to the equipment. The decision room may be single purpose (used for only one type of decision), or multipurpose.

The most complex decision rooms in service have been built by governmental units; in particular, in the Office of the President of the United States and in the U.S. Department of Defense. Many simpler decision rooms are found in commercial service. In one example of an outside service, a consulting firm provides a conference room where clients can visit to undergo a "decision analysis" process. Commercial systems are sold by several vendors, including IBM. Multipurpose decision rooms for experimental use exist at many universities.

Observations at the experimental facilities indicate that decision rooms are better used for large meetings (more than 6 or 8 people) than small meetings (3 or 4 people). In the case of small meetings, the technology adds little value. In large meetings, a phenomenon usually referred to as "human parallel processing" takes place. That is, by having people enter information into the meeting simultaneously through the computer, they are working in parallel rather than sequentially. Thus, for example, rather than each of 10 people in an hour's meeting having an average of 6 minutes to speak, 15 to 20 minutes of computer input by all participants provides much richer results more efficiently.

LOCAL AREA NETWORKS

Local area networks are the communication networks that tie individual computers together and allow sharing of programs and data among many workstations simultaneously. As the local in their name implies, these networks cover a small geographic area, such as a building or a contiguous work area for a group. They require both hardware and software. The hardware consists of cabling that interconnects the units and electronic boards that provide the communications interfaces among the computers. Typically, one unit on the network acts as a "file server," that is, a place where the programs and files to be shared are stored. Thus, servers usually have a large amount of hard disk storage capacity. Servers also provide printing and external communications interfaces that connect the network to common printers, to other networks, and/or mainframes. Networks may have anywhere from a few to hundreds of users connected to them. With a local network, the users can communicate with one another and with the file server.

The network software moves data between the user workstations and the file server, controls data access and file sharing, and manages security. It often includes electronic mail. Networks can be connected (in increasing order of speed and cost) with twisted pair wire, coaxial cable, or fiber optic cable. The interaction between the cabling method and the software determine the speed with which information can be moved across the network. The units in the network are connected in either a bus, ring, or star configuration. In a bus configuration, a single cable runs past all the workstations. A ring is like a bus, except that the two ends of the cable are hooked together. In a star, each workstation is connected directly to the server. A bus or ring requires less wiring (and hence less cost) than a star, whereas the star is easier to troubleshoot.

Applications that run on a network usually require special versions of the software to run successfully. This software achieves the needed data integrity. That is, applications must be set up so that if one user is accessing data, another user cannot change it at the same time. Otherwise, chaos results.

Local area networks are one solution for the data sharing problem. An alternative is the departmental computer. Here a time-sharing minicomputer is dedicated to a department for its own use. The choice of a departmental computer versus a local area network involves a tradeoff among costs and benefits. Local area networks can add costs of $1000 or more per workstation. Personal computers are more expensive than terminals. On the other hand, the minicomputer and the associated software costs are also significant. For an organization with a considerable investment in stand-alone workstations, local area networking provides a reasonable upgrade to connectivity. For a new installation, detailed cost/benefit analysis needs to be performed.

HUMAN FACTORS AND SOCIAL CONSIDERATIONS

The new technology changes the way people work. Not all of these changes are necessarily good or well designed. Increasing attention is being given to good ergonomic design in screens, lighting, chairs, and other elements of these technologies. Radiation hazards from screens are a matter of controversy. Telecommuting has been characterized by some as the "new sweatshop." Working in a computer/communications world requires higher technical skills and hence more education. Concern exists that some groups, such as minorities and the poor, will be even further disadvantaged in this new technology world.

In summary, increasing the technological content of work implies a need for taking human as well as technical factors into account.

QUESTIONS

1. What are some of the major applications of local area networks?
2. As these technologies evolve over time, some are likely to be combined with one another. What technologies are likely to be combined? Explain why. What additional benefits might be gained from such a merger of technologies?
3. What improvements in these technologies will have a larger impact on productivity: improved usability or more powerful technical performance? Explain your rationale.

2.2 Telecommunications: The Opportunity and the Challenge

PETER G. W. KEEN AND J. MICHAEL CUMMINS

Although the computer receives most of the attention, the new technologies are possible only because of the telecommunications that glue them together. Computers and business are two different cultures. Telecommunications is yet a third culture. In most cases, the people in the computer culture must learn the ideas and the jargon of the business culture because many business people have only a rudimentary understanding of computers. However, the telecommunications expert finds that he or she must learn about both the business and computer cultures because each of the two other groups is ignorant of what telecommunications is about. This section and some of the subsequent chapters will help you understand what the issues are and how telecommunications can be used to create substantial productivity gains.

The material in this section is taken from Chapter 1 of a textbook by Keen and Cummins. This book differs from most in the field in that it takes a business view of what telecommunications is about. This section covers some (but not all) the major topics in the field. It shows how technological change, particularly telecommunications change, is altering both the pace and nature of business. It shows how telecommunications has changed business so that it is now global. The section also discusses networks, the core telecommunications technology that ties people together.

Source: Networks in Action: Business Choices and Telecommunications Decisions. Belmont, CA: Wadsworth Publishing Company (1994), 2–27.

HOW TELECOMMUNICATIONS IS RESHAPING BUSINESS

Telecommunications is the electronic transmission of information, whether it be voice phone calls, computer data, photographs, music, documents, or television broadcasts. Transmission is the movement of signals along a communications link and is the core of telecommunications; it provides a connection between sender and receiver and manages the routing of the communication. The telephone system, for example, transmits your voice along a complex set of lines and exchanges. An automated teller machine transmits to your bank's computer your request to withdraw cash from your checking account. CBS transmits its programs to your television antenna.

As a business resource, telecommunications is already vital to the efficiency and effectiveness of public and private organizations and can only become more so in the coming decade. By making it practical to transmit information at high speeds, telecommunications has transformed the basics of entire industries. Obvious everyday examples are the automated teller machines (ATMs) that provide an electronic bank branch that stays open nearly 24 hours a day instead of a "bricks and mortar" one with much shorter hours, and the airline reservation systems that allow travel agents to book flights on hundreds of airlines in real-time. Real-time means that as soon as the booking is entered by hitting a key on a terminal that accesses the system, the relevant computer records, which may be thousands of miles away, are immediately updated. The word *terminal* was originally used for the devices that are the end points of the telecommunications service; today, personal computers, phones, bar code readers, and television sets are all forms of terminals. Figure 1 shows a very simplified diagram of an ATM network shared by many banks that allows you to use, say, your First American Bank of Transylvania bank card at an ATM in Boston, Little Rock, or Los Angeles. The telecommunications network handles the transmission of the transaction and the routing of it to your own bank's computer.

ATMs are so well established that we take them for granted and can easily overlook how they transformed the nature of banking in less than ten years. A less common but almost surely as far-reaching example of the transformation of an industry through telecommunications is the electronic claims processing systems that reduce the paper and time involved in handling five billion medical insurance claims a year in the United States, costing $8 billion to administer. These systems cut insurers' costs by 50% or more and speed up payment by weeks; it costs around $5 to process a paper claim but just 3–10 cents if the claim is sent directly to a computer that processes it and makes the payment electronically. It costs 29 cents just for the stamp on a mailed document. Industry experts have good empirical evidence to suggest that the savings from electronic claims processing will be $10 billion a year by 1995.

In July 1992, 15 health care providers announced a phased plan that will streamline the entire claims and payment process by 1996. All the technology is already in use and has been proven by leading health care insurers and hospitals. The plan is a response to a challenge by the head of the federal Department of Health and Human Services to ensure industrywide use of electronic claims processing and cut administrative costs, which amount to one-quarter of the $800 billion spent on health care in the United States each year. Figure 2 shows an electronic claims processing system that links doctors' offices to insurers, eliminating paper and delays.

EBS (electronic benefits system) may be as common an acronym in 1995 as ATM is today. It is also increasingly difficult to identify any industry whose *basics* are likely to remain unaffected by telecommunications.

Figure 1 The Power of Telecommunications: Bank Automated Teller Machines

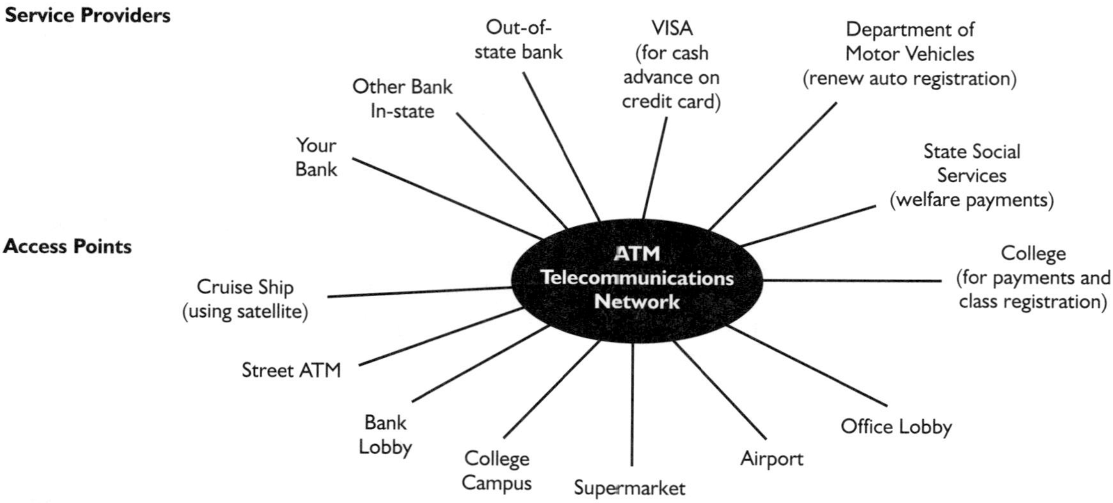

Note: The telecommunications network for a multibank national ATM network (such as Cirrus, Monec, or Most) allows you to use your bank's ATM card at any location that links to the network. The telecommunications network, although shown in the figure above, is something a user neither sees nor needs to know about; you merely insert your card and enter some information to get cash.

The diagram shows both some services that are available and some typical points of access to the network. Note that access need not occur from a bank and that services are not restricted to authorizing a withdrawal, making a deposit, or finding out your balance. The functions this telecommunications network provides are essential to users in that when the ATM network is "down," so is the bank.

THE EXPLOSIVE PACE OF TECHNOLOGICAL CHANGE

The rate of change and its effects in the technology of telecommunications is explosive. Four factors, which occur individually and/or in combination, pace the use of telecommunications as a business and social resource: 1. competition, 2. speed of transmission, 3. access tools, and 4. interoperability.

Competition

Until the 1980s, telecommunications historically was treated as a "natural monopoly," part of a nation's social infrastructure. A PTT—Poste Telegraphique et Telephonique, the generic term for the government-controlled agency that operates a nation's telecommunications monopoly—operated the telephone system, with limited licensing of other service providers in such areas as telex. The United States has never had a PTT, but AT&T was the equivalent; its chairman offered in the 1920s that it be subject to federal and state regulation in return for its being allowed to operate as a monopoly. When the United States first deregulated the long-distance phone industry in 1982, and the United Kingdom liberalized it (allowed partial competition) a few years later, they stood alone. In most countries, companies had no choice of equipment, services, or prices beyond what was offered by the PTT. The divestiture of AT&T in 1984, the outcome of an antitrust suit, created competition in long distance and split up AT&T's local phone services into seven regional phone monopolies.

Now, monopolies are the exception, not the norm. Sweden, the Netherlands, New Zealand, Spain, Mexico, Germany, and many other nations are recognizing the importance of low-cost, high-quality telecommunications in a world of global operations and competition. The result has been a drop in prices of, typically, 30% and a flood of

Figure 2 Electronic Insurance Claims Processing Network

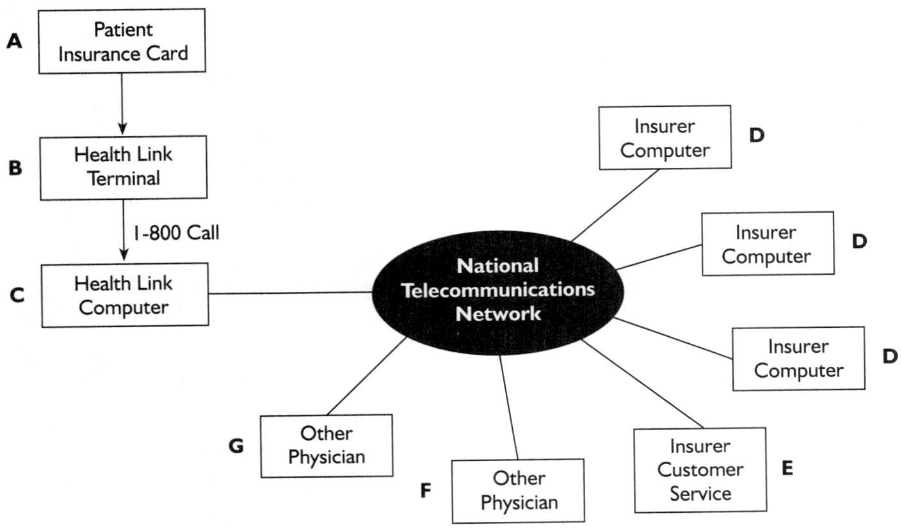

Note: Health Link was introduced in 1991 by a Princeton, New Jersey firm that offers thousands of doctors, insurers, and health providers a simple service that reduces paper, complexity, and time. *Computerworld* summarizes it as making "the process of filing insurance claims as paperless and speedy as using a credit card to buy a new suit and tie."

A doctor's office staff uses a small and unobtrusive computer terminal, almost identical to the device used to scan a credit card for authorization in a store. The card here is the patient's insurance card (A in the diagram). The terminal (B) dials a 1-800 number that connects it to the Health Link computer (C). A national telecommunications network processes the claim request, identifying any special information needed for the relevant insurer and formatting the data so that it can be directly processed through the network by the insurer's computer system (D). The insurer's system confirms coverage and immediately authorizes payment after the doctor has finished. If the patient is not covered for the treatment, the Health Link system automatically dials the insurer's customer service unit (E) for help. The physician may also refer the patient to other physicians on the patient's health plan (F, G).

Health Link has cut claims processing costs for its users by about half and processing time by two-thirds.

innovation in products and suppliers. For example, a U.S. leader in the use of one of the most advanced high-speed telecommunications transmission methods is a firm called WilTel, a subsidiary of an oil company that was set up to dispose of unused oil pipelines. Instead, it installed fiber-optic cables in them and became a long-distance telecommunications provider. Fiber optics send light signals along glass cables, thinner than a human hair, via laser beams that pulse signals at sub-sub-subsecond speeds. This permits high communications transmission rates and the corresponding ability to move volumes of information cheaply and rapidly, both of which were unimaginable ten years ago. WilTel preempted the traditional providers of long-distance computer-to-computer communications, such as AT&T, MCI, and Sprint, by exploiting its ownership of the pipelines and rights of way, fiber-optics technology, and new transmission methodology.

Competition in the long-distance telecommunications market has cut the average price of a phone call by over 40% in under 10 years. Lack of competition in the local market, which is three times bigger, has resulted in just a 10% cut. The competition between AT&T, MCI, and Sprint, the three major long-distance providers, has created a flood of discounts, new products such as personal 1–800 toll-free numbers, and aggressive technical innovation. New technology creates new competition. In 1991, for instance, more cellular phones were put into service than regular residential phone lines: 2.5 million versus 1.9 million.

Competition has at last come to the local phone companies. Cable television firms can offer phone call transmission, and phone companies can offer entertainment. A new race is on to capture customers and preempt the marketplace. Mobile communications, cable TV, fiber into the home, satellites, personal communication networks, and electronic publishing are a little like Hollywood in the 1920s; it is difficult to see who will win and where the marketplace will move. But the movement will be at least as fast in terms of technical innovation, products, and cost cutting as that in the long-distance telephone market between 1984 and 1993.

Speed of Transmission

The core of telecommunications is how quickly and cheaply information can be moved. A famous epigram by science fiction writer Arthur Clarke captures the progress of telecommunications in this area: he said that in any society there is a point at which technology is indistinguishable from magic. Much of what is happening in transmission in the 1990s is magic in this sense; almost nothing you can predict is implausible. The telecommunications planners of the 1970s hoped to achieve by the end of this century a transmission speed that has already been exceeded by a factor of almost 40,000 because of new telecommunications transmission media. Transmission media are the various types of cables and radio, and light waves that carry the transmission signal. Originally, telecommunications meant the movement of electrical signals along copper wires; now it also means the movement of light signals and radio signals. Fiber-optic cables occupy the same space as copper cables but carry information up to 100,000 times faster. Radio waves do not need any wires but move information via satellites 22,300 miles above the earth or through line-of-sight microwave relays. (Line of sight means that no physical blockage stands between sender and receiver.)

The same information may be sent over a variety of transmission media, all of which are equivalent in terms of the basic techniques of telecommunications. A phone call may travel by cable, or radio waves (with or without a satellite).

The main advantage of fiber optics is massive speed at low cost, but it requires installing new cables or replacing existing ones. The advantage of wireless communications is that it provides mobility; cellular phones can be used in cars, for example. The main impediments to deploying wireless communications are the high cost (over $100 million) of launching satellites, many unresolved issues of regulation and allocation of radio frequencies by the Federal Communications Commission (FCC), and the greater level of interference from noise, weather, and other sources.

The main advantage of the copper phone lines in use everywhere today is simply that they are in place. Replacing them with fiber optics is mainly a question of costs, and it will be at least three decades before every house has fiber. Moreover, the early 1990s have seen major innovations in transmission over existing phone lines. Your local phone company is racing to head off cable television providers and others to prevent them from capturing their core residential and business market; they are offering high-speed transmission that can even provide pay-per-view video with no replacement of the installed wires. One of the primary targets of the cable companies is large businesses and long-distance carriers that must currently route their calls via the local phone company; close to 50 cents of every dollar of revenue MCI gets from long distance is paid to the local exchange carrier as an access charge. This all adds up to guaranteed innovation and competition because fiber optics, satellite communication, and high-speed transmission through copper wire cut costs, make wider uses of telecommunications cost-effective, and create new uses. Consider, for example, the transmission of a document. Before long-distance competition cut phone rates, faxing a document was very expensive and took several minutes. Now, it takes just seconds along a standard phone line. However, it is still impractical to send a full-color photograph by fax. Fiber optics make it practical, as does satellite communication, but this capability is confined to use by businesses that lease the necessary high-speed transmission services. However, we are already seeing local phone companies offering this capability for an installation cost of under $500 and for only pennies per minute.

The National Research and Education Network, approved by President Bush in 1992, will be able to transmit in a single second 100,000 typed pages, or 1,000 high-resolution photographs, or 10,000 three-dimensional X-rays. In 1983, sending a database of a thousand figures, such as a spreadsheet or price list, took from five seconds (a special leased line) to a minute (over "POTS"—plain old telephone system). Today, it takes under a second over a leased data communications link and around 10 seconds on a standard "dial-up" phone line, at a cost that is at least a tenth of that just a decade ago. T1, the standard transmission link used by telephone companies and large businesses, can transmit this book in about five seconds. No end is in sight for continued acceleration of the pace of technical change in every area of telecommunications transmission.

Moving signals along a transmission link is one of the two basic elements of transmission. The other—routing the signal from originator to receiver—is needed because it is economically senseless to even think about providing a direct connection from every sending device to every receiving one. Instead, signals move through the network through a series of switches. As the term suggests, a switch is a device that receives a message and directs it to the next transmission link.

In the telecommunications industry, the national phone system is termed the "public switched network." Public means open to any subscriber, and switched means calls are routed from sender to receiver via equipment at nodes in the network. This process is depicted in Figure 3, which shows in highly simplified form how a phone call from Long Island to Los Angeles moves through AT&T's long-distance network.

Historically, the electrical signals transmitted along the phone line were processed by switches that were electromechanical devices, which have moving parts driven by electricity. Your phone call connected you to the local exchange, where a device responded to the signals and mechanically moved connectors to establish the circuit to complete the call. Now, more elements in the switching process are handled by computers, which operate at speeds measured in microseconds (millionths of a second), nanoseconds (billionths), and even picoseconds (trillionths). As a result, switches have become faster and cheaper, enabling them to handle the faster transmission media of fiber optics and wireless communications. Together transmission links and switching technology form the core of modern telecommunications—and its cost base as well. The investment in the public long-distance phone network is estimated at over $250 billion. Japan has calculated that it will cost at least $215 billion to upgrade its public system to fiber optics, a process that will take over 25 years.

Access Tools

Much of the innovation in the use of telecommunications stems from an increase in the places from and tools by which people can access services. Until the 1970s, the tools were limited to telephones, telex machines, radios, and very expensive and slow facsimile machines; now they include personal computers, mobile phones, paging devices, portable fax machines, bar code readers, cash registers, and even interactive television. POTS now offers push-button voice messaging, in which you select from a menu of options by pressing, say, a "1" or "8" in response to recorded voice instructions; a computer receives the instructions and responds to your selection. Personal computers can dial up over the phone line a host of information services. ATMs can be used to collect welfare payments or renew car registration and are even on aircraft carriers. Telecommunications is fundamentally about ease of access and convenience. Its major impact on society is that it removes time and location as barriers to service and coordination.

Interoperability

The more services and facilities to which an access tool can link, the more valuable and convenient it will be. The telephone is valuable because it is easy to use and can access other states and countries. Imagine what it would be like if you had to use twelve different types of phone—one

Figure 3 The National Phone System

Node	Approximate Number in U.S.
Central Offices	28,000
Toll Centers	1,300
Primary Centers	230
Sectional Centers	70
Regional Centers	12

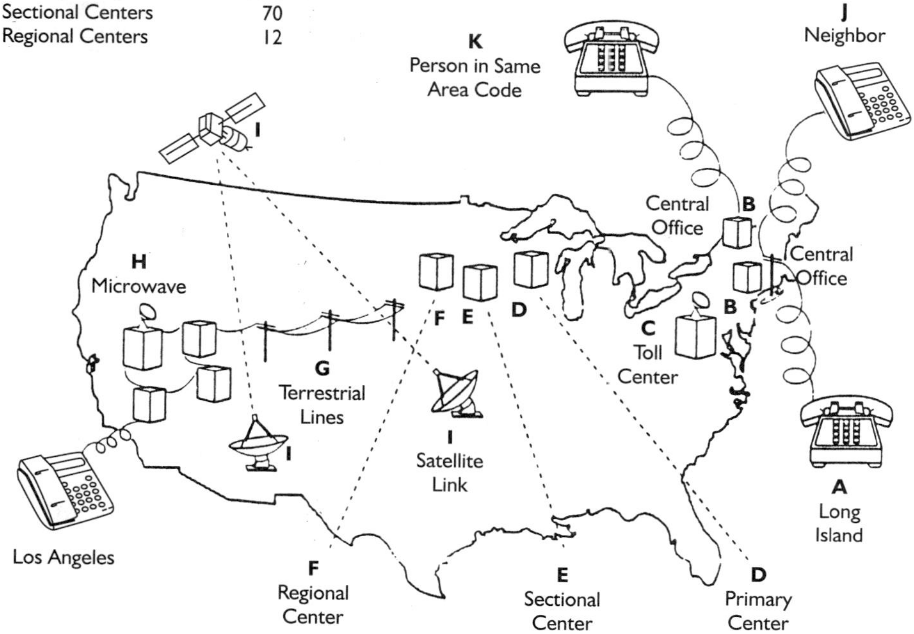

Note: In a call via AT&T from Long Island to Los Angeles, the telephone (A) is connected by copper wires to the first node, the local phone company's central office (B), which switches the call to a toll center (C). From there it is switched to a primary center (D), a sectional center (E), and a regional center (F), either along terrestrial cable trunk lines (G), microwave radio (H), or wireless satellite link (I). The centers concentrate traffic onto high-speed communication links, with the switches selecting the route on the basis of cost and congestion. A local phone call will be switched directly from the central office if the subscriber is connected to it—to a neighbor of the caller, for instance (J), or to another central office if the person called is within the same area code (K). Other switches and transmission links handle international traffic; these are not shown.

for connecting with AT&T customers, one for MCI, one for Sprint, one for calling the UK, and so on. What if your AT&T phone could call rotary phones only but not push-button ones?

Historically, the field of information technology—which comprises access tools, computers, telecommunications, and information stores—was dominated by exactly that level of incompatibility. Incompatibility means that this device cannot directly link to that device, this computer cannot use that software, or this personal computer cannot use that telecommunications facility. In 1981, the New York corporate treasurer's office of a leading multinational firm had 17 different computer terminals in daily use. One was for Citibank's electronic cash management system, one was for Chase's, another was for Midland Bank (of the United Kingdom), another was for Reuters foreign exchange quotes, still another was for foreign exchange trading, and so on. Even today, most such companies have a mass of incompatible computers and corresponding multiple and incompatible telecommunications systems.

Ending incompatibility is the major goal in the information technology field: providers, users, and the many consortia and committees are trying to work together to reduce incompatibility and increase what is often termed interoperability, the working together of the elements of a telecommunications resource. The wider term integration refers to the working together of all the elements of an information technology resource: hardware, software, and telecommunications.

The move toward interoperability is the fourth key force driving the explosive technological pace of innovation in telecommunications. The starting point was to interconnect countries' phone systems, each of which had different voltages, signals, and equipment. Now, you can routinely dial just about any country, and switching equipment will handle the needed conversions. The next step was to enable communication to, from, and between computers. Here, every major vendor developed its own telecommunications procedures and techniques, greatly limiting interconnection.

The situation in telecommunications in 1982, the pivotal date when the divestiture of AT&T was announced, was thus one in which the technology base for voice communications—spoken messages sent from and to telephones—was very separate from data communications—the transmission and processing of computer-generated information. One of the goals of industry leaders was voice/data integration—the ability to handle either type of traffic through the same telecommunications facilities. The business community of telecommunications users was more interested in open systems, ones that were not vendor-dependent, so that they could interconnect the many different types of computer and communications facilities they operated. Open systems are contrasted with proprietary systems, which require vendor-specific products.

The move toward integration of just about everything is now the driver of information technology. Integration poses massive technical challenges, and many vested interests are slowing it. That said, every year sees more and more progress in interoperability. This allows a firm like Wal-Mart to integrate its entire supply chain. Its stores' point-of-sale registers can link by satellite to its head office computers, which in turn link directly to suppliers' computers, which link to shippers, banks, insurers, and other parties.

Change as the Norm in Technology

These four forces are all moving rapidly. Competition ensures innovation in products and decrease in prices. Developments in transmission stimulate competition, drastically cut prices, and increase the type of information that can be transmitted and hence the range of practical uses of telecommunications. For instance, 1992 saw widespread pilot projects using teleradiology—the transmission of high-resolution images of X-rays from small hospitals to medical centers, where experts interpret them and review them over a telecommunications link with the relevant physician. This process was completely impractical as late as 1990; transmission speeds, let alone costs, blocked the opportunity.

Access tools open up new uses of telecommunications, mainly because adding a new service to an existing base can be very inexpensive. ATMs are becoming the access point for registering at college or handling state benefit payments. Telephones contact computers now, not just other phones. Finally, interoperability of information technology systems allows interconnection of business processes and eliminates many procedures, documents, locations, and people that can slow service and reduce coordination. That is why a few banks can now process a mortgage application over the phone in 15 minutes instead of 30–60 days, why Federal Express can tell a customer within half an hour where a package is along its complex delivery chain, and why electronic data interchange (EDI) has cut the number of staff in purchasing departments from hundreds to around 20. EDI is the transmission of electronic documents in a form that allows another firm's computer to convert the sender's message format (for, say, a purchase order) into its own formats.

NETWORKS: THE CORE TECHNOLOGY

The basic element of a telecommunications resource is a network—a set of devices that share a directory and can thus directly access each other. The directory is analogous to a standard telephone directory, except that it provides an address for each component of the network. If a device is not included in the directory, it is not part of the network. Directories must be updated when components and locations change, in the same way that if you wish to add a new phone to the public telephone system, you must have a phone number; having a handset is not enough.

The main distinctions between networks are public versus private and local area versus wide area. A public network is available to a wide range of subscribers on a pay-as-you-go basis: the most obvious and widely used example is the public phone system. A private network is one in which the transmission facility is leased by a firm for its own use. A local area network links personal computers and printers within the same building or building complex; a wide area network covers a broad geographic area.

Examples of business networks are a bank's wide area network that connects ATMs to central computers that handle check-cashing and deposits; a local area network that connects, say, the personal computers in a company's legal department; and an international private network in which a firm leases telecommunications services for its own (hence private) use.

There are many other types of networks, and a typical network may link thousands of devices and include thousands of additional components. Networks of connected networks can create new ways of doing business; leading retailers like Wal-Mart, Toys R Us, and Dillards, for instance, have linked their own stores' point-of-sale workstations and local (in-store) networks to their corporate cross-country network to provide up-to-the-hour data on stocks and purchases. The network links directly to suppliers for automated and increasingly paperless purchasing, invoicing, and payments through EDI. Many of the suppliers' networks in turn link to manufacturers' networks. This "quick response" technology has transformed the competitive dynamics of retailing. During the 1991 recession, when retailing slumped badly, Wal-Mart, the Gap, Limited, and Toys R Us reported sales gains of up to 30 percent. Their networked logistical systems are a major explanation of their success; they have better information, faster reaction times, and more coordinated distribution than their competitors.

Many of the most innovative and far-reaching business opportunities of telecommunications come from linking networks across companies and even industries. So, too, do many of the most complex technical problems. Generally, a specific network is built around a limited set of technical building blocks and is designed to meet a targeted range of needs and types of traffic. Adding new devices into the network is relatively simple if they fit into the original technical design. Linking networks of very different technical components is not so easy and may require special-purpose devices and software.

THE NEED FOR NEW MANAGEMENT PROCESSES

The mixture of technical innovation, competition, and variety of applications makes telecommunications (1) an important business resource, (2) a complex, specialized, and volatile technology, and (3) a field that is highly interdependent with the computers and information systems that process business transactions and access and manage the relevant databases. An ongoing problem organizations report is that few professionals or managers are comfortable in all three of these areas. Business experts do not understand a technology whose jargon makes computer terminology look simple. Telecommunications specialists are often unfamiliar with business needs and opportunities. Neither they nor their peers in information systems can be expected to keep up with each others' areas of expertise, and each is often ignorant of terms and key developments in the other's field. Organizations everywhere are talking about the need for people who can build a new dialogue that can bring these people together. This dialogue focuses on two main issues:

- *Business decisions:* the practical business and organizational opportunities of telecommunications.
- *Technical choices* of building blocks needed to create and operate the relevant telecommunications resource.

The telecommunications business decision sequence has four steps:

- *Identify the business opportunity.*
- *Define the business criteria* for the technical design of the telecommunications platform needed to turn opportunity into action.
- *Identify the network design priorities* and trade-offs between them.
- *Make the economic case* for the investment.

Table 1 summarizes this decision sequence.

INNOVATION AND GLOBALIZATION

Telecommunications and Business Innovation

Well-known technologies that indicate how deeply telecommunications affects us all in our daily lives are telephones, facsimile (fax) machines, automated teller machines, satellite dishes, credit card authorizations, and telemarketing. Table 2 lists some of the recent innovative uses of telecommunications to provide entirely new products, services, and ways of working, communicating, and operating.

Together with computers (most obviously personal computers), telecommunications changes the status quo. Just 20 years ago, many of the services listed in Table 2 bordered on fantasy. Today, it is routine for long-distance phone companies to send tens or even hundreds of thousands of phone calls simultaneously along a strand of optical fiber. (When AT&T's monopoly was ended in 1984, the number of calls that could be carried along a phone line simultaneously ranged from 50 to 600.)

Manufacturers today routinely handle orders and deliveries with no paper, delays, or people in the middle. Companies such as General Motors, Sears, and Wal-Mart require their suppliers to link to them through EDI instead of mail. Transactions that once took weeks now take days instead. The $25 billion that apparel retailers once spent annually on administration is being halved through the use of EDI and related quick response (QR) systems in which computers and telecommunications are used to streamline the entire chain of transactions, from point of sale to retailer to manufacturer to shipper to warehouse to store.

Routinely, the CEO of a company can know yesterday's sales to the penny and be alerted to events and trends now, instead of in several months, when it may be too late to do anything about them. According to the chief executive of Frito-Lay, one of the leaders in using hand-held computers and telecommunications to move information at the start and end of the business day to and from its field sales force and head office, the result was that for the first time in his career he had the information to act instead of react. Top retailers such as Wal-Mart, Toys R Us, and Dillards get information from stores every two hours via telecommunications. They can spot today's problems today and can spot a trend in under a week. The rest of the industry hears about the trend after it has already happened.[1]

Many industries across the world now literally depend on telecommunications operations. The international banking network SWIFT, which handles transfers of funds between international banks, electronically processes a trillion dollars of transactions *each day*. That is almost a billion dollars a minute! The city of London processes over $120 billion of foreign exchange deals a day; currency traders live by telecommunications. The computerized reservation systems for American Airlines and United Airlines each handle close to 2,000 transactions a second. Omaha, Nebraska, handles 100 million "800" calls a year, fully justifying its claim to be the 800-number capital of the United States.

These business innovations create both new

TABLE 1
The Telecommunications Business Decision Sequence

1. Identify the Business Opportunity
Run the business better
Gain an edge in existing markets
Create product and market innovation
2. Define the Business Criteria for the Telecommunications Platform
Reach: the locations/people/organizations to which we must link
Range: the variety of information and transactions we must be able to share directly and automatically across business functions and processes
Responsiveness: the level of service we guarantee, in terms of speed, reliability, and security
3. Identify the Network Design Priorities and Trade-Offs
Capability: the technical applications and services the platform provides
Flexibility: how easily it can accommodate change
Certification: how reliable it is and what service it guarantees
Cost: fixed, variable development, and operations
4. Make the Economic Case
Improve profit management and alerting systems
Improve traditional costs: labor, real estate, information technology
Provide premium quality without adding a cost premium
Provide premium customer service without adding a cost premium
Improve long-term business infrastructure costs; improve revenues

opportunities and new management challenges. When the optical fiber link is cut, thousands of phone calls die. When the manufacturer's purchasing systems cannot link to suppliers for half a day, production may stop dead. "Telecommunications" is synonymous with "service" in more and more aspects of both everyday business and our personal lives.

The New Gross Information Product

Telecommunications is a relatively new factor in international business competition. It is also a relatively new part of how we all live. The power of telecommunications in everyday life became especially apparent during the Gulf War, the attempted 1991 coup in Russia, and in the Tiananmen Square massacre in China. During World War II, people watched cinema newsreels that were a month old. During the Vietnam War, television news was only some 24 hours behind the events it showed. During the Gulf War, news was in real-time. Intelligence officers watching on CNN saw where Scud missiles landed and immediately either adjusted their aim or instituted countermeasures. The new tools of journalism included phones that fit in a suitcase, portable satellite earth stations, fax machines to send Arabic documents to experts for immediate translation, and high-resolution photographs transmitted via phone lines. None of this is "gee whiz" any more than is ordering a pizza by fax, or receiving 110 channels of TV via satellite, or being able to direct-dial Tokyo, Moscow, or Rio de Janeiro. We live in an age in which telecommunications is now the equivalent of electricity, cars, and airplanes 90 years ago.

TABLE 2
Eight Proven Telecommunications Innovations of the 1990s

1. Interactive Education by Satellite

Maine: 47th in percentage of high school graduates going on to college, 50th in adult education; ⅔ of population too far away from one of seven college campuses to attend classes. Maine's Interactive TV system has over 4,000 students, with many graduates and enrollments growing at 12% a year

Access tool: large-screen television sets in schools and even fire stations

Enabler: satellite transmission, with data compression reducing the amount of information to be sent

Impact: for less than the cost of a dormitory, Maine has created a whole new education infrastructure; major impact on women in rural Maine who dropped out of high school and have no job training

2. ATMs for Welfare Payments

Maryland: uses existing interbank ATM network (called Most) for food stamps and welfare payments via benefit card

Access tool: ATM and point of sale register in supermarkets (for processing food stamps electronically, eliminating physical coupons)

Enabler: in-place, low-cost ATM infrastructure; switching technology connecting to Maryland and Federal computers (for food stamps)

Impact: major cost savings for the state and for supermarkets handling food stamps; reductions in thefts and losses of checks

3. Tax Refunds Authorized on the Spot

Virginia: uses on-line systems to enter and check key tax form data, approve refund, and send message to central system to issue the check

Access tool: workstations in 40 tax offices, where citizens can bring in their returns and get requests for refund authorized on the spot; 1,800 other workstations in government offices

Enabler: low-cost private network

Impact: saves $50 million a year, with no staff increase in three years despite 7% in growth volume a year

4. A Mortgage in 15 Minutes

Citibank: replaces a paper-dominated process that took 30–60 days with a 15-minute session over the phone

Access tool: telephone (in real estate dealer's office or at home); workstations in Citibank with access to credit bureau, bank records, and so on

Enabler: 800 numbers; powerful personal computerlike workstations for online processing by customer service agent

Impact: a radical change in the basics of service for the mid-1990s

5. $1 Trillion a Day Traded in Futures and Options

Tokyo: trading in Chicago from Tokyo on 24-hour GLOBEX system

Access tool: personal computers

Enabler: low-cost global telecommunications links; satellite and high-speed fiber optics

Impact: complete globalization around the clock, even when the Chicago exchange is "closed"

6. Personal 800 Numbers

Sprint: offer of an 800 number for family use: students phoning home, elderly parents and so on; small business use, too

Access tool: standard phone

Enabler: falling costs of high-capacity wide area network transmission

Impact: over half the traffic on the long-distance phone network at peak hours is now 800 calls; allows small businesses to go national

(*continued*)

TABLE 2 (*continued*)

7. Paperless Aircraft Carrier

U.S. Navy: the Navy's paperless-ship project stores documents in image form

Access tool: personal computer

Enabler: compact disk storage, ultra-highspeed local area network

Impact: the 40 tons of basic documents needed for daily operations can be stored on a few dozen optical disks; time spent in funding documents is cut by 70%; frees up space for bunks, storage, weapons

8. Worldwide Delivery in 48 Hours with no Warehouses

Laura Ashley: this retailer, with 547 stores in 26 countries, has eliminated all its warehouses and uses Federal Express to handle all distribution, with a guarantee of delivery anywhere Laura Ashley operates within 48 hours

Access tools: personal computers in Laura Ashley locations that are linked to Federal Express's systems

Enabler: Federal Express worldwide network and use of mobile communications

Impact: goods can move straight from the supplier in, say, Hong Kong, to Vienna without having to go through Laura Ashley's UK distribution headquarters

Telecommunications has become so crucial to business operations that it is widely recognized as one of the main forces for economic growth of cities, states, and countries and for the competitive survival of companies. In the 1980s, world expenditures on telecommunications grew by 600 percent; even with the recession of the early 1990s, it is on pace to grow by another 500 percent in this decade. Across the world, there is a strong correlation between a country's per capita gross domestic product and the number of telephones per thousand population. That is a major explanation of why India launched its Mission Better Telecommunications in 1987 and will spend close to $40 billion by 2000 on its goal of doubling its entire communications systems in five years.

To be modern, a city or country must have modern telecommunications. The world is moving more and more business transactions on-line—that is, the data needed to answer a question or carry out a transaction can be accessed immediately by a computer, instead of having to be located in a file or a book and then made available. We are seeing the emergence of what may be termed the gross information product, and it dwarfs the gross domestic product. The total *physical* exports of the United States and Germany, the two world coleaders in trade (Japan is in third place), each amount to just over $400 billion a year. That is less than a week of London's international *electronic* trade in the form of foreign exchange, securities, funds transfers, and credit card transactions. It is well under half a day's electronic financial transactions handled in New York City, which amount to $1.5 trillion every day.

In international trade and finance, many cities are exploiting telecommunications as an explicit element of their economic policy. Singapore and Hong Kong, for instance, are using EDI to eliminate the masses of paper documents that add about 7% to the cost of international shipments. Government documents that once took several days to obtain and process are now electronically handled in under 15 minutes and routed to the relevant port, shipping company's office, or freight forwarder. EDI helped move Singapore from tenth to first in the world in terms of volume of shipments. Substituting telecommunications for paper is rapidly becoming a new cornerstone for business across the world.

Telecommunications and International Development

Telecommunications disparities among countries become disparities in both economic well-being and competitiveness. Sweden has 615 phone lines for every 1,000 people; the United States has 409, West Germany 403, and Japan 367. India has 4, China 5, and Mexico 47. Every 1% increase in the number of phones in a developing country is estimated to increase per capita income by 3%. The relationship is a causal one: Improve telecommunications and a business has new opportunities, but impede telecommunications and those opportunities are blocked. The *Financial Times* of May, 1990 commented that telecommunications traffic is now a far better measure for gauging economic activity than are more traditional metrics of imports and exports and gross national product.

The takeoff of the Pacific Rim is as much a reflection of its investments in telecommunications infrastructures as of reliance on cheap and very skilled labor; one without the other would not have led to growth in South Korea, Indonesia, Singapore, Hong Kong, and Taiwan at a rate that will double their GNPs in this decade. In order to attract trading services, Singapore set its rates for international phone calls at a third those of other Asian countries. Its 1987 National Information Technology Plan states: "The information communications system is the backbone of the information economy. We must continue to have the best information communication system in the world in order to maintain an advantage in the information age."

NOTE

1. "In December 1986, for example, Toys R Us tried out scooters—skateboards with handles—with a trial order of 10,000. They sold out in two days, a trend the computers immediately spotted and acted on. Last year, Toys R Us sold over 1 million scooters. When the computer conflicts with his own judgment, [the CEO] will go with the computer." *Forbes* (February 22, 1988).

QUESTIONS

1. What is the role of telecommunications in improving organizational productivity?
2. What are the barriers to more effective use of telecommunications in business applications?
3. What new business applications can be created through fiber optics technology? Satellite technology?

2.3

Waves of the Future

JOHN EDWARDS

John Edwards describes four of the newest of the technologies: personal digital assistants, wireless communications, multimedia, and virtual reality. It is not clear whether all or only some of these technologies will survive beyond the initial enthusiasm. For example, personal digital assistants—small, hand-held computers such as Apple's Newton—came on with a splash and then almost disappeared as the gap between promise and performance was not bridged.

Each technology brings with it a promise of increased productivity. Reducing the size of the computer so that it does not take up space and using it for messages has obvious implications on the factory floor. Wireless communications, on which PDAs and much else rely, make it possible to be anywhere, anytime, and not be tied to location by wire. Multimedia increases the ways in which we can communicate. Virtual reality allows us to simulate situations, such as conditions on the production floor or in the office without large capital investments.

Will these technologies make a difference? Only time will tell. By the time you read this section, one or more of these technologies may have taken off or been discarded.

FOLLOWING the computer industry is like spending a day at the beach. Every few minutes, a wave crashes in and deposits something new at your feet. One may toss a pretty seashell your way. The next may bring in a rusty nail or an unsavory glob. On very rare and lucky occasions, a wave may actually present the visitor with a pearl-laden oyster or a lost treasure.

The computer revolution's innumerable waves have borne such bounty as PCs, client/server networks, and laser printers. But they've also brought forth unsavory globs, like bubble memory, Apple Computer Inc.'s Lisa, and IBM Corp.'s Top View.

The latest batch of technologies has just arrived on the shore. It's too soon to tell if this particular breaker has deposited anything of lasting value, but the nice thing about waves—whether aquatic or technological—is that there's always another one on the way.

PERSONAL DIGITAL ASSISTANTS

Call them personal digital assistants (PDAs), portable communicators, or mobile assistants, these easy-to-carry handheld devices seem destined to make a big splash in business. The various models

Source: CIO Magazine (December 15, 1993–January 1, 1994): 58–61.

differ only slightly in design and concept; all merge pen input, handwriting recognition, personal-organization tools, and state-of-the-art mobile communications into a compact, easy-to-use package.

Apple's Newton Message Pad is the best-known PDA, but similar units have been developed or are being planned by such companies as Compaq Computer Corp., Tandy Corp., Casio Inc., and AT&T's EO Inc. subsidiary. By late 1994, a dozen or more PDA models should be on dealers' shelves.

Although nearly all industry observers agree that PDAs are a good idea and will be with us in one shape or another in the years ahead, the jury is still very much out on whether the first ones are anything more than expensive novelties. Many early MessagePad users have complained about the unit's slow and inaccurate handwriting recognition, hard-to-read screen, and limited storage capabilities. Users have also griped about the lack of decent PDA software, claiming that the few available applications are mostly warmed-over PC products that fail to fully exploit the PDA concept of fast, on-the-spot information.

PDA supporters are quick to point out that users' expectations are too high and that first-generation assistants like the MessagePad should be viewed as harbingers, works-in-progress of an important technology. The supporters draw comparisons between PDAs and early personal computers, most of which allowed users to do little more than balance a checkbook and play a few text-based games.

In any event, market researchers seem convinced that PDAs are here to stay. International Data Corp., of Framingham, Massachusetts, says PDA shipments will grow to between 2 million and 5 million units by 1998, with sales topping $3.5 billion in five years. BIS Strategic Decisions, of Norwell, Massachusetts, forecasts that the market will grow to 2.3 million units by 1998.

WIRELESS COMMUNICATIONS

Wireless technologies are an important part of the budding PDA field. New-generation wireless systems are also making an impact on local area networking and person-to-person voice communications.

In the PDA area, plug-in radio receivers currently allow users to receive brief onscreen messages transmitted via nationwide paging services. Next-generation radio modems will enable users to both receive and transmit text, video, sound, and fax files. In the years ahead, PDA users will be able to dial up local area networks and an array of online information and transaction services from virtually any point on Earth. Ericsson GE Mobile Communications and Motorola Inc. are poised to become major players in the radio-modem field with units based on the PCMCIA standard for plug-in PC peripherals.

Several development projects are focusing on new ways of connecting users by radio. These ventures are grouped under the umbrella classification of personal communications services (PCS). PCS is based on a simple premise and a lofty promise: to give users data and voice connectivity anywhere, anytime.

Ventures and companies, including ARDIS Co., EMBARC (Electronic Mail Broadcast to a Roaming Computer), RAM Mobile Data Corp., McCaw Cellular Communications Inc., Pacific Telecom Inc., and SkyTel Corp., are looking for technologies that will make wireless communications fast, easy, and cheap. Motorola's Iridium project is attracting particular interest; the venture aims to create a global wireless network using 66 satellites that would link up wireless users around the planet.

Another budding PCS technology is a communications infrastructure called the personal communications network (PCN). This system, based on microcell wireless technology, is designed to provide lower-cost wireless telephone coverage in metropolitan areas. The technology's cells, which are more-closely spaced than current cellular-telephone cells, will allow smaller and more-power-efficient radio transmitters. With PCN, many analysts see wireless PDAs and portable telephones eventually becoming as commonplace as wallets and pocketbooks.

The PCS market got a shot in the arm last summer when the Federal Communications Commission approved the allocation of radio frequencies in the 1850MHz-to-1990MHz range to companies with PCS ventures. The channels will be auctioned to the highest bidders in a projected four-year process that's expected to place about $4 billion to $7 billion into the government's depleted coffers.

As with PDAs, market researchers see boom times ahead for wireless technologies. Market Intelligence Research Corp., of Mountain View, California, reports that wireless personal-communications products and services will balloon from $70 million in 1992 to $1.9 billion in 1996 and more than $13 billion by the end of the decade; New York–based Frost & Sullivan Inc. concurs.

MULTIMEDIA

Although multimedia is usually associated with home-oriented applications such as computer-based encyclopedias and learning programs, the technology also has a wide range of business applications. It is already making inroads in training, where products combining text, graphics, video, and sound are helping users acquire and retain information and skills faster and more effectively. Well-designed multimedia applications combine audiovisuals drawn from videotape or laserdisc with computer intelligence to produce a learning experience that can quickly adapt to the needs of the user.

The most-conclusive study of multimedia training to date was conducted by the U.S. Department of Defense. The agency, which has an annual training budget of about $18 billion, found that the technology is generally more effective and less costly than conventional instruction. It also determined that multimedia could be used to teach students effectively in a variety of training environments.

For example, in manufacturing, multimedia can be called upon to help workers learn without risk about intricate and hazardous tasks. In all businesses, multimedia pop quizzes allow managers to assess a trainee's knowledge and steer instruction to most-needed areas. The technology is also useful in social-behavior training, as with programs that simulate sexual-harassment and minority-discrimination scenarios.

Compared with traditional classroom instruction, multimedia training can be cheaper, more convenient, and less intimidating to employees. A product like the $4,995 Authorware Professional from Macromedia Inc., of San Francisco, allows nontechnical users to construct an icon-based interface that closely resembles a book, complete with multimedia "pages." The $3,395 Quest Multimedia Authoring System from Allen Communication Inc., of Salt Lake City, lets users create interactive multimedia courses featuring a wide range of visual and aural effects.

Multimedia is also making strong gains in other business areas. Marketing departments are installing multimedia-based kiosks that work as on-site demonstration and purchase centers, and management is discovering the benefits of multimedia-based presentations and business conferencing.

Even traditionally straitlaced business applications are getting into the multimedia act. For example, Lotus Development Corp., of Cambridge, Massachusetts, is in the process of embedding multimedia capabilities into its core products. In March 1992, the company announced Multimedia SmartHelp for Lotus 1-2-3, a CD-ROM that offers sound, graphics, and video learning capabilities. Another product, Lotus Sound, enables users to link and embed audio messages into business documents. Users can create and play back digitized audio files that can be attached to any Windows OLE-enabled application.

One barrier to more widespread acceptance of multimedia in business is the technology's cost and evolving nature. Fully configured multimedia PCs are priced higher than plain PCs, while peripherals such as CD-ROM drives and add-in audio boards range from $300 to $800 each. Another key drawback is the fact that multimedia products add more weight to already sagging and overburdened networks.

Network throughput carries the greatest load when huge packets of full-motion video images and more traditional data contend for limited bandwidth. A single frame of video requires 150Kbps of throughput. With full-motion video running at a rate of 30 frames per second, the strain on a network can be enormous. To run multimedia without compromise requires added disk storage, more-muscular processors, faster networks, and advanced software, most of which remain unavailable and cost-prohibitive. Even when higher-bandwidth and increased-storage multimedia building blocks arrive, these extras will still be necessary.

Still, expectations for multimedia technology are running high. According to Market Intelligence, multimedia sales worldwide will almost quintuple over the next five years, going from $5 billion in 1993 to $24 billion by 1998.

VIRTUAL REALITY

Anyone who has watched TV or picked up a news magazine in the past few months knows something about virtual reality. Unfortunately, since most media reports have concentrated on the technology's "gee whiz" aspects, all most people know about it is that you must put on a funny helmet and ride up and down various computer-generated escalators.

But virtual reality actually has a broad array of practical business applications. Major airlines and the military have used aircraft simulators for decades—a type of virtual reality that combines computer-generated effects with real-world equipment and surroundings. The big advantage offered by the emerging generation of fully digitized virtual-reality systems is lower cost, which will eventually make the technology available to many more businesses.

Some of the potential applications are genuinely mind-boggling. For investment managers, virtual reality could be used to quickly uncover undervalued stocks. In one possible system, the user would wander through a field of 3-D blocks, with blue blocks representing rising stocks and red ones denoting falling issues. The view would allow the investor to instantly see how individual stocks are doing in comparison with their competitors—an edge that could mean quick profits in fast-paced financial markets.

Real-estate agents and conference planners could use the technology to scout properties "on-site" without leaving their offices. For companies that deal with hazardous substances and technologies, it could be used to train workers in a danger-free setting (just as the airlines and nuclear-power industries currently do with their high-cost simulators).

For most businesses, practical applications lie at least several years away. But a number of software publishers are now offering virtual-reality toolkits and simulators that can help businesses become acquainted with the technology.

Sausalito, California, vendor Autodesk Inc. recently released the Cyberspace Developer Kit—a sophisticated toolkit for writing virtual-reality programs. The company predicts that the kit's programming capabilities and relatively low $2,495 price tag will accelerate the use of virtual-reality applications across industry lines. Applications are currently being developed by the software's users in the business, science, education, and entertainment markets. The programs themselves offer "worlds" that users enter using anything from simple keystroke commands to full glove-and-helmet interaction. The kit provides built-in networking capabilities and runs on any 386-based PC or higher.

Businesses looking for a bargain-basement taste of virtual reality can check out Virtus VR, published by Virtus Corp., of Cary, N.C. The $99 package allows users to explore, alter, and invent realistic 3-D worlds on their PC screens. Virtus VR comes with five prefabricated worlds: the White House (including the Oval Office), the Hindenburg zeppelin, a "mystery house," a futuristic ocean-floor research/apartment complex, and Dallas's Dealey Plaza, site of the Kennedy assassination.

The latter world features President and Mrs. Kennedy in their limousine riding by the infamous grassy knoll before a scanned image of the Texas School Book Depository. Users can conduct their own investigations of the assassination by viewing the knoll from the perspective of the limo or by "flying up" to the Depository and peeking through the fateful window.

Pretty strong stuff. But if reliving national tragedies isn't your thing, you can always enter the software's ocean-floor world and watch the waves pass overhead. You may even spot a new technology.

QUESTIONS

1. Pick one of the four technologies described in this section. Can it be used to improve productivity in a manufacturing organization? Explain why or why not.
2. Pick one of the four technologies described in this section. Can it be used to improve productivity in a service organization? Explain why or why not.
3. Which of the four technologies described in the article are most likely to have a lasting impact on the way we will work in the future? Explain why.

2.4 Image Processing Systems: Users Put Benefits—and Costs—in Focus

R. KELLY RAINER, JR., HOUSTON H. CARR, CHARLES A. SNYDER, AND GLENDA DENEEN

When very large amounts of data must be handled, computers come to the fore. A typical case is the insurer who must keep complete files on a policy and all the claims that are made against it. For a large insurer this requirement meant that literally hundreds of thousands of file folders had to be stored in warehouses. In the past, when a claim came in, people had to find it in the warehouse and bring it out for an adjuster to handle. Naturally, some files were never found and others took weeks to locate. Furthermore, paper is heavy. It took large, specially reinforced warehouses to store all the material. Productivity in such an environment is naturally low.

Imaging refers to the use of magnetically or optically stored digital equivalents of paper. A typical application that you, as a consumer, may have seen are the copies of your American Express vouchers that accompany your monthly bill. As Rainer and his co-authors point out, simply reducing paper to electronic form is not enough. A full system has to be created which takes the paper, converts it to electronic form, and then allows its retrieval and use for business processes.

Imaging is not cheap. It is a classic case of economies of scale. Unless you have a large enough volume of transactions, the investment and operating costs are prohibitive. In this article, you will find a description of the technology, of the costs, and of the benefits. You will find that if the volume justifies it, not only does imaging increase efficiency, it also changes the way you can serve the customer.

Source: Information Systems Management (Summer 1992): 30–37.

ORGANIZATIONS can usually benefit from implementing new technologies in two ways. First, the technology may result in a relatively long-lasting strategic advantage to the business. Second, the technology may result in increased efficiencies and lower costs. The popular press and industry trade publications have typically regarded image processing systems as contributing only to a reduction in paper. Discussions of the benefits of image processing (also known as imaging) systems have mostly centered on efficiency gains and often have overlooked strategic gains.

The driving force behind the development of image processing systems has been the desire to reduce the need for paper. Organizations continue to generate a large amount of paper despite their use of computers. In fact, the use of paper negates some of the advantages possible with computer-based systems. Paper is neither addressable nor readable by computer-based applications. Image processing systems may ultimately improve an organization's data manipulation methods by bringing the so-called paperless office closer to reality.

ALTERNATIVES TO PAPER

The simplest and most direct method of storing data is to file the source document (i.e., the paper on which the data was originally placed). Organizations have long realized that paper is expensive to create, handle, store, retrieve, and return to storage. In addition, paper is fragile, bulky, and subject to chemical deterioration. For these reasons, many businesses have sought alternatives to paper storage.

Paper reduction alone may not be sufficient justification for an image processing system, however. An organization may also need to examine alternatives and complements to image processing technology—for example, micrographics, electronic data interchange (EDI), E-mail, and corporate videotext. An organization may decide that an image processing system is not suitable for its particular application or that a combination of technologies is necessary for the efficient and cost-effective handling of documents. For example, a company may use EDI for common business transactions, E-mail for most internal transactions, microfilm for rarely referenced documents, corporate videotext as a data base enhancement and image processing for applications that require quick access to pictures of the original documents.

Micrographics

Micrographics—the use of microfilm, microfiche, or computer output on microfilm (COM) to record pictures of documents—is an alternative to paper for the simple archival storage of documents. When black-and-white 35-mm and 16-mm film were first used as storage media for documents, there was no way to directly address (i.e., retrieve) stored images. With time, systems were developed that could easily and quickly access these media; however, micrographic material remains neither readable nor addressable by computers.

Microfilm has some advantages over image processing systems, however. Microfilm requires smaller financial outlays and has lower operating and maintenance costs and faster input time. An organization may choose microfilm or microfiche when the document is rarely referenced or the quick access time affordable by image processing is not needed.

ELECTRONIC DATA INTERCHANGE

EDI eliminates paper altogether. Whereas image processing technology scans an original document into the system, with EDI the document originates in the source computer as an electronic record. The electronic record is a replacement for the original source document, not a means of storage. EDI is therefore a complementary technology, not an alternative, to image processing systems.

E-MAIL

For internal communications (e.g., memos and letters), E-mail can help an organization reduce its total paper output. As with EDI, E-mail can eliminate paper completely because the communications are electronic. E-mail is not a means of storage and therefore can be used in conjunction with image processing systems.

VIDEOTEXT

Corporate videotext refers to a company data base system that has the capability to manage data, text, and graphics. Image processing fits closely with corporate videotext because documents may be digitized and stored in the data base for easy access by company personnel. The capability provided by corporate videotext is generally limited and designated to be an extension of a particular field (e.g., a picture of a person in a human resource data base).

A STUDY OF IMAGING SYSTEMS IN USE

In general, image processing applications are effective when large amounts of paper can be eliminated. These systems are useful when data is nontextual and cannot be keyed into a computer-based system. In addition, image processing systems are practical when document processing requires frequent access or retrieval by many different people.

To investigate these current organizational applications of image processing systems, an empirical study was performed by Auburn University. A two-part questionnaire was mailed to a random sample of 210 geographically dispersed companies chosen from the subscribers to *Information Center* magazine. The first part of the questionnaire defined an image processing system. The definition was important because these systems represent a new technology, so a precise definition may not be universally accepted among academicians and practitioners.

The second part of the questionnaire was used to gather demographic data on each organization. It also asked for data concerning the development, operation, costs, and benefits of image processing systems. A pretest of the instrument yielded some suggested changes that were incorporated into the final survey instrument.

Of the 210 questionnaires mailed, 27 were returned as undeliverable. Of the 183 questionnaires that were delivered, 50 were completed and returned, for a response rate of 27.3%. Fourteen of these organizations indicated that they had implemented an image processing system; the 36 others had not. The percentages used in this article to describe current practices are based on the 14 businesses with systems in place. Because the respondents did not answer every question in some cases, the percentages are based on the number of responses received. All the issues discussed in the remainder of this article are presented in light of survey data from the 14 organizations with image processing systems.

The Demographics

Organizations with large processing systems represent a variety of industries. The organizations surveyed have 6,530 employees on average and list total corporate assets of $2.55 billion. The respondents report that their image processing systems have been in operation an average of 14.7 months.

Thirty-six of the 50 responding organizations (72%) did not have image processing systems; of these, 23 are investigating such systems, 12 are planning for one, and one business is testing a pilot system. When asked why image processing technology was not implemented, the reasons given reflected financial (84%), technological (28%), project support (16%), and legal (4%) considerations. (These percentages do not total 100% because some respondents gave more than one reason.)

The chief financial considerations were high cost, budget constraints, and other budget priorities. Technological reasons that were cited were that the organization's current infrastructure cannot handle the storage and transmission load, the technology is considered risky, and the organization lacked suitable applications. The lack of a sponsor, labor resources, or user support were other factors inhibiting image processing projects. On the legal front, image processing technology has not been proved to meet legal requirements. One respondent stated that hard copies of the organization's records must be stored to meet federal and state regulations.

Image Processing Systems Planning

The respondents using image processing systems reported that the planning phase lasted 9.1 months, on average. The actual implementation process took an average of 4.3 months after the planning stage. Ten organizations indicated that they had senior management support for their systems, and 12 had a champion for the project.

When asked which storage media were employed before the imaging system was implemented, eight respondents said microfiche, six said microfilm, and four used computer output on microfilm. Four organizations reported the use of multiple methods of storage. When asked about the extent to which the image processing systems replaced the technology previously used, the respondents were asked to answer on a scale from 1 (not at all) to 7 (completely). Microfilm replacement averaged 4.8, microfiche averaged 4.1, and computer-output micrographics averaged 1.3. These findings suggest that image processing systems are being employed in conjunction with technology already present in organizations.

Three explanations are plausible. First, companies may be using previous technologies as backup. Second, image processing systems are a new technology, so companies may not have the confidence to fully convert to them. Third, it may be too expensive to completely convert to image processing systems.

Systems Structure and Operation

Image processing is an emerging technology that converts documents into digital form for storage on optical disk, magnetic disk, or magnetic tape for use with an organization's computer systems. Using such a system, an organization can capture, store, retrieve, display, process, distribute, and manage business information in digital form.

A typical image processing system contains five functional components:

- A document scanner
- A document storage processor
- Storage devices and media
- Workstations
- Plotters and printers

All of the organizations indicated they used document scanners, and 10 used document storage processors. Most of the organizations (13) had dedicated printers attached to their image processing systems, and one organization reported the use of a plotter.

Image processing systems may be either standalone or integrated with the organization's existing information system. Although most currently installed image processing systems are standalone, the trend is toward larger, integrated systems (i.e., workstations and networks) that rely on standard architectures and components. In this study, nine respondents reported that they have standalone image processing systems; five have integrated systems.

DIGITIZATION AND COMPRESSION

Most systems use page readers and scanners to convert a paper-based image to a magnetically or optically stored, digitized equivalent. This stored image is a series of dark and light points. Optical character recognition (OCR) scanners detect and store text characters. Text characters are readable by the computer and can be manipulated by it. Stored dark and light points are not intelligible to the computer. When the stored image is attached to a machine-readable record, the record can be used for computer processing, but the image cannot.

The job of scanning existing documents into the imaging system was done in-house, according to nine of the respondents; three other organizations subcontracted this task. Two companies indicated they did not scan existing documents into their systems. Eight of the respondents did not destroy documents after scanning; the other six said they did destroy documents after scanning.

The scanner creates a digital representation of the original document so that it can be stored on magnetic or optical media. This digital image, called a bit map, is an exact representation of the paper image. After the image has been digitized, it is compressed to reduce required storage space. Compression removes extraneous information and condenses redundant patterns of pixels, the basic elements of an image, into a kind of "coded electronic shorthand."[1] The compression process does not alter the content of the paper source.

INDEXING

For a paper or imaged document to be stored so that it can be retrieved, it must be indexed. Paper documents in file cabinets have identifiers, and imaged documents in computer-based storage areas have record numbers that are similar to data base records. For example, the United Services Automobile Association (USAA), a large insurance company that relies heavily on image processing, identifies its documents, either paper or image, by a special member number.

STORAGE MEDIA

After a document has been scanned, compressed, and indexed, the digital image is sent to a host, usually a minicomputer or mainframe computer, called the document storage processor (DSP). The DSP controls the movement of images in and out of storage and routes information.

The choice of storage medium—either magnetic disk, magnetic tape, or optical disk—usually depends on the access time required, the frequency of access, and cost. Magnetic disk storage has traditionally provided the fastest access time, followed by magnetic tape and then optical storage. However, advances in optical media have improved response times and lowered costs so that optical disks now compare favorably with magnetic media.

Because most imaged documents are archival in nature, storing them on such optical media as WORM (write-once/read-many) or CD-ROM (compact disk/read-only memory) devices makes sense. Because both media are read-only, they are nonalterable. Eight respondents said they employed magnetic disk as the storage medium in their imaging systems, eight use optical media, and two use magnetic tape (the responses total more than 14 because four of the respondents indicated they use multiple storage media).

A user accesses the document through a microcomputer or workstation terminal. The workstation allows the user to access, retrieve, and view the document. Printers and plotters are attached to the image processing system so that a hard copy of the imaged document can be created as required. The companies surveyed linked, on average, 14 workstations to their imaging systems.

Costs of Image Processing

Costs associated with imaging include system costs, employee costs, storage costs, transmission costs, and back-file conversion costs. All relevant costs should be considered before an organization decides to purchase and install an image processing system. If only the hardware and software costs involved are considered, the actual cost to the organization will be underestimated.

SYSTEM COSTS

Imaging system costs vary, depending on whether the system is standalone or integrated. Findings from this study support the common claim that standalone systems are less expensive than integrated systems. In this study, the organizations with standalone systems reported an average cost of $52,000. Those with integrated systems stated that their imaging systems had an average cost of $1.65 million.

EMPLOYEE COSTS

The costs of skilled workers include their salary, training, and retraining in new technologies. It has been estimated that the cost of adding a new knowledge worker to an organization is $38,000 for the first year, with annual costs of approximately $19,000 to maintain that knowledge worker in an image-enabled environment.[2]

If these figures are typical, organizations may choose to retrain existing employees rather than hire and train new ones.

STORAGE COSTS

Storing a document as an image requires computer and peripheral storage space. For example, after applying data compression techniques, an 8½-inch by 11-inch page still occupies 50K bytes or more of storage. Potential benefits, but also storage costs, increase with the widespread use of imaged documents.

TRANSMISSION COSTS

Not only do images take up computer storage space, but their transmission costs are also greater than those of simpler electronic data. One estimate states that the cost of transmitting a typescript page electronically as an image is about six times higher than the cost of sending a similar page in the form of computer text.

BACK-FILE CONVERSION COSTS

After an organization implements an imaging system, it must scan the necessary documents to convert them to images. This conversion costs both time and money depending on the condition of the documents, how much indexing is needed, and whether the scanning is done by a third party. Estimates for back-file conversion are approximately $0.25 to $1.50 per page.

The survey respondents noted which costs they expressly considered during the planning process. Some considered more than one cost. For example, five of the respondents said employee costs were carefully considered, six considered storage costs, four considered transmission costs, and five considered back-file conversion costs. No cost categories were omitted from the survey, according to the respondents.

BENEFITS OF IMAGING

Information technology can affect an organization's competitive advantage by lowering costs or enhancing differentiation in its products or services. Imaging systems may reduce storage space needs, labor costs, communications expenses, and mailing costs. There is less reliance on copying equipment as well as improved document repair. The average total cost savings reported by survey respondents was $324,000.

REDUCED STORAGE SPACE REQUIREMENTS

Converting documents to images reduces the space needed to 0.5% of that required for paper-based documents. Because most of an organization's information is still on paper, considerable space must be allocated to document storage, which incurs rent, protection, maintenance, and insurance costs. If the space requirement is eliminated, as in the case of a warehouse that is no longer leased, the cost avoidance is direct and obvious. If the space is reallocated, the cost may be shifted to a revenue-generating activity that can support it directly.

REDUCED LABOR COSTS

The reduction in labor costs results not only from the reduced employee time required to retrieve documents but from a reduction in the number of staff members required. According to a 1990 report, "corporations spend an average of $1.50 to retrieve a filed document; a misplaced document runs about $120 in lost employee time."[3]

Image processing systems have resulted in staff reductions in some businesses. For example, North American Van Lines, Inc., which installed an imaging system to handle customer billing, has been able to trim clerical staff by 20% because of the reduced paper handling. Another organization has estimated that it will not expand in personnel for at least two years because of the implementation of an imaging system. These estimates suggest that image processing has a significant cost savings potential through staff reduction and avoided or delayed staff expansion.

REDUCED COMMUNICATIONS EXPENSES

Because documents can be kept online with an imaging system, a company may be able to cut its communications expenses. For example, the number of customer call-backs is reduced because service personnel have immediate access to relevant and historical information on customer accounts with an online imaging system. Furthermore, customer perception of the organization's service levels may be enhanced.

REDUCED MAILING COSTS

Imaging can reduce mailing costs associated with returning such documents as checks and credit slips. Even if a company must return a hard copy to customers, it can store original documents as images, reduce the size of these images, and print several images per page on lighter-weight paper, thereby reducing postage costs. American Express Co. provides its customers with compact images of receipts rather than the receipts themselves.

REDUCED COPYING COSTS

Image processing systems reduce the need for document copies, the number of copy machines, and employee time spent copying paper documents. One Exxon study found that the average document in the company's headquarters office was copied 40 times; 15 of these copies were filed away in permanent storage. Imaging reduces the need for excess paper.

IMPROVED DOCUMENT REPAIR

Image processing systems protect documents from damage that can result from handling or aging. Documents that are damaged before they are scanned into the system can be restored to their original condition with image-enhancement techniques that can remove stains or bring out hard-to-read areas.

Enhancing Product or Service Differentiation

Companies also use imaging technology to differentiate their product or service from the competition's, thereby raising barriers to market entry of competing products. Imaging affords a company benefits, ranging from faster access time to improved work control, that can help preserve competitive advantage.

IMPROVED DOCUMENT ACCESS TIME

An important benefit of image processing systems is that they reduce the time it takes to retrieve documents. With documents stored online, in media directly retrievable by the computer, an employee can quickly access the documents through a workstation instead of searching through file cabinets. The USAA reports access times with its imaging system of five to 20 seconds, compared with the several minutes needed to physically retrieve and replace documents. This improved access time provides other competitive benefits as well—for example, increased productivity and improved customer service, each of which is discussed later.

FASTER TRANSACTION PROCESSING

Faster transaction processing means less time is needed to complete work on a document and is a direct result of the information's being available online. Documents are immediately available to all interested parties and can be routed to different parts of the organization without the need for human intervention. The time savings are significant. At one insurance company, an imaging system has helped cut the time needed to underwrite a new liability policy from 33 days to five.

ABILITY TO HANDLE LARGER TRANSACTION VOLUMES

An image processing system can streamline an organization's processes so that work can be done more efficiently and the company can handle a much larger volume of transactions. For example, the McDonnell Douglas Corp. is reportedly handling three times the volume it had been handling in accounts payable before the implementation of its imaging system, and it is able to do so without a corresponding increase in staff or facilities.

IMPROVED CUSTOMER SERVICE

Customer service is a critical component of an organization's success. The quick access time afforded through use of imaging can improve the quality of customer service. That fewer calls are put on hold leaves customers with the favorable impression that the business is being run competently. With better response times, telephone support may replace mail as the preferred form of customer support.

REDUCTIONS IN ERRORS

The Department of Veterans Affairs in Washington DC has reduced its mispayment rate from 3% to almost zero after installing an image processing system because of the access it provides to complete and correct documentation. One way to control errors is with work-flow software. With such software, users receive all the information needed for processing and the information is presented in the correct sequence. Errors that result from insufficient documentation or processing can be reduced through the combined use of work-flow software and an imaging system.

IMPROVED WORK CONTROL

With work-flow software, managers can also monitor operations, work can be assigned to those employees who are most skilled in a specific area, files can be automatically routed from one department to another, and the current work status of a file can be obtained.

INCREASED PRODUCTIVITY

Although productivity is difficult to measure, it appears that an organization can increase its productivity through the implementation of an imaging system. A Yankee Group survey revealed that "in 99% of [image processing] installations, workers productivity improved by 20% or more."[4] This increased productivity may be the result of workers having easy access to documents, or it may be because the imaging system eliminates the problem of documents being out of file. Increased productivity may also result from the use of hypertext documents. Although documents are generally stored as graphic images and are not machine readable, hypertext techniques are possible if the document is attached to a record in a data base. Keywords in the data base can then be hypertext-linked to related images.

One reason for the benefit of increased productivity is that time savings for employees result in cost savings for the business. USAA insurance company has estimated that after installing its imaging system, increased worker productivity has provided a yearly savings of approximately $5 million in annual benefits and overhead costs for employees.

Table 1 lists the potential benefits of image processing systems, as discussed in this article, along with the importance rating each benefit received from the survey respondents. The benefits are ranked from 1 (not at all important) to 7 (very important).

The benefits of enhanced differentiation have an average importance score of 5.64; benefits related to reducing costs average 3.97. These findings suggest that businesses implementing imaging systems place more importance on differentiation benefits than on cost benefits.

CRITICAL SUCCESS FACTORS FOR IMPLEMENTATION

According to the popular and trade press, there are eight critical success factors for the implementation of an image processing system. Organizations need to:

- Create a knowledge base about image processing.
- Develop executive sponsors.
- Redesign work-flow processes.
- Implement a pilot project linked to the business strategy.
- Integrate image processing with existing information technology infrastructures.
- Delineate the conversion process.
- Join industry groups that are exploring imaging technology.
- Ensure cooperation between technical and business managers.

These eight critical success factors (CSFs) for the successful implementation of an image process-

TABLE 1
Potential Benefits of Image Processing

System Benefits	*Importance Ratings**
Lowering Costs	
Reduced Storage Space Requirements	4.60
Reduced Labor Costs	5.30
Reduced Communications Expenses	3.40
Reduced Mailing Costs	3.20
Reduced Copying Costs	3.70
Reduced Copying Equipment	3.70
Improved Document Repair Capability	3.90
Enhancing Differentiation	
Improved Document Access Time	6.40
Faster Transaction Processing	6.00
Increased Productivity	6.00
Improved Customer Service	5.80
Improved Work Control	5.30
Larger Transaction Volume	5.10
Reduced Errors	4.90

Note: *Ratings are on a scale from 1 (not at all important) to 7 (very important).

ing system may also apply to other information technologies. Therefore, survey respondents were asked to provide an open-ended list of imaging system CSFs. Table 2 shows the issues they indicated and the number of respondents who answered.

Respondents added the importance of technology reliability and the relationship between users and image processing systems as critical success factors. The relationship between users and the system (e.g., ease of use, availability to users, and training) reflects the fact that imaging is an emerging technology and therefore may not yet have sophisticated user interfaces.

TABLE 2
Critical Success Factors Listed by Respondents

Critical Success Factors	*Number of Respondents*
Hardware and Software Reliability	5
Ease of Use	5
Availability to Users	5
Training of Users	4
Integration with Existing Hardware and Software	4
Management Support and Commitment	3
Business Orientation	2
Redesign of Work Processes	2
Planning	1
Costs	1

CONCLUSION

The implementation of image processing is often initially viewed as an efficiency measure because it reduces the cost of storing and using documents. The reduced cost is the combined result of the compression of storage space and the elimination of personnel. Reduced cost is frequently the basis for installing expensive equipment, software, and procedures and provides the tangible benefit against which the cost of installation is gauged. Yet these obvious savings may not be nearly as significant as the ability to provide improved customer service levels, greater responsiveness, and an image of enhanced organizational effectiveness.

Imaging systems are not as yet well documented and need further academic investigation of their planning, implementation, and evaluation. Case studies to verify the critical success factors reported in this article are needed to assist organizations hoping to implement their own imaging systems. In addition, more rigorous studies to verify the benefits and costs are needed. Issues companies must better understand include:

- How to measure costs.
- How to quantitatively measure system benefits.
- What cost/benefit methods to use to evaluate image processing technology.
- What implementation methodologies to use.
- The successes and failures companies have experienced with imaging systems.

NOTES

1. "Imaging: The Big Picture," *Solutions* 11: 5 (1990): 22–23.

2. May, T.A., "Fixing Your Image Problem," *CIO* 3 (November 1989): 68–73.
3. "Imaging: The Big Picture," 22–23.
4. Fisher, M.J., "Digging Out with Image Technology," *Datamation* 35 (April 15, 1989): 18–26.

QUESTIONS

1. What key technologies are used in image processing systems?
2. What type of industries are most likely to benefit from image processing systems? Explain how.
3. Do you think image processing systems will replace paper in most organizations employing more than 100 people? Why or why not?

2.5 Bar Coding Making Inroads in Manufacturing

BILL McILVAINE

All of us are familiar with the bar codes on our groceries that speed checkout. Bar coding on retail products is so universal that when President Bush marveled about it, his level of knowledge made headlines. In this short article, McIlvaine explains how bar coding is now used in manufacturing as a way to keep track of parts and finished products. The main advantage of bar codes is that they eliminate the extra time and the errors that result when information is keyed into the computer over and over. It is McIlvaine's contention that the capabilities of bar codes will increase over time, making it possible to remove more and more paper from the manufacturing and logistics processes.

FEW recent technologies have gained such widespread acceptance in manufacturing as bar coding. On the shop floor and in the warehouse, bar coding has become a standard means of acquiring more timely and more accurate data through the elimination of manual data entry systems—which also lowers its cost to the user.

The market for bar code products such as scanners, printers, software, and media is still growing at 15% to 20% annually, according to market research firm Venture Development Corp. (Natick, MA). Bar code equipment accounts for 70% ($2.4 billion) of the total auto ID industry, which also includes radio frequency, machine vision, magnetic stripe, voice recognition, and other technologies. Europe is the next market frontier for bar code manufacturers, as sales of bar code systems in Europe are expected to triple over the next five years, according to Market Intelligence Research Corp. (Mountain View, CA).

But despite the healthy picture for the bar code industry, the revolution is far from over. Ac-

Source: Managing Automation (September 1992): 24–26.

tivity is taking place on both the cutting edge and the trailing edge of the industry that will enhance the capabilities of bar coding while bringing its benefits to more and more users.

On the most basic level, AIM USA Inc. (Pittsburgh, PA), the auto ID industry trade association, is producing its first bar code primer, aimed at corporate executives who still are not familiar with the technology, its applications, and advantages.

"We're finding companies forming all the time, and this information is aimed at them," explains Bert Moore, technical communications specialist at AIM USA. "As companies reorganize, people may be moved to new locations and new replacements may be brought in. The person who was the auto identification 'guru' may have been replaced by someone who doesn't know that much about auto ID. So there is a tremendous amount of reeducating to do."

Linda Barber, auto ID executive at Xerox Corp.'s Printing Systems Div. (El Segundo, CA), is in charge of drafting the primer, which is being compiled from various sources of information within AIM USA. "Industry looks to AIM USA as a leader and a primary source of information about bar coding," she explains, "and this publication will provide a single, definitive source—something from beginning to end."

The need for a primer is there, she says, because the bar code industry still has room to grow, and new users are coming into the market every day.

"There is definitely room to grow," explains Dave Cohen, director of marketing for Accu-Sort Systems Inc. (Telford, PA) which markets bar-code-based sortation systems to large manufacturers. "Many of the larger companies have implemented extensive applications and that has opened the door to even more applications to get more accurate and timely data flowing through the manufacturing enterprise."

Among manufacturing industries, the one that is leading the growth in bar coding is semiconductors, where individual chips and packages are being bar coded. This is the result primarily of increased automation in the electronics manufacturing process and also from the need to track smaller lot sizes, down to individual wafers. Use of bar coding is also increasing in areas such as chemicals, aerospace, and metals processing. For pharmaceutical makers, bar codes are not just convenient but necessary to validate the manufacturing and packaging processes.

"Manufacturers believe bar coding will help them achieve management objectives in total quality management, just-in-time inventory controls, and quick-response order processing," states Venture Development's recent report on the bar code industry.

But future growth is likely to be fueled by standardization of bar code symbols, which encourage more companies and industries to adopt the technology. The Universal Product Code (UPC), for instance, has grown beyond the grocery store into retailing and other industries. Interleaved 2 of 5 is the symbology of choice in warehousing and manufacturing for product-tracking applications. Another example is Codabar, which is used extensively in retail price-labeling.

"There is a great deal of efficiency to be gained for the U.S. economy by standardizing the movement of products between trading partners," says David J. Collins of the Data Capture Institute (Duxbury, MA), a market research group. "Bar coded messages are used as the fundamental medium of exchange."

Bar codes originally were nothing more than object identifiers, and most of their applications remain in simple automatic identification. But those codes are beginning to carry even more information. They are becoming full-fledged tracking devices, carrying in-process manufacturing data and shipping notices, for instance.

But a standard bar code symbology is just the beginning. Standard symbologies are merely frameworks, which can be modified or extended as applications dictate.

The UPC, to take one example, is nothing more than a product identifier: the standard identification number for a manufacturer's jars of peanut butter, for instance. It doesn't represent a particular jar of peanut butter. This is primary identification as defined by the Uniform Code Council. Secondary identification is to identify the individual units of the product: manufacturing history, dimensions, weight, or other characteristics.

THE UCC Shipping Container Code was established in 1987, based on Interleaved 2 of 5. Its symbol specifications are defined for containers of both fixed and variable contents. But it was not flexible enough to handle shipping containers with non-standard product mixes: jars of peanut butter and cans of applesauce, for instance. Thus the Uniform Code Council, the organization that administers the UPC as well as the UCC Shipping Container Code, developed the UCC/EAN-128 Serial Shipping Container Code, based on Code 128. This code, published in 1989, includes a seven-digit manufacturer ID number instead of the six-digit UCC Shipping Container Code number and a nine-digit number assigned by the manufacturer which uniquely identifies each carton and the contents in it.

"The code was also designed to be used in conjunction with EDI to allow trading partners to provide advance shipping notification, to verify shipment, and to detect misshipments at the loading dock," explains Collins.

Yet even this new code had limitations. It still could not hold all the information that manufacturers found that they wanted to include. "The format of the labels needed to be standardized to contain more information. Manufacturers wanted to code ship-to locations, sell-by dates, production dates, packaging dates, lot numbers, and more," says Collins.

And every user of the code would not necessarily need to scan every item in it during every application. A system was needed by which "only the information required for that operation should be used," says Collins.

Therefore, in 1991 the UCC and the International Article Numbering Association (EAN) jointly developed a definition of new "application identifiers," and published their guidelines. Application identifiers extend auto ID capabilities to manufacturers that need more specific data coded on their bar code labels—for instance, perishable-foods producers that need to convey sell-by dates, or wallpaper manufacturers that sell by both lot number and stock-keeping unit (SKU).

Collins says one of the major advantages of application identifiers is the ability to combine several data fields in one, eliminating the need for multiple bar codes. "When all the data is concatenated into one symbol, the host computer can easily isolate the correct data fields," he says.

"The fundamental message is that this is a serialized label for a serialized package. There's a vendor number and a serial number. You know exactly what's coming and you can process it faster, move it around, and also pay the supplier faster," continues Collins.

Collins also emphasizes that it is universal—a truly global communication system that all trading partners in all countries recognize and use. "It's not an America-only system that works here and nowhere else," he says.

Such standardization is crucial to bar coding's future growth. "If organizations are willing to adopt standards, it will make more people want to use the technology," says Cohen of Accu-Sort.

Such developments as application identifiers improve the functionality of existing bar codes like Interleaved 2 of 5 and Code 39. But the effort to squeeze still more information into less space in bar codes has led to denser, two-dimensional codes. Code 49, developed by Intermec Corp. (Lynwood, WA) is a stacked code of bars and spaces in horizontal rows. Symbol Technologies Inc. (Bohemia, NY) has introduced PDF 417, a matrix-style code that compresses up to 1,750 characters per symbol. Code 49 can work with conventional bar code scanners and modified software, but PDF 417 requires special equipment.

New codes and standards have an effect on

the bar code printer manufacturers, depending on the particular technologies that each manufacturer uses. The advantage in speed, density, and print quality tends to go to non-impact technologies such as laser and thermal/thermal transfer. The ability to print very fine lines—more than 300 by 300 dots per inch, or 90,000 spots—is an advantage of non-impact print technology. As bar codes get to 600 by 600 dpi (360,000 spots) the advantage goes up further.

"The trend in bar codes is definitely toward putting as much information into the codes as possible," says Drew Smith, director of marketing at printer manufacturer CIE America Inc. (Irving, CA). "That lends itself to non-impact technologies." As a result, he says, CIE plans to move into thermal printing markets because the new symbologies will make obsolete most applications for its impact serial and line printers.

"We're going to thermal bar code printing for the print quality. Not that impact printing can't handle that, but the perception is that non-impact is better," says Smith.

"One of the advantages of a laser printer is the ability to handle a brand new font," says a printer executive who asked not to be named. "It's not costly and you can keep up with the characters. Therefore, it's easier to cost-justify."

"Additional check characters such as identifiers can be handled in software," states Ann Marie Cook, director of marketing at Weber Marking Systems Inc. (Arlington Heights, IL), a maker of laser and thermal/thermal transfer bar code label printers.

"With the new 2-D symbologies, you have to make the changes in firmware," continues Cook. She states that the demand for the 2-D symbologies has been slow to develop but she expects it to pick up in the next few years.

Through the 1980s, scanning and printing hardware technology helped the bar code market advance. New bar code standards will require more sophisticated capabilities from the equipment.

"I think there will be more laser diode scanners," predicts Cohen of Accu-Sort. "They're solid state, smaller, and offer more OEM opportunities. CCD (charge-coupled device) readers are also beginning to appear. They're limited in the area they cover and the size of the scanner package—it's a contact technology, but we'll see non-contact CCD readers soon."

Looking at the progress of bar codes so far, Collins of the Data Capture Institute believes that manufacturing—and industry in general—has made a breakthrough.

"I look at how we used to run around looking for scraps of paper—did it arrive, did we ship the right thing back?—and how we've given the computer the power to do all this for us now." Collins concludes, "when I see how far we've come, in five years we'll look back at today and think, 'How did we ever get along without it?' "

QUESTIONS

1. What are the advantages of bar coding?
2. What are the barriers to wider use of bar codes? Describe both technical and management issues.

3. Suppose a new technology made it possible to apply bar codes to individuals when they were born. What advantages would be gained? What societal problems would be encountered? Do the gains outweigh the problems or vice versa? Explain.

2.6 Try Before You Buy: Virtually Real Merchandising

PAUL BANDROWSKI

Virtual reality is a technology of the future. Yes, there are examples today in which people use gloves and spectacles and go into specially constructed rooms. There are commercial arcades where, for a small sum, you can experience the phenomenon. However, these simple examples merely provide a taste of what true commercialization will bring.

Virtual reality refers to the concept of using a computer environment to simulate life experiences. At one level, there is no need to travel to the mountains to get cold and wet as you experience skiing. At a more sophisticated level, virtual reality stands poised to change the nature of commerce. In Bandrowski's article, he examines the idea of using virtual reality to learn about store layouts before building them and populating them with products, not after. The idea is intriguing because it is possible to evaluate a broad range of possibilities in the comfort of the office without having to invest in building the real thing.

In the long haul, we can expect to see virtual reality come to the home, moving marketing two steps forward from seeing products on television. The first step is the ability in the coming interactive television to look at the products we want to see, not just the products that appear on a particular home shopping program. The next step, implied by virtual reality, is to make it possible to try products out much as we would in a store.

At present, virtual reality is expensive. Hence it can be expected to be adopted first in industry. The gains in productivity come from the reduced elapsed time examining alternatives and from the ability to spot mistakes before they happen. As prices go down, moving virtual reality to the home market becomes just a short step away.

Source: Corporate Computing (May 1993): 209–211.

PICTURE yourself as the CEO of a chain of retail department stores. Business is good, but not growing very fast. Marketing says a fresh image will attract new customers. You agree; moving toys down to a child's eye level, offering more self-service counters, or even redesigning the storefront, logo, and sign could mean new business. Yet you hesitate. Will it be worth it? How can you be sure you'll get what you want without spending millions? This scenario is played out repeatedly in the retail industry.

The traditional solution is to design and build a prototype store. If it doesn't fly, the cycle may be repeated several times until you arrive at a successful solution. Designers, architects, and builders make money each time. You don't.

Now visualize yourself walking into a prototype store, turning left, and frowning. The colors are all wrong, the new logo looks like the international warning symbol for toxic waste, and the self-service counters you imagined sprinkled around the store are huddled in one corner. You pick up a pair of shoes and notice the labeling is all wrong. Off come the goggles, off comes the glove. Your marketers and designers go back to the electronic drawing board, your company saves millions, and you keep your job.

Using 3-D modeling, visualization, and CAD tools, marketing and design departments can sit down at a computer and virtually build an entire store with everything from walls to cash registers. Some companies go as far as placing products such as shirts and bicycles on store racks with their proper labels, packaging color, and product size.

NEW PERSPECTIVES

"So what's the big deal?" you ask. "We've had CAD and computerized drawing tools for years."

Most design tools today are 2-D and, for a few lucky individuals, 3-D. With virtual reality, the designs are no longer just images; they become real. New interactive modeling packages allow a person, via computer, to enter the design and walk through the model.

Using a mouse, other pointing device, or in some cases your actual body action, you can manipulate the visual orientation of the model. For example, when you turn left, the image on the screen changes to what would be seen if you had actually turned left.

Just think: You'll know exactly what a new store will look like because you have virtually been there. Some products even allow you to define through whose eyes you view the virtual environment.

Look at your new store through the eyes of an eight-year-old child. Are the shelves too low or too high? Is the placement of a particular product on target for impulse or preplanned purchases?

Some companies are also using virtual modeling in product design. In this application, marketing works with the product design team to create anything from a new car to a can opener. Once the new product has been designed within a modeling system, you can actually use the product. At first sight, it is amazing. You can virtually pick up the can opener, pick up a can, and virtually open that can. Although you are not actually using your hands to turn the opener or hit the on/off button, the use of input devices such as a mouse or an electronic glove gives you a realistic simulation of that action. Once you have a working product, the marketing department can let customers test it and suggest changes before it goes into production. In the auto industry, for example, consumers could take a virtual spin in a concept car, critiquing looks, placement of controls, and so forth.

Remember, all this test marketing happens before any product is even built. Although developing virtual product models can be anything but cheap, it is usually an order of magnitude less costly than building clay or plastic models or one-of-a-kind prototypes.

Northrop now uses virtual reality to test airframe assembly parts in a 3-D virtual world. The assembly modeling application is directly linked to Northrop's CAD system and allows engineers to manipulate objects as if they were handling them on the shop floor. The virtual reality system is designed to save millions in development costs by eliminating the need for full-scale mock-ups.

Boeing is also evaluating virtual reality for use in designing and testing new commercial aircraft. Boeing believes that virtual reality will help development teams address human-factor issues in preproduction designs. In the future, Boeing's Advanced Technology Center plans to extend the use of virtual reality technology to the assembly and maintenance of aircraft.

Matsushita Electric Corp., a Japanese kitchen-appliance manufacturer, also tests products via virtual reality. The company recently implemented a system that allows it to display thousands of kitchen products in a computer-generated environment. Shoppers can enter the virtual world and use the products.

A national women's-apparel retailer with more than 900 stores found it could save time and money by using the 3-D virtual-reality drawing package Virtus Walkthrough for designing store layouts and storefronts. Previously, it used 2-D photos and layouts in the design of the stores, which meant actually going to the location from New York and photographing the store. The company sped up the marketing strategy process by not having to rely on time-consuming travel and multistep layouts. The Macintosh-based package allows the store designers to change fixtures at will using the program's library of stock architecture fixtures.

Not all applications of virtual reality deal with consumers and manufacturing. A unique application now under development lets financial market traders experience life in the pits before ever making a trade. It's just like the real thing: Greenhorn traders signal trades and the virtual reality environment reacts to the frantic waving and bellowing. A brokerage firm can save millions of dollars by eliminating the mistakes made by new staff. The FBI is also putting SWAT operations through different scenarios with virtual reality. The technology is used to test recruits' reactions in tense situations.

Although virtual reality offers businesses a wide variety of benefits, don't expect it on every desktop anytime soon. Cost, even for large corporations, is one of the biggest barriers to wider use of virtual reality technology. Development time can also be exorbitant.

Yet for applications like prototyping a new design for retail chains, the cost-benefit analysis is a breeze. The cost of developing a virtual store doesn't approach the outlay for building and rebuilding a real store. For other applications, you may have to weigh costs more carefully. Virtual modeling requires very powerful CPUs, disk drives, and video displays to present picture-quality images in real time. Even then the resolution is barely adequate. In addition, there are no standards for hardware or software.

How quickly can we expect organizations to employ virtual reality technology as a strategic component of their business strategy? Market acceptance of virtual reality will be evolutionary rather than revolutionary. This will result in continued delays in introducing products because they will go through the existing, time-consuming route of prototype creation, testing, and prototype revision.

SIDE EFFECTS

Scientists and businesses alike are concerned by the effects that prolonged exposure to a virtual world could have on human physiology. This may seem ridiculous now, but as these products continue to mature and become more realistic, users may require a recovery period following exposure—to regain their balance, for example.

While experts don't agree on when, or if, virtual reality will be a household word, most say that by the mid-1990s virtual reality applications will be common enough to buy off the shelf, like word processing software today.

QUESTIONS

1. How does virtual reality differ from simulation technologies of the past?
2. What are the major limitations of virtual reality at this time?
3. Explain how you would go about justifying virtual reality for a specific business application (e.g., marketing a home, manufacturing a bicycle, designing an insurance claims payment).

2.7

Internet: A New Business Building Block

MARY CRONIN

The Internet, long the private province of academics worldwide, has arrived on the business scene. The number of users increased ten times between 1990 and 1994. In a few short years, the Internet became a preferred way for many companies to communicate. For example, the editors of this book used the Internet to stay in contact with the book's publisher. With the Internet, telephone tag is eliminated. Written records of communications are available instantly. Small companies can compete in the global arena.

The Internet is a major case example of how the new technologies transform the nature of business. The article by Cronin explores the nature of these transformations. It shows, for example, how the Internet is one way businesses can compete in time (see Section 1.5).

BEFORE 1990, the Internet network and the corporate MIS department seemed a world apart. Mainframes dominated the computer room, client/server networking was just a dream, and the only business conducted on the Internet involved federally funded research.

But times have changed.

Commercial use of the Internet has moved from a novelty to a necessity for many companies. During the past three years, the world's largest research and educational network has become one of the fastest-growing avenues for businesses to share information. The number of Internet host computers increased from about 300,000 in 1990 to more than 2 million in 1993. During that time, businesses have registered more than a half million unique computer addresses with the Internet.

Today there are Internet connections in more than 100 countries, and each year the number of hosts on the network and the traffic it carries doubles. To support further growth, the Internet community is already working on a new version of the Internet Protocol (IP) used to carry the network's traffic.

It often seems like everyone in business is on the network.

"Most of our customers are now on the Internet, says Pete Ciufetti, director of corporate development for SilverPlatter Information Inc., Norwood, Massachusetts. SilverPlatter develops CD-ROM databases, and Ciufetti is one of the company's most avid Internet users.

Business users say that relying on the Internet makes sense technically and strategically. As businesses downsize and distribute their applications, many are adopting the same technologies and protocols that have been used successfully on the Internet for years—the Unix operating system and TCP/IP.

Source: Communications Week (April 18, 1994): WP41–47.

In addition, today's global economy and emphasis on strategic partnerships requires worldwide electronic communications between business colleagues and customers—a need the Internet can fill.

The Internet's wealth of technical and business information, meanwhile, offers new opportunities for marketing products and services. Businesses can download public domain software from the network, communicate with vendors and even hire employees.

"The Internet isn't just for corporations like IBM," says Howard Funk, an independent consultant and former IBM executive who recently stepped down as interim executive director of the Internet Society. "It can make any size company a competitive player in the global arena, if managers know how to use its resources effectively."

But along with instant access to millions of potential customers, business partners, and technical expertise, the Internet also poses some challenges for businesses.

For network managers, melding the Internet with the corporate network does not happen overnight.

And while Internet access may broaden the wealth of information available to employees, it also raises other concerns, including security. For many managers, concerns over security have led to the construction of electronic fire walls between their internal data networks and the Internet. In most instances, these fire walls are created by using a gateway to provide dedicated access to the Internet.

INTEGRATION CHALLENGES

For other uses, the notion of supporting integrated access to yet another network is the big challenge. The prospect of integrating the Internet's electronic-mail service with multiple internal E-mail systems that don't even communicate seems overwhelming.

Suzanne Johnson, an information technology manager at Intel Corp., Santa Clara, California, says that before the Internet was widely available to Intel users, the company strung together a number of different E-mail networks with individual gateways. It was not uncommon for messages flowing from one E-mail system to another to experience noticeable delays.

Eventually, an Intel senior manager realized it was faster to communicate via the Internet to colleagues outside the company than it was to send E-mail internally. This led to a redesign of Intel's internal network to support TCP/IP.

Today, Intel's different E-mail systems communicate through gateways based on the Simple Mail Transfer Protocol, the TCP/IP E-mail protocol.

The Internet connection has "made people more aware of the importance of network infrastructure within the company," Johnson says.

Adopting TCP/IP as a model for network integration, however, does not mean companies are substituting the Internet for their internal, private networks.

The Internet puts an increasing number of applications at a business user's fingertips, such as global E-mail and data exchange capabilities. It offers access to files and information on everything from international trade opportunities to economic conditions around the world.

One way to get a handle on this intricate weave of Internet resources is through the array of navigational aids offered on the Internet. These software tools are especially helpful to commercial users who lack Unix expertise.

Gopher, World Wide Web, Wide Area Information Servers, and Mosaic are navigational tools intended to provide a user-friendly front-end to the global network. They work with menu-driven commands or graphical icons that let a novice sign onto a remote computer, transfer files and locate information.

Even the largest multinational companies are using these tools to compete in the global market.

Hermann Tai, a computer applications staff specialist at Motorola Inc., Schaumburg, Illinois, credits the company's Internet connections with improving external communications and speeding up development projects.

Every day, in fact, Motorola employees log about 10,000 E-mail messages on the Internet. In addition, employees post questions to some of the Internet's thousands of technical discussion groups.

Employees use navigational tools like Gopher to locate information they need and to gain access to software code.

Motorola is connected to the Internet by an internal T1 connection to CICnet, an Internet service provider, based in Ann Arbor, Michigan. For $12,000—just a fraction of its multimillion-dollar networking and systems budget—it travels the Internet daily.

ONE SIZE FITS ALL

Smaller companies, too, say they benefit from connecting to the Internet. Terry Brainerd Chadwick, president of research company InfoQuest Information Services, Portland, Oregon, uses Gopher and other tools to scan the Internet for instant foreign trade information.

"The Internet is the best place to connect with international sources and trade opportunities," says Chadwick, whose company specializes in gathering information from the Internet. "I use it to match up my clients with overseas customers."

Chadwick also taps into federal sources like the Economic Bulletin Board at the Department of Commerce, which provides daily updates of the financial, manufacturing, and trade data that her small-business clients need to compete globally.

Using Internet discussion groups dedicated to particular countries, Chadwick can put her clients in touch with sales opportunities in Eastern Europe, China, Latin America, and other parts of the world. To do this, she uses a dial-up connection to The Well, a San Francisco–based Internet service.

FARMER'S MARKET, INTERNET STYLE

One of the most intriguing but controversial trends in business use of the Internet lies with vendors using the network to market their products and services.

"In [one] week, I've had four calls from businesses interested in reaching customers through the Internet, including a plumbing supply company," Chadwick says. "Sometimes I worry about marketing overshadowing the information-sharing aspects of the network."

The Internet's acceptable-use policy applies to users who send traffic over the National Science Foundation's NSFNet backbone, the largest backbone on the Internet. Under the policy, only educational and research traffic can flow over this federally subsidized piece of the Internet. Businesses with commercial Internet connections—which are not on the NSFNet—have no such restrictions.

Nevertheless, the Internet community discourages a hard-sell approach to customers, and many Internet providers do not permit unsolicited advertising, the electronic equivalent of "junk mail."

Gopher servers provide Internet-connected companies with a new model for marketing their products and services that fits in well with the culture that has grown up around the Internet. The Gopher software, developed by the University of Minnesota, organizes large information files and makes them available on Internet-accessible file servers.

Tom Caldwell, manager of information services at Rockwell International Corp.'s CMC Network Products Division, Santa Barbara, California, runs a Gopher server at CMC to reach potential buyers for the company's IP routers and other products.

Every week, hundreds of customers browse through product information, company press releases, and new technical developments, he says. The server can also connect users to other Internet services or Gopher services to get more information about a particular topic.

"Gopher is a cost-effective way to attract new customers and keep old ones," Caldwell says.

CMC plans in the "near future" to add multimedia features to its server using the Internet-provided Mosaic software tool. With Mosaic, the textual information now residing on the server will be augmented with photos, diagrams, and voice descriptions about CMC's products.

CMC also runs its own Internet File Transfer Protocol server that gives its customers direct access to new software releases and upgrades, as well as files of Internet standards specifications.

These enhanced services for customers cost the company very little in staff time and hardware. Since CMC uses the Internet as its primary link to the rest of the Rockwell organization, the $10,000 annual service fee it pays for access to the network is already built into its network budget. CMC accesses the Internet through the University of California–Santa Barbara, which is connected to the CERFNet regional provider.

To support the additional traffic generated by customer and marketing use of the servers, CMC simply had to upgrade its routers with support for the Point-to-Point Protocol. Its 56-kilobit-per-second network access remained the same.

CMC engineers and technical staff remain the heaviest users of the Internet, Caldwell says. Information gleaned from the technical discussion groups on the network helps them in their development and design of CMC products. "Our staff keeps current with technical developments in networking and security and shares its expertise with other Internet users," Caldwell says.

The most popular discussion groups among CMC users are those dealing with the Simple Network Management Protocol and security, Caldwell says. CMC engineers also answer customer queries through the discussion groups, offering immediate advice and fixes for problems.

While marketing via the Internet is important for high-tech companies, offering Internet customer support is crucial.

FASTER PROBLEM RESOLUTION

More and more customers expect to be able to reach vendors over the Internet to ask questions and get technical problems resolved. Businesses that put a priority on computer and network system performance contend that speeding up problem resolution over the Internet makes the cost of the connection well worth it.

For example, data from J. P. Morgan, the New York–based banking and financial services company, flows from its private corporate network to traders, economic planners, and clients 24 hours a day. A system malfunction requires immediate attention—something that was hard to get before J. P. Morgan joined the Internet.

"Without the Internet, it was difficult to get consistent support and fast problem resolution from vendors," says David Spector, vice president for technology services at J. P. Morgan.

Before the Internet, J. P. Morgan's systems and technical support staff sent problem reports to vendors by overnight courier. By messaging vendors over the Internet, turnaround time for resolving problems is now measured in hours rather than days.

VENDORS BENEFIT, TOO

This networked support system is just as beneficial to vendors. It lets technicians troubleshoot customer problems and send off software without leaving the building.

At network hardware manufacturer Xylogics Inc., Burlington, Massachusetts, a dedicated 56-Kbps Internet connection is used mainly for customer support. Most of Xylogics' customers are connected to the Internet and prefer contacting the company over the network with questions and requests, says Doug Mildram, administrator for Unix and network support.

Xylogics engineers can respond to them immediately using the Internet, and engineers sometimes resolve problems by actually connecting to the customer site over the network. This special attention makes customers more willing to serve as beta sites for new equipment and software releases, Xylogics officials say.

For other companies, the cost savings from using free, public-domain software more than justifies the cost of their Internet connection.

Using a dial-up Internet connection, automo-

PIECING TOGETHER THE FACE OF THE INTERNET WITH MOSAIC

▪ ▪ ▪ ▪ ▪

The days of associating the Internet with complex Unix commands and text interfaces soon may be over.

Enter Mosaic, the emerging multimedia interface to the Internet. Mosaic is browsing software that sits on top of Internet access tools to let users easily locate and retrieve text, image, and sound files on thousands of Mosaic-based servers on the Internet.

"Mosaic is the interface for which nontechnical users have been waiting," says Dan Lynch, founder and chairman of Interop Co., the sponsor of the Interop trade show. "It makes accessing Internet resources much more intuitive—a long-standing need. That's why Mosaic applications are spreading so quickly."

In one application, students at the University of Hawaii in Honolulu, as well as outside Internet users, can retrieve campus maps, photographs, and course information using Mosaic.

Mosaic is being developed by the National Center for Supercomputer Applications at the University of Illinois at Urbana-Champaign. Last November, the NCSA made Mosaic available for X-Window System, Microsoft Corp. Windows, and Apple Computer Inc. Macintosh systems. Since then, more than 200,000 Internet users have downloaded the software from the NCSA; countless others have acquired it from other sources.

A COMBINATION TOOL

Mosaic combines browsing, navigation, and delivery of information. Users move from file to file simply by clicking on a highlighted word or image within the document and can create a personal menu of the Internet resources they use most, says Joseph Hardin, an associate director at NCSA and head of its software group.

There are both business and educational applications for Mosaic. "Mosaic is the first easy-to-use interface to the Internet that puts the capabilities of Gopher, World Wide Web, WAIS [Wide Area Information Server], and File Transfer Protocol all together on one screen," Hardin says. "Companies can use it internally to distribute information, bundle it with products for the public, or use it as a platform to develop new Internet applications."

Hardin says Mosaic users generate many of the transactions made from the World Wide Web server maintained by the NCSA.

The Mosaic client and server software is free for personal use in educational and commercial sites. It is available from the NCSA server at FTP.NCSA.UIUC.EDU.

To license Mosaic for a commercial application, contact NCSA at INFO@NCSA.UIUC.EDU.

One of the first commercial Mosaic products is Internet-in-a-Box, from Spry Inc., Seattle, and O'Reilly & Associates, Sebastopol, California. The software—which comes with a Mosaic client package called Global Network Navigator—is designed to provide easy desktop access to the Internet. It is priced at $149.

tive die maker Greenville Tool and Die Co. has downloaded public-domain software that would cost thousands of dollars to purchase commercially.

Steven Meinhardt, a systems manager for the Greenville, Michigan, company, estimates that the software he gets from the Internet is worth about 10 times the $250 monthly connection charge to the network.

Software developers at Viewlogic Systems Inc., Marlborough, Massachusetts, download compilers, debuggers, and Unix code to help them write software application products. Recycling already developed code saves developers time in building new products, says Steve Caissie, Viewlogic's Unix system administrator.

One of Caissie's favorite finds on the Internet is a public-domain program that synchronizes the clocks in Viewlogic's computer system. "I wouldn't have written it myself, and the company probably wouldn't have spent money for it, but it makes my job a lot easier," he says.

The bottom line is that using public-domain software from the Internet speeds up Viewlogic's operations. "If we didn't have Internet access, our jobs would be a lot more difficult," Caissie says.

The Internet also gives companies access to strategic information about their competitors and even links them to potential business partners.

Recently, the Securities and Exchange Commission's database was added to the Internet. The database includes SEC filings, such as corporate earnings statements, annual reports, and financial analyses for all publicly held companies in the United States. This information, typically repackaged and sold by commercial information providers, is free to Internet users.

Other federal government resources are also available to businesses over the network. The U.S. Patent and Trademark Office offers a weekly update on new patent titles and numbers that is distributed to subscribers via E-mail upon request to PATENTS-REQUEST@WORLD.STD.COM.

The White House and federal agencies like the Environmental Protection Agency, the Food and Drug Administration, the National Technical Information Service, and the Justice Department provide Internet access to outside users. The Library of Congress even provides access to its card catalog and other references.

The range of applications and resources available on the Internet has lured thousands of U.S. businesses to join it in the 1990s. Corporations from Apple Computer Inc. to Xerox Corp. use the Internet for business. This flurry of commercial activity has not only changed the face of the network; it has spawned a new generation of user-friendly tools and resources.

The Internet's mix of cultures may sometimes clash, but most traditional research users are getting used to business—and even marketing—on the network. Advertising may still irritate some old-school Internet users, but the network has become a major clearinghouse for finding product data, electronic publications, and catalogs.

There is always the risk, however, that the Internet could become a distraction rather than a productivity tool. It's tempting, for example, for users to take detours of one sort or another on their Internet path—a mental picnic on the grassy banks of a recreational discussion group, perhaps, or an E-mail conversation. It is difficult to monitor what users are actually doing on the Internet, just as it is difficult to monitor telephone use.

But sometimes even side treks on the Internet can help business. Users at Xylogics, for example, frequently dial into the Internet in the evenings from their home computers. They come back the next day, Xylogics officials say, with suggestions or information that can be put to use within the company.

QUESTIONS

1. What is the main reason for commercial interest in Internet?
2. What are the advantages of using Internet over using a company-owned network? What are the disadvantages?
3. What additional Internet services do you envision in the future?

3 Office Productivity

EARLY applications of office systems focused on improving individual office worker productivity by replicating existing tasks more efficiently. The latest technologies allow office workers to do things they were unable to do before. These technologies provide instant access to large amounts of information. They allow people to communicate and share information with co-workers, customers, and suppliers, not only locally, within a single office, but worldwide. As a result, office work does not have to take place in the traditional office. It can be performed anywhere, anytime.

We are living in a world of software as well as hardware inventions. When combined, they enable office work to be performed in ways that could not be done before. For example, combining the laptop computer with specialized software for sales people increases the productivity of the sales force. Electronic data interchange software allows companies to carry out all the paperwork in a transaction that includes placing an order, receiving an order, billing, and payment.

This chapter describes office applications that are well beyond word processing, spreadsheets, and E-mail. They range from groupware in meetings and on the desk, to the idea of not coming into an office at all, to providing managers with executive information on demand. Not all the technologies are described in this chapter. Additional technologies will be found in subsequent chapters where their use is described.

3.1 Group Decision Support Systems

PAUL GRAY AND JAY F. NUNAMAKER

Peter Drucker, who wrote Section 1.1, forecast that office work in the 1990s and beyond would be done much more by teams and groups than by individuals. One of the great inventions to come out of business schools was the development of Group Decision Support Systems which facilitate this move toward groups.

As described by Gray and Nunamaker, a large amount of research has focused on group support systems and these systems are moving into the commercial mainstream. The basic idea is quite simple technologically. Each person in a meeting is provided with a personal computer and software that allows the group to work together on the same problem at the same time. Although most of the studies involved people in the same room connected through a local area network, the concept works just as well when people are in different locations as long as they are connected by a network.

The major productivity gain from group support systems comes from the concept of "parallel input." That is, in a conventional meeting, people speak one at a time according to Roberts Rules of Order. In an hour's meeting with ten people, on the average each person receives 6 minutes of "air time" and spends 54 minutes listening to others. In a group support system, because people can type information into the computer at the same time, it is possible for everyone in a meeting to participate simultaneously. It is, of course, true that in a meeting with group support, such parallel input occurs only some of the time. Nonetheless, the effect is to significantly increase the amount of information obtained.

INTRODUCTION

At the First International Conference on Decision Support Systems, Peter G. W. Keen (1981) pointed out that the fundamental model of DDS—the lonely decision maker striding down the hall at high noon to make a decision—is true only in rare cases. In real organizations, be they public or private, Japanese, European, or American, most decisions are taken only after extensive

Source: Decision Support Systems: Putting Theory into Practice (3rd edition) edited by R. H. Sprague, Jr. and H. J. Watson (Englewood Cliffs, N.J.: Prentice Hall, 1993): 309–326.

consultation. Although it is possible on occasion for decision makers to go counter to the consensus of their organization, this is not a viable long-term position for them. Decision makers out of tune with the people in their organization either depart the organization or the organization undergoes massive personnel turnover.

In this reading, we consider group decision support systems, usually referred to as GDSS. This is an emerging subfield within DSS in which there has been a marked increase in activity since the mid-1980s. The evolution of GDSS has been rapid. To a first approximation, the evolution can be characterized in terms of a chronology having some very short time intervals (Ahituv, 1985).

1982–85	Survey papers and research agendas (e.g., Kraemer et al., 1983; Huber, 1982; DeSanctis and Gallupe, 1986).
1981–83	Initial papers describing group decision support systems (e.g., Gray, 1981; Huber, 1982; Konsynski, 1982).
1982–86	Initial experimentation and experimental results (Kull, 1982; Gray, 1983; Lewis, 1983; Gallupe, 1986).
1987–	Building of advanced facilities in universities (DeSanctis, 1987; Gray, 1986; Mantei, 1988; Nunamaker, 1991 and 1989; Stefik, 1982), and formal experimentation (e.g., Applegate et al., 1987; Nunamaker, 1987; Watson, 1988; Zigurs, 1989).
1988–	Commercial versions of hardware introduced by IBM (McGoff, Hunt, Vogel, Nunamaker, 1990) and other vendors and commercial versions of software introduced by Ventana Corporation of Tucson, Arizona, and others.

As indicated by this chronology, GDSS was in a laboratory stage throughout the 1980s. In the past, when these systems were installed in industry and government, they often behaved like shooting stars. They were put in by one senior executive and used during his or her tenure. However, as soon as that individual was replaced, the system was dismantled or fell into disuse. The classic "not-invented here" syndrome held sway. As we move into the 1990s, the systems have become commercial in the United States. The move to commercialization became possible because we found out:

- How to use these systems effectively.
- How to design systems so that people (particularly middle-aged executives) do not require any significant training.

There is growing evidence that group decision support systems are able to improve both the efficiency and the effectiveness of organizational group processes.

THE NATURE OF GROUP DECISION MEETINGS

Although most business organizations are hierarchical, decision making in an environment involving choices among alternatives and assessment of risks is usually a shared process. Face-to-face meetings among groups of senior executives (or boards of directors) are an essential element of reaching a consensus. The group may be involved in a decision or in a decision-related task such as creating a short list of acceptable alternatives or creating a recommendation for approval at a higher level. These group meetings are characterized by the following activities and processes.

- The meetings are a joint activity, engaged in by a group of people of equal or near equal status,[1] typically five to 20 or more individuals.
- The activity, as well as its outputs, is intellectual in nature.
- The product depends in an essential way on the knowledge, opinions, and judgments of its participants.
- Differences in opinion are settled either by

fiat by the ranking person present or, more often, by negotiation or arbitration. The results lead to action within the organization.

Another way of looking at group decision meetings is in terms of what groups do. Specifically, groups

- Retrieve (or generate) information
- Share information among members
- Draft policies and procedures
- Use information to reach consensus or decision

DEFINITION

Definitions of GDSS have been offered by both Huber, and DeSanctis and Gallupe. These definitions are more than adequate for our purpose here:

- "A GDSS consists of a set of software, hardware, and language components and procedures that support a group of people engaged in a decision-related meeting" (Huber, 1982).
- "An interactive, computer-based system which facilitates solution of unstructured problems by a set of decision makers working together as a group" (DeSanctis and Gallupe, 1985).

An important point to note is that the group using the GDSS may not make the ultimate decision. It may be creating and/or reviewing alternatives to be submitted as a short list to the next level in the organizational hierarchy.

GDSS TECHNOLOGY

Although almost all group decision meetings today are face to face, technology is starting to be applied to make it possible for participants to be separated in space and/or time. These electronic meeting systems include computer conferences (either online or extended in time) and audio and video teleconferences. DeSanctis and Gallupe (1985) discuss the four combinations of proximity and separation in space and in time. In this reading, we will concentrate on one form of electronic meeting system, the "decision room" in which the participants are in the same room at the same time.

The motivation for creating a decision room comes from the observation that in almost all organizations, office automation had resulted in terminals being ubiquitous in work areas. However, as soon as one stepped into a conference room, the only technology available was a telephone, and high technology was a speakerphone. The personal and online computer capabilities used routinely elsewhere were not available.

In a typical decision room, terminals (or personal computers) are provided at some or all of the seating positions at a conference table. Input to these terminals is by keyboard, mouse, touchscreen, bit pad, or some combination of these devices. Participants can do "private work" at their individual displays. One member of the group (either the group leader or a staff member acting as a "chauffeur") operates the software needed to create the "public" display (e.g., a projection TV) that can be seen by everyone. In some rooms, multiple public screens are provided where one screen is used for the current discussion and the others for reference or slowly changing information.

Because decision rooms are designed for senior managers, they tend to have an "executive feel" to them. Even the experimental laboratories have plush carpeting and quality furnishings.

The computer and communications equipment in the decision room typically includes file servers to act as dedicated storage, local-area networks to interconnect the terminals and the servers, and connection to a central computer (usually a minicomputer) and to peripheral equipment including printers to provide hard copy and (sometimes) electronic blackboards. Whiteboards, overhead and slide projectors, and other audiovisual support are also provided.

GDSS SOFTWARE

GDSS provide software to support the individual and to support the group. To allow each individual to do private work, the usual collection of text and file creation, graphics, spreadsheet, database, and help routines is provided at the individual workstations. For the group as a whole, in addition to providing information retrieval and display, a GDSS provides software for summarizing group opinion. Thus, for example, the public screen can be used to present a cumulative list of all suggestions (such as from a nominal group technique session) or to show the aggregated results of voting and ranking or ratings of alternatives. Votes and preferences can be either identified by individuals or aggregated so that individuals need not expose themselves if they hold views contrary to the consensus or to those of the senior person present. This last feature allows managers to obtain a truer set of advice, because people need not fear retribution if they do not follow the prevailing group opinion.

As discussed later, introducing GDSS hardware and software into a conference changes the content of the discussions. By being able to do private work, a participant can examine an alternative (e.g., a "what-if") on his or her private screen and, if it is good, send it to the public screen for discussion. However, if the alternative is poor, it can be quietly buried without embarrassment. The recording of information on the public screen reduces the redundancy of the conversation. People do not keep bringing up the same ideas over and over if they are displayed. As a result, communications focuses on what participants know and on the rationale that led them to hold their views. Recording of information also provides group memory by recording all electronic comments.

GDSS COMMUNICATIONS

A GDSS must have a communications base as well as the model base, data base, and interface required in conventional DSS (Bui and Jarke, 1986). This is particularly true for a GDSS distributed in time and space. However, even for a decision room where everyone is present at the same time, communications links are required. For example, a participant may want to send you a message "for your eyes only." These links provide electronic mail among participants, access to remote computers, and the ability to send information from a workstation to the public screen via the chauffeur.

GDSS MEETING STYLES

GDSS software and hardware in full-service facilities are sufficiently flexible that they can be used in any of the three styles:

- Chauffeured
- Supported
- Interactive

As shown in Table 1, in the chauffeured style, only one person uses the software, either a group member or the meeting leader. A workstation is connected to the public screen to provide the electronic version of the traditional blackboard. The group discusses the issues verbally, with the public screen used as a group memory to record and structure information.

The supported style is similar to a chauffeured style, but differs in that each member has access to a computer workstation that provides a parallel, anonymous, electronic communication channel with a group memory. The meeting proceeds using a mixture of verbal and electronic interactions. Each member is able to enter items that appear on the public screen.

In the interactive style, the parallel, anonymous, electronic communications channel with group memory is used for almost all communications. Virtually no one speaks. Although a public screen may be provided, the group memory is typically too large to fit on a screen and thus is maintained so that all members can access it electronically at their workstations.

TABLE 1
GDSS Meeting Styles

Chauffeured	*Supported*	*Interactive*
One person enters group information	All group members can enter comments	All group members can enter comments
Public screen can provide group memory	Public screen can provide group memory	Group memory accessible via workstations
Verbal communication predominates	Both verbal and electronic communications	Electronic communication predominates

In practice, GDSS meetings involve use of two or all three of these styles.

A TYPOLOGY OF GDSS FACILITIES

Table 2 lists representative GDSS facilities. Such facilities can be characterized by the delivery mode and the range of tasks supported. Delivery modes include the following.

1. *Permanent installations at the user's site.* Here a conference room is instrumented with terminals and dedicated to GDSS use. For such a system to be successful, it must be used frequently. Otherwise it is not cost-effective and falls victim to one of the periodic cost-cutting efforts in organizations. Many permanent facilities currently are located at universities. Two corporate installations (COLAB at Xerox Palo Alto Research Center (Stefik et al., 1987) and Electronic Data Systems (Mantei, 1988) are developmental systems being used for research.
2. *Portable installations brought to the user's site on an on-call basis.* The equipment and the services of support staff to run the meeting are rented from a vendor. The skills of the vendor's staff in facilitating meetings is often as important as the equipment in such arrangements.
3. *Permanent installation at the vendor's site.* Here the group travels to the vendor site to hold its meeting. The vendor supplies the software, hardware, and support staff for a fee. The support staff usually acts as the chauffeur. Such a system is offered by Decisions and Designs, Inc. Some universities also offer this service.
4. *Facilities designed and sold by commercial firms.* IBM installed over 50 group decision support centers in its own facilities by 1991 and began marketing these facilities to other companies. Two organizations in England, ICL (a computer manufacturer) and Metapraxis (a consulting firm and software house), designed and developed permanent GDSS facilities that they installed at company locations.

The range of tasks supported can be from one or a few to a "full service" GDSS. An example of the former are the rooms provided by Decision and Designs and SUNY-Albany that support primarily interactive decision-analysis sessions that allow users to create decision trees and utility functions and in which the system does the Bayesian statistics. A full-service GDSS contains a broad range of software and the ability to support a diversity of tasks from financial decision making to crisis management to personnel selection to project review to long-range planning and forecasting.

RELATION OF GDSS TO DSS

GDSS can be viewed as subsuming conventional DSS within it. That is, the concepts of model base, database, and human interface (see Sprague and Carlson, 1982) all apply. Thus, as group size

TABLE 2
Selected Group Decision Support System Facilities

University-Based Facilities	
University of Arizona	Queens University (Canada)
The Claremont Graduate School	National University of Singapore
University of Georgia	San Diego State University
Universitat Hohenheim (Germany)	State University of New York at Albany
University of Indiana	Western Washington University
London School of Economics	
University of Minnesota	
Corporate-Based Facilities	
Electronic Data Systems	IBM Corporation
Execucom Systems Corporation	BellSouth
Xerox Palo Alto Research Center	Greyhound Financial Corporation
Commercial Facility Vendors	
IBM Corporation	ICL
Metapraxis	Ventana Corporation
For-Hire Commercial Systems—Permanent Installations	
Decisions & Design, Inc.	
For-Hire Commercial Systems—Portable	
Applied Future Inc.	Perceptronics, Inc.
K. R. Hammond	Wilson Learning Systems

shrinks to one, a GDSS reduces to a DSS. Conversely, in moving from a DSS to a GDSS, some new requirements are introduced.

1. The addition of communications capabilities.
2. Enhancement of the model base to provide voting, ranking, rating, etc., for developing consensus.
3. Greater system reliability.
4. Enhanced physical facilities.
5. Increased setup before use of the system.

The first two have already been discussed. System reliability must be much greater for a GDSS than for a DSS. If a GDSS fails, many people are affected, not just one. Because these people are high-level executives and well-paid, there is a much greater loss, both in terms of financial costs and trusts in the system. A GDSS requires much more setup before the system can be used because both the people and the facility must be scheduled, an agenda must be prepared, participants must be able to prepare for the meeting by seeing its data files and models, etc., and, if necessary, create any additional knowledge bases that are needed.

A GDSS requires capital investment in physical facilities. If the GDSS is located in a decision room, the room has to be elegantly furnished and have the feel of the executive conference room that it is. A GDSS also requires much more display and communications hardware.

APPLICATIONS

GDSS have been used principally for tasks involving idea generation, planning, competitive analysis, and consensus building. For example, Greyhound Financial Corporation used its GDSS for developing a mission statement, strategy formulation, evaluations of senior managers, and information systems planning.

GDSS applications cover a wide range. For example, Gray and Borovits (1983) discuss the use of gaming in GDSS environments, with special emphasis on the use of gaming for risk assessment and crisis management. They point out the following.

- Gaming is a form of training that improves both the intuition and skills of a manager in facing real-life situations, particularly high-stress, high-stakes situations such as crises and difficult negotiations.
- The experience and data gained from extensive gaming of a particular situation under laboratory conditions provide a basis for understanding how to cope with "surprise" and with outcomes of low probability.

- Gaming involves highly complex situations in which decision makers respond in novel and unanticipated ways. The iterative nature of the design process for GDSS thus makes it possible to use the results of gaming to increase the ability to handle complex conditions and cope with low-probability events.
- Gaming can assist senior managers both in understanding what consensus exists within their organization and in creating consensus. As such, it provides a basis for decision making that takes into account the internal viewpoint of the organization.

Other applications of GDSS include:

- Supporting negotiations. For example, in international situations, two or more groups that speak different languages and have different cultural backgrounds are negotiating a contract.
- Supporting business teams involved in design work (the Xerox Palo Alto Research Center facility was used primarily in this way), quality control reviews, and such relatively new tasks as reengineering and concurrent engineering.
- Supporting visual decisions such as selecting packaging for a new product.

WHAT HAS BEEN LEARNED

Experimentation with the use of decision rooms has been undertaken since the mid-1980s. A number of papers have appeared (e.g., those of Applegate, 1987; Gallupe, 1986; Gray, 1983; Grohowski, 1990; Kull, 1982; Lewis, 1983; Nunamaker, 1991; Watson, 1988; and Zigurs, 1989) that report observations, field research, action research, and laboratory experiments. These papers indicate the high potential for the contribution of computer and communication-based mediation, facilitation, and support in creating effective group decision support. The following summary of what has been learned thus far is based on these and other papers and our own observations.

The research results have underscored the need to review and examine the theories and hypotheses on group decision making that have been developed in the past. There is reason to believe that many previously held assumptions about the conduct of group deliberation are subject to review in the new electronically based forum. Factors such as speed, anonymity, recording of group processes, voting, participation levels, group size, and other facilitated activities change the group environment significantly.

Analysis of the available data shows that using decision rooms can improve group work in many situations because they:

- Enable all participants to work simultaneously (human parallel processing), thereby promoting broader input into the meeting process and reducing dominance of the meeting by a few people.
- Provide equal opportunity for participation (through anonymity).
- Trigger additional points to be made.
- Enable larger group meetings that can effectively bring more information, knowledge, and skills to bear on the task.
- Provide process structure to help focus the group on key issues and discourage irrelevant digressions and nonproductive behaviors.
- Support the development of an organizational memory from meeting to meeting.

In the following sections, we describe these and other findings in more detail.

Group Size and Composition

Groups numbering from three to 20 or more of differing composition have used GDSS facilities to accomplish a variety of tasks. One finding is clear: Individual satisfaction increases with the size of the group. Computer support assists groups in building toward a consensus. Larger groups appreciate the inherent structuring that keeps the group from becoming bogged down or subject to domination by personalities. Small groups find that the fixed overhead associated with using the computer-based systems eats up the gains from using the system. Small groups are less likely than large groups to conclude that the computer-aided support is more effective or efficient than an unstructured face-to-face meeting. The various electronic brainstorming approaches do not work effectively with groups of less than four. Such techniques are more effective for groups of eight or more persons. Many other techniques, such as Stakeholder Identification and Assumption Surfacing, also increase in satisfaction with group size. The implication of this finding is that different tools are required for different group sizes. Work toward this end is currently underway.

The increase in satisfaction with group size is due, in part, to "human parallel processing." That is, in many situations, participants are entering information into the computer simultaneously. They are functioning in parallel rather than in sequence. In a typical meeting that follows, say, Roberts Rules of Order, verbal input is sequential. As a result, if a group of 10 meets for an hour, each participant has the floor for only six minutes on the average. In parallel processing, the group can finish in 20 minutes and each individual (even slow typists!) has made a larger individual contribution.

Anonymity

The anonymity facilitated through use of nominal group techniques (Van de Ven and Delbecq, 1971) and other tools for electronic brainstorming is a positive factor in encouraging broad-based participation. Anonymity is important when sensitive issues being discussed can easily be confounded with personalities in the group. Anonymity also provides a sense of equality and encourages participation by all members in the group, independent of perceived status. Problems of "group think," pressures for conformity, and dominance of the group by strong personalities or particularly forceful speakers are minimized even though the participants are face to face. Group members can contribute without the personal attention and anxiety associated with gaining the floor and being the focus of a particular comment or issue. As a result, issues are discussed more candidly.

Anonymity does tend to heighten conflict within the group because members tend to become more blunt and assertive in their written comments and often are not as polite as when speaking face to face. (However, people are more accepting of written criticism than verbal criticism because it does not carry with it the same level of embarrassment. Written comments become more polite with age and rank in the organization.) Further, in any written medium, the richness of voice inflections and facial expressions is lost, which can lead to misunderstanding. Occasional face-to-face discussions, as well as breaks and social time, are important as issues become more politically charged and sensitive.

Facility Design

The lighting and physical organization of the decision room affect outcomes. Better results are obtained when the facility has aesthetic appeal and provides a comfortable, familiar setting for executives. Carpeting, wall coverings, executive-style furniture, and quality acoustics provide an atmosphere that is well suited to long sessions over a number of days.

Adequate lighting control and arrangement of lights is needed to assure that results are legible on both the public and the individual workstation screens. For example, front screen projector images are "washed out" by fluorescent lighting and can reduce the effectiveness of a meeting.

The decision room should be full-service in order to meet the needs of groups that differ in size and task. The decision room should be designed so to accommodate a range of group sizes and be able to support tasks that range from passive to active. Breakout rooms located adjacent to the decision room that provide computer-based support make it possible to divide a large group into smaller groups and are useful for changing the environment during a meeting.

Multiple Public Screens

More than one public screen increases group productivity. Not all information can be displayed adequately in the standard 25-line-by-80-character format of current computer screens. Windowing on a single screen allows presenting multiple sets of data but further reduces the amount of information that is shown about individual items. In a multiple-screen setup, the group can view both the current information being discussed and reference information (e.g., sales trends, financials) at the same time. They can see the existing version and the proposed alternative side by side. In rank ordering, they can see the unordered and ordered items simultaneously.

Knowledge Bases and Databases

The documentation of meeting activities, the creation of working papers, and the recording of decisions and commitments are particularly useful by-products of GDSS. These outputs are provided without detracting from meeting activities. File servers handle the knowledge bases and databases, facilitate coordination and management of input from individual decision makers, and serve as "organizational memory" from session to session. The file server functions as a knowledge-base repository and provides access to organizational data that are relevant to a particular meeting. A key to effective use of GDSS in supporting planning is the continuity from planning session to planning session provided by an ongoing, expanding knowledge base that is integrated through the output from software tools. This continuity and integration provide the opportunity for analysis from multiple perspectives.

Communication Network Speed

Users become impatient if they must wait more than 1 or 2 seconds for a screen. Experience has shown that users expect to receive subsecond response for all activities. A wide-bandwidth local area network (LAN) is needed to maintain these high levels of network response.

Fixed versus Customized Tools

In planning, groups usually start with idea generation, followed by the development of alternatives, and conclude by converging on a course of action through forming a consensus. A group can go through this process in either of two modes.

1. It can create a customized methodology from the set of available tools.
2. It can follow a standard sequence for using the tools.

Some groups prefer to adopt a standard methodology, whereas others feel that their needs are very different from anyone else's and therefore prefer to generate the sequence. Both approaches have given excellent results.

Software Design: Ease of Use and User Friendly

The best GDSS software helps rather than frustrates individual users. It supports a continuum of modes of working ranging from electronically based, self-directed participation to facilitator-directed discussion. However, a minimum amount of instruction and direction is still required. In the system designed by the University of Arizona, for example, it takes less than five minutes to explain how to use each software tool. Efforts to increase software ease of use are particularly worthwhile. Techniques that use color, overlays, windowing, consistent interfaces, and on-demand help screens all help the user (particularly the novice and the computerphobic user) to master the software. One or two group members who have difficulty with the software can affect the productivity of the entire group.

Consistency in the dialog interface protocols permits effective dialog management, ease of introduction of participation to new support tools, and ease of tool building. Common keystroke assignments, window layouts, use of color, messaging, and icon semantics facilitate this dialog management.

Screen sharing across, and among, participants opens new opportunities for particular decision tasks. Activities such as local editing of shared screens, help and monitoring activities, and personal messaging create alternative forms of communication. Using the keyboard as an input device has not proven to be an inhibitor of active participation.

Satisfaction

Individual satisfaction is reflected in user reports on the positive aspects of the group decision-making process in a computer-based support environment. Group participants conclude that they are not blocked out of the group and, as a result, they support the group solution with increased confidence.

THE FUTURE OF GDSS

The foregoing discussions were based on the assumption that GDSS is an emerging technology that will become pervasive in many organizations. Certainly, the commercialization of GDSS by IBM and others is an important indicator of its long-term success. People meeting in a group need to have a free flow of ideas and the technology—with its typing, pointing of mice, and other mechanical interruptions—can act as a barrier and a delayer rather than a facilitator of thought. Certainly, the present technology could be made even more user-friendly and transparent. New technologies (such as pen-based systems and voice recognition) should help in this regard. One of the dangers in the development of GDSS is that, like many other computer-based solutions offered previously, it may overpromise and underperform in its early stages, resulting in user expectations that are too high for what GDSS can deliver. The research programs and the commercial experience now underway should increase the long-term viability of GDSS.

What can we expect? Group decision support systems today use relatively standard computer, communications, and display technology. Physically, existing decision rooms look very much like conventional conference rooms with terminals and projection screens added. This is to be expected. When a new technology is introduced into an existing situation, it tends to look like and be used in the same way as the technology it replaces. The early automobile looked like and was used in the same way and over the same roads as the carriage, with only the horse replaced by a motor. GDSS provides an enhanced environment for performing existing tasks and will, in the short term, be used the way conference rooms are used now. As understanding of GDSS develops, the nature of group decision making itself can be expected to change. The idea of human parallel processing is such a change.

One potential new direction comes from the changes going on in the workplace. As large companies decentralize more and more and as computer and communications costs continue to decrease relative to transportation costs, we can expect that many people will "telecommute" (e.g., Nilles, 1976) to meetings just as they telecommute to work. That is, they will stay where they are and will be brought into the decision conference electronically—through video and computer conference. It is also possible to use these technologies to call experts on a particular subject into the meeting, obtain their advice, and let them go without their ever leaving their place of work.

Another potential direction of change comes from the expert/knowledge-based system realm. As expert systems become more pervasive, one or more of them will be brought into meetings to assist in deliberations. In effect, they become the $(n + 1)$st person in a meeting of n people. Their role can range from retrieving information to synthesis of new alternatives to helping resolve conflicts of opinions.

The present approach of using the technology to mechanize group processes such as voting, Delphi, and nominal group techniques is relatively crude and rudimentary. We can anticipate that new ways of gaining group interaction and group consensus will be developed that take advantage of the capabilities offered by GDSS.

We are at a very early and a very exciting point in GDSS. The level of activity is building as firms start to explore the possibilities. The 1990s should bring additional innovation and lead to maturity.

NOTES

1. However, the use of group systems has proven to be effective as a way of holding meetings whose members cross many levels of an organization.

REFERENCES

Ahituv, N. Personal communication, 1985.

Applegate, L. M., et al. "Knowledge Management in Organizational Planning," *Journal of Management Information Systems* 3 (1987): 20–38.

Bui, T., and Jarke, M. "Communications Requirements for Group Decision Support Systems," *Proceedings of the 19th Hawaii International Conference on Systems Science,* Honolulu, January 1986, 524–533.

DeSanctis, G., and Gallupe, R. B. "Group Decision Support Systems: A New Frontier," *Data Base* (Winter 1985): 3–10.

DeSanctis, G., and Gallupe, R. B. "A Foundation for the Study of Group Decision Support Systems," *Management Science* (May 1987): 589–609.

Gallupe, R. B. "Experimental Research into Group Decision Support Systems: Practical Issues and Problems," *Proceedings of the 19th Hawaii International Conference on Systems Science,* Honolulu, January 1986, 513–523.

Gray, P., et al. "The SMU Decision Room Project," *Transactions of the First International Conference on Decision Support Systems,* Atlanta, June 1981, 122–129.

Gray, P. "Initial Observations from the Decision Room Project," *Transactions of the Third International Conference on Decision Support Systems,* Boston, June 1983, 135–138.

Gray, P., and Borovits, I. "Gaming and Group DSS," *Transactions of the Sixth International Conference on Decision Support Systems,* Washington, D.C., April 1986, 165–175.

Grohowski, R. C., McGoff, C., Vogel, D., Martz, B., and Nunamaker, J. F. "Implementing Electronic Meeting Systems at IBM: Lessons Learned and Success Factors," *MIS Quarterly* (December 1990): 369–384.

Huber, G. P. "Group Decision Support Systems as Aids in the Use of Structured Group Management Techniques," *Transactions of the Second International Conference on Decision Support Systems,* San Francisco, June 1982, 96–108.

Huber, G. P. "Issues in the Design of Group Decision Support Systems," *MIS Quarterly,* 8 (1984): 195–204.

Keen, P. G. W. "Remarks" Closing Plenary Session, *First International Conference on Decision Support Systems,* Atlanta, June 1981.

Konsynski, B. R., and Nunamaker, J. F. "Plexsys: A System Development System," in J. Daniel Couger et al., *Advanced System Development/Feasibility Techniques.* New York: John Wiley, 1982.

Kraemer, K., et al. "Computer-Based Systems for Group Decision Support," Working Paper, University of California at Irvine, Public Policy Research Organization, December 1983.

Kull, D. "Group Decisions: Can a Computer Help?" *Computer Decisions,* 14 (1982): 14.

Lewis, L. F., II "FACILITATOR: A Microcomputer Decision Support System for Small Groups." Ann Arbor, Mich.: University Microfilms, 1983.

Mantei, M. M. "Capturing the Capture Lab Concepts: A Case Study in the Design of Computer Supported Meeting Environments," Research Paper 030988, Center for Machine Intelligence, Electronic Data Systems Corporation, Dallas, TX., 1988.

Mason, R. O., and Mitroff, I. I. *Challenging Strategic Planning Assumptions.* New York: John Wiley, 1981.

McGoff, C., Hunt, A., Vogel, D. R., and Nunamaker, J. F. "IBM's Experience with Group Systems," *Interfaces* (1990): 39–52.

Nilles, J. M., et al. *Telecommunications–Transportation Tradeoff: Options for Tomorrow.* New York: John Wiley, 1976.

Nunamaker, J. F., Applegate, L. M., and Konsynski, B. R. "Facilitating Group Creativity: Experience with a Group Decision Support System," *Journal of Management Information Systems,* 3 (1987): 5–19.

Nunamaker, J. F., Dennis, A. R., Valacich, J. S., Vogel, D. R., and George, J. F. "Electronic Meeting Systems to Support Group Work," *Communications of the ACM,* 34 (1991): 40–61.

Nunamaker, J. F., Vogel, D. R., and Konsynski, B. R. "Interaction of Task and Technology to Support Large Groups," *Decision Support Systems International Journal,* 5 (1989): 139–152.

Nunamaker, J. F., Dennis, A. R., Valacich, J. S., and Vogel, D. R. "Information Technology and Negotiating Groups: Generating Options for Mutual Gain," *Management Science* (October 1991): 1325–1346.

Sprague, R. H., Jr., and Carlson, E. D. *Building Effective Decision Support Systems.* Englewood Cliffs, N.J.: Prentice Hall, 1982.

Stefik, M., et al. "Beyond the Chalkboard: Computer Support for Collaboration and Problem Solving in Meetings," *Communications of the ACM,* 30 (1987): 32–47.

Van de Ven, A. H., and Delbecq, A. L. "Nominal Versus Interacting Group Processes for Committee Decision Making Effectiveness," *Academy of Management Journal,* 14 (1971): 203–212.

Watson, R., DeSanctis, G. L., and Poole, M. S. "Using a GDSS to Facilitate Group Consensus: Some Intended and Unintended Consequences," *MIS Quarterly,* 12 (1988): 463–478.

Zigurs, I. "Interaction Analysis in GDSS Research: Description of Experience and Some Recommendations," *Decision Support Systems International Journal,* 5 (1989): 233–241.

QUESTIONS

1. How are group decision systems different from the more traditional decision support systems? How are they alike?
2. What are the major advantages of group decision support systems over group meetings without such a system?
3. How can GDSS enhance organizational productivity?
4. Pick a specific organizational decision-making process. Describe how you would apply group decision support systems in this situation. What would be the benefits of this application?

3.2 Groupware Goes Boom

DAVID KIRKPATRICK

Groupware comes in many flavors. One of them is a major product from Lotus, called Lotus Notes. Lotus Notes is designed for people working in different places at different times. It allows people to put information into the system at any time. A major feature of Lotus Notes is that it is organized by topics and hence allows people to comment on specific topics and keeps that information together. People can share information of all kinds and keep track of it. An unusual feature of Notes is that a copy of the entire database is kept at each computer that acts as a network server (the central node of each local area network). The result is that information is available quickly to everyone. Notes may be best suited for organizations that are highly information intensive with a culture of cooperation rather than competition.

Kirkpatrick points out that by the beginning of 1994, over 300 products bore the groupware label. These products fall into meeting software (discussed in Section 3.1), work flow software designed to speed processes, scheduling software, and basic groupware, of which Notes is an example. Installing Notes is more than adding a technology. As you will see in Kirkpatrick's article, it changes the nature of the organization and how people work.

WHEN Rick Richardson arrives at his office each morning, he gets a cup of coffee and turns on his PC. On an average day he receives 20 to 25 E-mail messages—not surprising, since he's in charge of all the work that Price Waterhouse, the accounting and consulting firm, does with New York City banks. Then he clicks the mouse to see news articles on topics he has given the computer standing orders to track. He has already done more with his PC than most American executives, but what happens next makes him truly exceptional.

Price Waterhouse maintains electronic bulletin boards on more than 1,000 different subjects, accessible to as many as 18,000 employees in 22 countries. Richardson calls up a board that holds messages about the firm's financial services business. A Washington staffer reports on an accounting regulation change; Richardson adds a comment for all to see. A manager in Miami needs Spanish-speaking consultants for an assignment in Buenos Aires; Richardson zaps him a name by E-mail.

Finally, he checks a board called PW Alert,

Source: Fortune (December 27, 1993): 99–106.

where employees post general information and queries. One message, from Dublin, asks if anyone has experience auditing a cheese plant processing 100 million gallons of milk a year. Messages are listed by title, so Richardson can skip that one and any others he's not interested in. But he reads carefully a query from Chicago soliciting CFO candidates for a client. Perhaps he'll call the sender later and suggest a name.

From the moment he turned on his PC, Richardson has been using Notes, a Lotus Development software product that is helping define the future of corporate computing. Notes is the most popular entry in a fast-growing category called groupware. While most PC software has been written for people working alone, these programs make it easier for groups to work together. That couldn't be more timely. Just when the theorists of business-process reengineering are telling companies to flatten hierarchies and get people working in teams, along comes the software that—properly used—can help make it all happen.

The new groupware tools are so powerful, in fact, that they virtually compel companies to reengineer themselves. By giving workers well below the top of the organizational pyramid access to information previously unavailable or restricted to upper levels of management, groupware spreads power far more widely than before. Many who have grown comfortable dwelling near the pyramid's apex are understandably distressed by the way groupware disrupts old-style hierarchies. Says a consultant who helps clients figure out how to use the software: "Fortune 500 companies are having a hell of a time with groupware. There's a tremendous amount of what I call antibody immune response."

Like E-mail, groupware became possible when companies started linking PCs into networks. But while E-mail works fine for sending a message to a specific person or group—communicating one-to-one or one-to-many—groupware allows a new kind of communication: many-to-many. Groupware can improve efficiency, because it keeps workers continually apprised of what colleagues are thinking and doing. To be sure, it also creates even more data for the already-besieged to digest. Some complain that groupware only adds to information overload.

More than 300 products today carry the groupware label. The most important fit into four main categories.

- Basic groupware, which Lotus Notes has all to itself, combines a sophisticated messaging system with a giant database containing work records and memos. It changes the way information flows in an organization. It creates a kind of corporate on-line service, an Internet of the office. Unlike plain E-mail, Notes doesn't require you to figure out who needs to know a fact or hear an idea. Instead, you simply forward your memo to the appropriate bulletin board. Like Rick Richardson of Price Waterhouse, anyone who needs to know about a subject will check there and find it. Hundreds of other companies besides Price Waterhouse have logged on, including Andersen Consulting, Compaq Computer, Chase Manhattan, General Motors, Monsanto, Reebok, and Texaco. Since it unveiled Notes in 1989, Lotus has sold over 500,000 copies to more than 2,500 companies at around $300 for each PC that uses it.
- Workflow software is designed to remake and streamline business processes, especially in paper-clogged bureaucracies. It helps workers understand the steps that make up a particular process and allows them to redesign those steps. It also routes work automatically from employee to employee. The best-established company in the field is Action Technologies of Alameda, California, whose ActionWorkflow System sells for about $16,000 for 20 users. Another major player: NCR, with Process-IT (cost: $19,900 for 20 users). Specialized tools for routing business forms are sold by Reach Software of Sunnyvale,

WINNING THE JOB WITH GROUPWARE

■ ■ ■ ■ ■

Here's how Price Waterhouse put together a proposal in four days and won a multimillion-dollar consulting contract by using Lotus's Notes software for groups. On Thursday a Price Waterhouse executive learned that a major securities firm was about to award a big consulting contract to help develop a complex new trading operation. Price Waterhouse was invited to bid, but there was a hitch: The proposals were due Monday. A Price Waterhouse competitor had been working on its own bid for weeks.

The four Price Waterhouse executives who were needed to write the proposal were in three different states. But they were able to work together using Notes, which allowed them to conduct a four-way dialogue onscreen. They also extracted key components of the proposal from various databases on Notes. From one, they pulled résumés of the Price Waterhouse experts from around the world who could be assigned to the job. From another, they borrowed passages from similar successful proposals.

As the draft evolved, each of the four modified it or made comments. Notes kept track of the changes. Other executives looked at the proposal via Notes over the weekend.

The proposal was ready Monday, and Price Waterhouse won the deal. Its competitor didn't even meet the deadline. A year later, the client hired Price Waterhouse to audit the new operation. That contract will probably last for years.

California; its Reach WorkMAN costs $5,980 for 20 users.

- Meeting software allows participants in face-to-face or videoconference gatherings to "talk" simultaneously by typing on PC keyboards. Because people read faster than they speak, and don't have to wait for others to finish talking, the software can dramatically speed progress toward consensus. It also helps ensure that everyone gets a chance to take part. The most prominent developer is Ventana Corp. of Tucson, which sells its GroupSystems V for $25,000 for 20 users. A 1991 Boeing study found that Ventana's software could shorten the time it takes groups to complete projects by as much as 90%.
- Scheduling software uses a network to coordinate colleagues' electronic datebooks and figure out when they can all get together. Several companies sell versions of it; at a recent groupware conference in San Jose, an expert panel preferred Network Scheduler 3 ($1,995 for 25 users) from Powercore of Manteno, Illinois.

Now that all these pioneering companies have proved that groupware works and customers want it, virtually every major software outfit has proclaimed its intent to jump in. Both Microsoft and Oracle Systems, the big database company, say they will soon release software that rivals the organization-spanning abilities of Notes.

None of this comes cheap. Companies with large Notes installations calculate that besides the software itself, hardware, programming, telephone connections, and servicing runs $200 to $500 a year per employee, in addition to the cost of the network. Dell Computer, with 1,125 people on Notes, spends $45,000 a month just in phone charges to distribute Notes to its offices worldwide. There's more: Lotus estimates that for every $1 customers pay for the software, they spend another $3 for consultants, trainers, and additional software to help them make optimal use of it.

Just because it's expensive and tough to do right doesn't mean you can ignore groupware. If your company isn't using it, your competitors probably are, or soon will be. In some businesses—information technology consulting, for example—so many players use Notes that it has become almost a cost of entry. The battle-scarred companies that have had the gumption to go to groupware swear all the disruption is worth it. Before you try it yourself, though, consider these hard-won lessons:

GROUPWARE CHANGES BUSINESS PROCESSES

Bankers Trust is pushing the limits of groupware as hard as any company, and its efforts show how powerful a tool groupware can be for business-process reengineering. The bank manages assets for big financial institutions, but with large amounts of money in various types of investments, clients occasionally detect errors (or think they do).

When these questions arose in the past, clerks in New York City often hauled out big boxes of papers, searched through them, consulted various departments, and then faxed or mailed documents to a servicing and records center, now in Nashville, for double-checking and further research. Someone there faxed back a reply, perhaps an annotated copy of the original fax—now sometimes too faint to be faxed on to the client with an explanation. The entire history of a case was kept in a manila file folder. Typically it took three to five days to complete an inquiry.

To automate all that, Bankers turned to workflow software from Action Technologies. The bank used Lotus Notes to store electronic file folders accessible to anyone on the Bankers computer network. To get started with Action's software, a team spent two weeks analyzing the steps in resolving customer queries. It created what Action calls a workflow analysis diagram that broke out the roles of the various people involved: account administrators who receive the query and manage the process, the people who actually run the money, the Nashville staff, and supervisors who approve the final response. Five people had the system up and running within seven months for 125 users, in Nashville and at two locations in New York.

Now the bank guarantees clients a maximum three-day turnaround on queries and completes many within a day. Notes organizes cases by the status of the client's query, its position in the workflow, or the person responsible. If someone is sick, others can easily pick up the work. One office of account administrators was able to reduce its staff from 13 to nine because it no longer needed the researchers who hauled out boxes and kept manila folders. It's now clear exactly who is doing what and where each query stands. Says Jeremy Keisman, a technology project manager in Bankers Trust's global assets group: "There used to be a black hole: 'It went to Nashville.' Now people take responsibility. There are no more mysteries."

MORE INFORMATION MEANS WORKERS MUST TAKE MORE RESPONSIBILITY

"Groupware will change the way we evaluate people," says Mike Blum, who leads technology services for Coopers & Lybrand Consulting, an arm of the accounting firm. "It's like what's been happening on the factory floor. Workers take responsibility for getting the car out the door. The same kind of thing will happen in the office."

Few companies use groupware more ambitiously to push authority down the organization chart than CKS Partners, a 65-person advertising agency in Campbell, California, that had revenues of about $12 million in 1993. CKS does what it likes to call "integrated marketing"—product design, packaging, and in-store displays as well as conventional advertising, mostly for high-tech clients like Apple Computer, National Semiconductor, Nextel Communications, and Tektronix. CKS sees technology as a key to everything it does; in 1993 the firm spent $18,000 per employee on hardware and software. Says CEO Mark Kvamme: "You cannot do true integrated marketing without a group collaborating through the use of technology. We can call on premier experts in a variety of disciplines, whether they are in-house or not, and get them all working together onscreen."

The agency paid a programmer $75,000 over the past three years to write a custom groupware application it dubbed "the electronic job jacket." The software keeps track of each employee's exact contribution to every job. It also maintains records of budgets, contracts, purchase orders, personnel, and contacts at subcontractors and suppliers. Before work starts on a job, managers estimate the hours various employees will need to spend on it and create a time budget for each one. Workers then record the hours they put in.

That might seem draconian, and Kvamme a major control freak, but for one thing: Each quarter, CKS pays between 15% and 20% of its profits to employees. The jacket allows the company to calculate exactly how much it made from each assignment and how much each person contributed. The firm pays bonuses accordingly. Says Kvamme: "When people screw up and lose money on a job, they've got to face their co-workers and take responsibility." Kvamme finds this especially salutary for creative people like designers and copywriters, who he says suffer from "creeping elegance": The more time they have, the more incremental but often unnecessary improvements they'll think up.

COMPANIES WITH FLEXIBLE CORPORATE CULTURES DO BEST

"Groupware is a tool that catalyzes organizational change whether you want it or not," says David Coleman, who edits a newsletter called *GroupTalk* and organizes conferences on the subject. Information is power, and groupware disperses it much more widely than before. Says Jeffrey Held, a senior technologist with Ernst & Young's consulting group: "People acquire power in an organization by knowing things that others don't. Unstructured information-sharing runs contrary to the culture of many organizations." Held estimates that only about 10% of companies adopting groupware are prepared for the organizational ferment it unleashes.

Teams may not behave as managers expect, especially when the teams are using groupware. For example, although Boeing conducted rigorous experiments two years ago that found meeting software could cut the time some projects take by 90%, the aircraft giant is not using it at all today.

The company will not explain in detail why it dumped what seemed a successful tool. A spokesman says groupware got dropped during a cost-cutting effort, because in comparison with other technology investments it offered less payback. In addition, he says, "some people didn't like it." Groupware experts outside Boeing speculate that managers may not have enjoyed finding themselves in an electronic spotlight where decisions that had once been their sole province were now fair game for comment and change by everyone. Possibly Boeing decided that if its old-style bosses were uncomfortable, it didn't matter that the decision process was dramatically faster. To encourage acceptance of groupware and give workers a stake in it, Ernst & Young advises clients with similar problems to find ways to reward teams, not individuals, for success.

GROUPWARE WORKS BEST TO SOLVE A SPECIFIC BUSINESS PROBLEM

Too many companies get into groupware without clear aims. Says *GroupTalk* editor Coleman: "A lot of people go out and buy Notes and then try to find a problem to solve with it. That's bass-ackwards. Start with the problem, not the technology."

Tom Thomas, head of information systems at Dell Computer, warns: "If you don't define what you want to do with Notes and how you want to use it, you can get yourself into a bit of a jam." He should know. Almost four years ago, in Dell's heady fast-growth days, engineers designing PCs brought in Notes to help them keep track of the progress of various projects. Other departments decided to try it too. They began setting up their own Notes databases.

Problem was, some of the information users began sharing was purely social. People used Notes to discuss movies, or what they would do after work. Instead of saving time, Notes was causing people to waste it. So Dell management imposed strict rules on how Notes was to be used and took it away from some employees. The company still relies on Notes extensively, but only where it can be justified for a specific task.

One such assignment: keeping track of the status of all design, marketing, and manufacturing work on different kinds of products. Employees anywhere in the world can follow and comment on each other's contributions. Among the 500 users are manufacturing managers, who come on-line often to order parts, plan staffing, and learn about new products. Says Tom Holman, program manager for advanced systems: "I don't really know how we'd run a global design and manufacturing operation without Notes." Before Notes, Holman's staff communicated by E-mail or conference calls, but the process sometimes broke down—for example, when engineers in Ireland misunderstood the Texas twang of designers at Dell's Austin headquarters.

GROUPWARE SELDOM SUCCEEDS WITHOUT TOP-LEVEL SUPPORT

Price Waterhouse was fortunate to have one of the world's great technology cheerleaders at the helm of its Notes effort. Sheldon Laube, the firm's chief technologist, has a techie's brain and a salesman's personality. He got his top-level job—one of the company's 20 most senior—in 1989 with explicit instructions to make the firm's technology state-of-the-art. When he found Notes, he recognized its potential for tackling one of Price Waterhouse's key challenges: capturing and storing information. Laube pushed, cajoled, joked, and lobbied for it. Richardson says Price Waterhouse would never have won one recent multimillion-dollar job if it hadn't been able to use Notes to put together a proposal quickly.

Groupware challenges conventional notions of proper public presentation. Comments in databases are generally more informal than the paper memos and letters they replace. For a company to use groupware successfully, employees must feel they won't be judged by how they write or how well they have thought through what they say. Says Clint Alston, who heads Ernst & Young's technology services: "It takes pretty confident people to put information that may be half-baked, or just the best they have at a given moment, into a fishbowl where anybody else can react to it." Says one disgruntled manager of a large national media company that recently started using Notes: "People are putting in poorly thought out stream-of-consciousness stuff, and it takes hours to wade through."

This manager finds Notes contributes mightily to his feeling of information overload: "People say if I'd just turn on Notes I'd know what's going on. And I say 'Yeah, and I'd also know about 400,000 other irrelevant things that simply waste my time.' Ideally I'd take half an hour to peruse Notes each day. But I never have a half hour of uninterrupted time." His solution: Keep Notes switched off and have his secretary occasionally print out new comments that seem germane.

Groupware may prompt business people to start developing new ways of dealing with information, says Tom Malone. He directs a research center at MIT's Sloan School of Management that studies groupware and networked computing. Malone recalls a time not long ago when people read every letter and returned every phone call that came to their offices. "But in a world where hundreds and thousands of communications enter the office every day," he says, "our expectation that we should carefully digest it all can make us feel overwhelmed and terribly inadequate."

Malone believes people should accept that they can't deal with every last bit of data. Some new software applications (but not yet Notes) already include so-called electronic agents to help workers filter and sort through information and find exactly what they need. Price Waterhouse deals with this potential problem by assigning people to keep Notes databases well organized and clear of confusing or irrelevant material. Dell, on the other hand, leaves maintenance up to the users of each database, and some end up looking "like garbage dumps," in the words of one senior Dell techie. Lotus says it will put electronic agents into the next version of Notes, available by the end of 1994. They will alert you if, say, your boss adds a comment to something in a database, so you'll know to look for it.

There's no turning back. Networking is making computers more and more interpersonal, but most American workers and managers still don't think electronically. Groupware can not only help bridge their conceptual gap but also give them power they never had before.

QUESTIONS

1. What are the main components of groupware? How do they function?
2. How can groupware improve organizational productivity? Give specific examples.
3. What organizational changes would be necessary to obtain the full benefits of groupware?

3.3 Who Needs an Office?

CLINTON WILDER

Peter Drucker[1] tells us that the organization of the future will be much flatter than the organization of the past. The traditional control span of five to seven people can increase as the information needed to manage more people becomes available from the computer. More people reporting to one individual reduces the number of management layers and improves productivity. Furthermore, people will spend more of their time working in teams than they do today.

The technology makes it possible for people to work together even though they are separated in space and time. Back in 1973, a team from the University of Southern California came up with the concept of *telecommuting*[2]—of moving work to people rather than people to work (see Section 2.1 for a brief description of telecommuting). The telecommuting concept has been carried to its logical extreme by VeriFone Inc., the largest supplier of electronic credit card verification equipment. This highly successful, world-wide organization operates almost exclusively in a telecommuting mode. It uses time zones to its advantage, handing over rush jobs such as business proposals from one team to the next, so that these jobs are handled on a round-the-clock basis without overtime.

The VeriFone case is an extreme example. Yet more and more companies are adopting the new technologies such as telecommuting to gain competitive advantage.

HATIM Tiabji's least favorite term is "corporate headquarters." The chairman, chief executive officer, and president of VeriFone Inc. expects his employees to conduct meetings from any location with the aid of a corporate electronic-mail network. "Our corporate model is the blueberry pancake," says Tiabji. "All the berries are the same size. All locations are created equal."

VeriFone, the world's largest supplier of electronic credit-card verification equipment, does have a base in Redwood City, California—if only to comply with Securities and Exchange Com-

[1]Peter F. Drucker, "The Coming of the New Organization," *Harvard Business Review* (January/February 1988).

[2]J. M. Nilles, R. Carlson, P. Gray, and F. Hanneman, *Telecommuting: Options for Tomorrow* (New York: John Wiley & Sons, 1976).

Source: Informationweek (June 6, 1994): 60–62.

mission requirements. But according to Tiabji, the workplace is where the employee is. Call it the mobile work force, the networked organization, or the virtual corporation, VeriFone is already living the model that management gurus such as Tom Peters have laid out for the next century. A VeriFone employee can operate from a hotel room in Tiabji's native India, an airport lounge in Boston, or a pay phone in Athens.

VeriFone believes it is a business necessity to be dispersed, nimble, and responsive. Although it dominates the field of transaction systems automation, the company is up against formidable competitors such as electronics giant Schlumberger Canada Ltd. VeriFone must move fast to close deals and develop new products. Folio, a mobile wireless card swiper VeriFone unveiled in May, took just nine months to develop.

The anywhere/anytime culture works only because VeriFone's 34 locations in nine countries are hooked to the company's Digital VAXmail system over private leased lines. Employees also exchange mail and documents over the Internet. Nearly all of VeriFone's 1,800 employees are equipped with a 386 or 486 PC, Apple Macintosh, or similarly powered laptop. New employees get a laptop before they get a desk. There are no executive secretaries. Paper memos are verboten.

BOTTOM-LINE PROOF

Does VeriFone's philosophy work? The bottom line seems to indicate yes: The company controls more than a 60% share of the U.S. market for credit-card swipers and associated software and has sold more than three million systems in 70 countries since its founding in 1981. Sales have grown by more than 25% a year since 1988, and the company posted $259 million in revenue and more than $21 million in profits last year.

Star Bank in Cincinnati is one of 500 customers, suppliers, and consultants linked directly to VeriFone's corporate E-mail system. Mary Karwisch, assistant VP and manager of merchant operations, dials into VeriFone through CompuServe on an IBM ThinkPad and places orders, exchanges messages with sales staff, and downloads articles about the credit verification process that VeriFone posts several times a week from commercial online services. "If I need a price for a merchant, I send an E-mail request and get it back either that afternoon or the next morning at the latest," says Karwisch, who orders verification equipment for 4,000 merchants and agent banks. "It's nice to get a fast response to a message instead of having it sit on someone's mainframe somewhere."

NOT STATE OF THE ART

VeriFone doesn't have any mainframes, but you won't find a leading-edge client-server architecture in place either. VeriFone runs on Digital VAXs—3000s, 4000s, and 6000s, all running VMS—clustered in various configurations in its Northern California "headquarters" and in Honolulu; Taipei; Bangalore, India; and Costa Mesa, California. PC users in all 34 facilities are interconnected with Novell NetWare 3.12 and the TCP/IP networking standard. VeriFone's "groupware" is simply VAXmail.

RevWatch, the company's customized financial system, provides managers with up-to-the-minute data on worldwide orders and sales. VeriFone provides a few power users, mostly financial analysts, with Pentium PCs from Gateway, Dell, and Compaq.

How can a company that sells information technology products and that relies so heavily on networking and electronic mail get by without the latest and greatest? "We believe in Ford Fairlanes and Chevy Impalas," says Will Pape, VeriFone's senior VP and chief information officer. "If you're just using E-mail, there's no reason to have a Pentium. You don't need a Ferrari to drive to the supermarket, and you don't need the latest wireless device to enter the information age."

Indeed, VeriFone spends just 2.5% of its revenue on IT, including telecommunications. That's about half the U.S. electronics industry average of 4.9%.

No one is more "virtual" in this corporation

than Pape, whose own base of operations is his home just outside Santa Fe, New Mexico. Known as the "sagebrush CIO" by other VeriFone executives, Pape can be heard on his speakerphone in New Mexico giving a visitor in California a software applications demo. Where's the application? On a PC at VeriFone's Redwood City headquarters.

VeriFone is dispersed and mobile primarily to be close to its customers. But it also wants to be located in what Harvard Business School professor Michael Porter likes to call a "center of excellence," or location known for its intellectual or capital resources.

VeriFone's smart-card operations, for instance, are based in Paris because France is generally recognized as the leader in smart-card technology. VeriFone has a software development group in Bangalore, where many U.S. companies are starting to tap applications development talent. And it builds its products in Southern California and Taiwan, both manufacturing centers.

VeriFone uses its corporate VAXmail network for 24-hour operations. Managers pass projects from one continent to another in a constant round of activity. The company wants to tap the expertise of its work force to meet a business demand.

LOADED FOR BEAR

Tiabji is fond of one story involving a sales rep who was trying to close a deal with a major bank in Greece. The negotiations hit a snag when the rep didn't have some information the bank needed. He asked the bank for a 24-hour grace period so he could find the answer. The bank's management agreed, skeptical that the rep could amass enough material in such a short time from the confines of a foreign hotel room. The rep posted an urgent request on the VeriFone VAXmail network to the company's worldwide sales force from his laptop, asking if anyone had ever faced a similar scenario. He got back 26 responses overnight, went into the bank the next day, and closed the sale. "To say that he was loaded for bear is an understatement," Tiabji boasts.

"It's not a matter of technology, it's a matter of attitude," says Roger Bertman, VeriFone's VP and general manager of marketing. A veteran marketing executive of technology stalwarts Ungermann-Bass and Rolm, Bertman confesses that he had never used a PC before joining VeriFone two years ago. "I'd heard VeriFone was different," he says. "But I was shocked to find that there were no secretaries here."

Tiabji admits that VeriFone's culture can be jarring to new employees, but that's part of its mission. "Every time I bring in a new person, I have to change his or her mindset," he says. Executives who think using a keyboard is beneath them needn't seek employment at VeriFone.

QUESTIONS

1. What technologies is VeriFone using for operating in "real-time and on-line" style?
2. How does information technology and the operating style enhance organizational productivity at VeriFone?
3. For what types of organizations can this style of operation be imitated? For what types of organizations could it not be imitated? Explain why or why not.

3.4 Executive Information Technology: A Strategic Necessity at Motorola Codex

CHRIS CARROLL AND CHRIS LARKIN

Executive Information Systems (EIS) are typically data retrieval systems operated by and for the use of senior managers. The data that senior managers need is different from that gathered and put into the computer by the typical accounting function. Senior managers do not care about individual transactions made at specific points in time. Rather, they need to know where the business stands now and where it is going. Much of the data in executive information systems tend to be time series information that shows history and forecasts. The data in commercially available executive information systems tend to be presented in fairly glitzy graphical form. What is intriguing is the number of corporations that have and use such information systems.

Carroll and Larkin present a case study that describes a particular EIS, the one at Motorola Codex, a subsidiary of Motorola, Inc. They argue that the EIS they built is a strategic necessity, that it is an information system used to stay even with competitors. They argue further that Motorola Codex' EIS helps managers improve product and process quality and, ultimately, customer service.

In reading this article skip over the heavy use of jargon and product references if the language and names are strange to you. Rather, focus on the companywide process involved in creating and implementing a major system for senior management. The implementation task is not easy. However, if done right, it can have major positive impact on the business and its profitability.

THE successful application of executive information systems (EISs) in established businesses requires cultural change and superior implementation efforts. There must also be a personal commitment from members of senior management to help define their information needs and to actually use the system after it has been developed. If senior managers don't take advantage of the data placed at their fingertips, the tremendous effort that goes into providing that data

Source: Information Systems Management (Summer 1992): 21–29.

on the desktop is a waste of corporate resources.

Executives need quick access to data on demand, and they must be able to view the data as accurate, summary in nature, and of strategic importance to the decisions they must make about how to run the business. This article describes how a subsidiary of a Fortune-50 high-technology company determined the need for an EIS, developed the system, and implemented a program to support its primary business objectives.

THE COMPANY AND ITS CHANGING STRATEGY

Motorola Codex has been in the data communications business for nearly 30 years and has sparked many networking industry standards. Headquartered in Mansfield, Massachusetts, the company has sales and service offices in more than 50 countries and 3,500 employees worldwide.

Motorola Codex, which is a subsidiary of Motorola, Inc., was established in 1962 as a supplier of coding techniques. Eventually the company developed modern technologies and, over time, helped build the modem industry, selling primarily component products that offered a fast economic payback compared with telephone company alternatives.

Following divestiture, however, business networking needs became more complex and innovations came much more quickly. More corporate networks began using mixed facilities and traffic types. In many companies, the network stopped being regarded as a tactical business operation and instead came to be viewed as a strategic business weapon. Airlines, banking and financial services, and retail corporations acknowledged that their network strategy reflected their ability to compete in the market.

Investments in networking rose, and purchase decisions escalated in importance as the dollar values increased. Motorola Codex's traditional customers, data communications managers, were now influencing purchasing decisions that were often made at the boardroom level. At the same time, large computer companies and carriers began to compete for those purchasing decisions, and in many cases, they were better positioned than Motorola Codex to distribute their solutions because of their relationships to senior executives in customer companies.

Customers were purchasing not just boxes but networking solutions. They were buying not only products but the capabilities of companies that could support the network. Quality became a key criterion as businesses recognized that downtime from equipment failure was unacceptable. Innovative and fast-moving small companies developed niche markets; carriers and computer companies acquired independent communications equipment manufacturers.

For Motorola Codex, all these developments meant that the company needed to increase its understanding of customer needs, market segmentation, and market dynamics. It meant accelerated product life cycles for product development. It meant a move from hardware-based systems to software-based systems so the company could continue to supply customers with rapid innovations.

Because ways of doing business changed so radically through the late 1980s, Motorola Codex adopted a strategy to transform itself from primarily a component supplier to a networking systems company. Professional services have become as critical to success as the networking expertise developed over 30 years. Whereas the company used to spend 36 months developing a product that would have a five-year life, today it spends 12 months developing a product that will have a life of three years. This change required total commitment to quality and a segmentation of the business into several smaller businesses, each run by a general manager and each requiring detailed management information.

MAPPING INFORMATION SYSTEMS TO STRATEGY

The corporate infrastructure needed to reflect the change in business strategy. Not only did Motorola Codex have to reevaluate its information systems, it needed to reexamine its buildings and grounds, its human resources, and even its management team. As the company began implementing changes in its corporate infrastructure, it found that the information needed to propel this business transition was inadequate to achieve business objectives.

Company executives recognized that in many ways they had become like the cobbler's children. For example, since the mid-1980s, Motorola Codex had sold backbone networks on the principle that the network was a strategic business asset. Executives cited examples, created models, kept up with information technology developments coming out of Harvard and MIT, promoted the value of corporate information, and evaluated the work done by such leading consulting firms as the Index Group and McKinsey. In short, the company's executives were selling information as a competitive advantage but weren't implementing information technology well enough in their own business.

Systems Improvement a Primary Business Objective

Motorola Codex made the improvement of its systems a primary business objective and hired Perot Systems to integrate the corporate computing environment. Together, they developed the Global Enterprise Management (GEM) program. The program's mission was to create an integrated global information environment that positions Motorola Codex to compete effectively in the marketplace.

The system's benefits would be providing managers with easy access to critical data, including profit and loss information, and fast access to summary-level data. To start, the company needed a consistent model of its business environment and an efficient means of extracting data from existing systems.

The GEM program had three clear objectives. The first was to reduce the time needed to make critical strategic, operational, and tactical decisions. The second objective was to have a consistent view on critical success factors among the management team. The final objective was to improve financial performance through the use of information technology.

The financial performance objective targets and supports higher growth and higher margins. Executives wanted to better manage low-growth and low-margin products and services and to retain customers and market share by constantly improving quality.

Cycle time is an issue affecting all three objectives. Reduced cycle time is a major corporate objective of Motorola, and Motorola Codex is mapping all its administrative processes in order to find ways to shorten these processes. The EIS improves the availability of business indicators so that decisions can be made earlier, which in turn facilitates compliance with audit and government requirements.

The EIS and the Malcolm Baldrige Award

In 1989, Motorola, Inc., received the first Malcolm Baldrige National Quality Award. Motorola had championed quality for more than a decade, and Motorola Codex products were regarded by many industry surveys as the highest-quality networking products in the world. Because continuous improvement is critical to total quality management programs, management viewed the EIS implementation as critical to tracking and monitoring quality statistics.

Surprisingly, after winning the Baldrige award, Motorola management heard from customers that although they were greatly satisfied with product quality, they were less satisfied with the company's order-processing and contract administration procedures. If the company didn't improve its billing system, customers indicated that they'd eventually switch to another vendor, one that would be easier to do business with. The reality that process quality is as critical as product quality hit home, and efforts began to adjust the company's infrastructure to provide managers with more data on each of these areas.

The company continues to move to a shorter, more accurate order-entry process, improving the availability of marketing support data and making data easier to access and manipulate. The company is also improving the availability of warranty data as well as the availability of installed base information. The company is also increasing its efforts to set up sales incentive plans and is moving toward quicker and more accurate invoicing to the customer.

The EIS and the Global Network

A sophisticated wide area network allows employees to communicate online from the desktop with colleagues around the world. Motorola Codex is a Macintosh shop, with nearly 3,000 Macs connected in every office in the United States and nearly 30 foreign countries. The Macs are networked through Apple Talk LANs connected by gateways to a global network. In the United States, LANs are connected over Motorola's fast-packet wide area network with high-speed dial connections to smaller field offices.

International subsidiaries and distributors are connected through either dedicated or dial circuits, and a central Mac-based E-mail system connects every user on the network and provides fast and timely file transfer. The company has realized the tremendous productivity benefits of the easy-to-use Macintosh without experiencing the communications restrictions usually associated with this computing platform.

Making Strategic Use of Information

Decision support systems and executive information systems are critical to achieving the company's business objectives. Together with the Global Enterprise Management program, decision support systems provide the tools needed to drive Motorola Codex's business transition.

Although the company's management has often sold customers on the strategic applications of information, most managers had little personal exposure to technology. Most senior executives had Mac SE-class machines on their desktops, primarily for communicating to marketing and international managers and to the field sales organization through electronic mail. Senior executives relied largely on lower-level management and staff analysts to gather data from discrete sources and manually generate an analysis of issues. However, these systems were inefficient, required too much staff time, and did not provide enough information.

Senior management needed to gain a consistent and comprehensive view of the business and to use this view as a means of monitoring operating results. They wanted a model that could be used to predict the patterns for earnings, margins, and revenues that might result from varying product, organizational, or financial strategies.

Tools include multidimensional data models that give access to a consistent data base of key business variables. The data models array bookings, revenues, costs, margins, and quality metrics across the business dimensions of product, distribution channels, customers, geography, and time.

The company contracted with Decision Support Technology of Cambridge, Massachusetts, for an executive decision support system named the Codex Information Window (CIW). Senior management accesses the information through a graphical, Windows-like interface. Unfortunately, because none of the vendors in the market could create a system that is based on the Macintosh interface, the system has the look and feel of an IBM-like system without the pull-down menus executives had been using.

In addition to these business models, the system provides executive support functions (e.g., an electronic briefing book, exception reporting, and drill-down capabilities). These features highlight the selected key factors in the product and distribution organizations, providing managers with a tool for accessing timely operating results and identifying areas that need action.

Two decision support applications—Comshare System W and Commander—were customized to implement the system. The models run on a 3-node VAX cluster. The Commander application was used to build the briefing book features of CIW, and the interface is the EXECUVIEW application. The Mac IIs reside on the desktops of senior management, and they are used to poll the VAX cluster weekly for updates. Similarly, the VAX system polls the IBM mainframe systems for transaction information. There are now about a dozen users of the Codex Information Window, and the company plans to eventually have as users a maximum of about 80 executives located in its Massachusetts offices in Canton and Mansfield, in U.S. field management offices, and in international subsidiaries.

SYSTEM GOALS

Although Motorola Codex needed an enterprise decision system that reflected its business, it also needed a tool that could be useful to some mid-level managers throughout the company who could either take action on the data supplied or use the data to create more customized models of an aspect of the business. For example, the business development analyst has the application on a desktop. The market research department has access through the analyst to this unit for preparing custom reports or trend or environmental analyses. Although the vision was for senior management to use the tools (executive acceptance is vital to realizing the value of the EIS), the company acknowledged from the beginning that the user may not always be a senior executive but may be a staff manager.

Effective Management Control

One of the first goals was to reduce administrative and overhead costs by improving management control over order cycle time. Delays in the order-processing cycle, at any stage, cost money by delaying the receipt of revenues, accumulating additional overhead costs, and reducing the level of customer satisfaction.

Since implementing the system, the company has made significant progress in manufacturing; cycle times have been reduced from about 20 days to five days. The company is now looking to CIW to help reduce order-processing cycle time by providing indicators of order efficiency. For example, the CIW will inform users of the average time necessary to complete orders for each product category in a given time period and in comparison with previous periods.

Variances of cycle times outside acceptable bounds as well as trends that help in planning future products can be identified and acted on. Gaining effective control of revenue-generating assets is a critical priority. The system already has the limited ability to identify exactly what equipment is leased, when the leases expire, the yield on leased equipment, and the market or book value of equipment returned from leases.

Because a significant portion of Motorola Codex's business is in leasing data communications devices, actively managing this resource is critical. CIW improves this control by providing timely information that lets managers know the revenue performance of leased assets over time. This allows management to access the standard margin on leased assets to highlight particularly profitable or unprofitable products. Another objective is improving the effectiveness of distribution channels by highlighting critical variances in product or distribution channel sales performance.

CIW provides access to key indicators of market performance for each of the company's major product groups and product lines. Management conducts performance measurements on a gross margin basis at the product group level. CIW includes certain leading indicators of performance—for example, the volume of product returns or the variances in actual unit sales versus marketing forecasts. This allows management to read markets much more quickly than in the past and with more accuracy.

Another objective is improving the responsiveness of manufacturing to variances in the current and future product mix from the mix forecast. The Motorola Codex product portfolio is so broad that the company faces a major element of business risk by investing in the materials, labor, and equipment necessary to support the production of each monthly manufacturing plan.

Management accepts that actual order volumes can never match those forecast and that there will always be instances of either overbuying materials or insufficient capacity to meet customer demands. Previously, these variances in demand became apparent only when orders reached the factory, even though weeks may have passed since the orders were booked.

The CIW allows management to fine-tune manufacturing decision making according to actual volumes of orders. By providing access to the number of units on order—by product and period—manufacturing management can plan short-term material and labor requirements. Through this process, the company hopes to realize lower material costs and improved customer satisfaction over the long term.

COMPETITIVE ADVANTAGE THROUGH INFORMATION

Motorola Codex executives believe that by investing in the Global Enterprise Management system and the Codex Information Window the company can gain an advantage in the marketplace through the use of information technology. Motorola Codex has been using the system since early 1991 and it is not yet fully populated with data, particularly historical information. Overall, the program has been on schedule with no surprises in the implementation.

CIW was implemented so that senior management can readily access information critical to making fast and informed decisions. The single area in which the company may realize the most long-term benefit is in quality. On the basis of these business objectives and a complex and changing business environment, the Codex Information Window provides a window into corporate data and serves as more of a corporatewide information vehicle than an executive support system. (The word *executive* was not allowed in the name of the system because of its elitist connotation). The exceptional information infrastructure that is available—the AppleTalk network—plays a major role in streamlining the flow of critical information.

THE DESIGN EFFORT

The design effort employed the Decision Support Planning and Analysis (DSPA) methodology. This methodology focuses on the extraction of data requirements from the user's point of view and frees the user from the laborious deciphering of data flow diagrams and long sessions in requirements validation. The user community expresses its needs in terms of key decisions made on operational and strategic levels, and the design team provides rapid feedback by developing a demonstration prototype along with the underlying data scheme that answers those needs.

Structured Interviews

The CIW design group interviewed more than 80 people involved in the corporate information flow process. The selection of interviewees was made from the data provider, data analyst, and ultimate data user levels. The interviews allowed the designers to listen to the interviewees' opinion of their impact on the decision-making process and what each interviewee thought was both critical or irrelevant to the current information flow process. Interviewees' data needs could be attained by inspecting their answers to such questions as:

- Where is information gathering most difficult?
- Who are the internal customers of your information?
- What questions do you ask most frequently each day?
- Which questions are hardest to answer?
- After a two-week vacation, what is the first thing you'd want to know upon your return? Who would you ask?

The interviews allow the users to express their needs in terms of their daily business activities, rather than in terms of electronic data flow. The result is a set of interview data that represents root cause information flow and quality problems. The decision analysis phase of the requirements analysis is a method for deriving a focus for the system design from critical issues revealed during the interview process. This focus can then be inspected for the impact on data availability, technical feasibility, and other issues concerned with answering some basic questions—for example:

- Who are my most important customers?
- What were last week's domestic bookings?
- What were the cycle times for each fulfillment site last year?
- What were factory defects per unit for each device type?

Knowing that the system's objective is to answer these questions provides a consistent focus from requirements analysis through adaptive design and development into successive releases and rollouts.

The Codex Information Window was developed using adaptive design and development that involves incremental, rapid prototyping. As the system's content evolves, so do the system's capabilities. In this way, users are given a platform they can react to (the prototype). Furthermore, they feel that they have helped mold the system. The result of this process is a system that they will be more inclined to use.

As the vision of CIW became increasingly defined, a prototype took shape. The strategy was to initially focus on the critical content of the system while representing the capability needed by the core group of users. As the architecture of the system was defined in conjunction with the requirements analysis, an evolution of prototyping environments occurred.

The initial presentations of the system were made with Bridge, a microcomputer-based batch language developed by Softbridge Microsystems of Cambridge, Massachusetts. The architecture required a shift to a Macintosh platform, so the prototype was quickly reengineered in HyperCard. A further migration occurred to Supercard to more closely resemble the emerging functions of the end system. The final prototype was developed within the environment that was actually rolled out.

Weekly, the prototype was shown to a different subset of users within the organization (e.g., marketing, product management, senior management, quality assurance, and manufacturing). The input from each weekly team session was engineered in the prototype. The team meetings were used to tear down and rebuild the prototype as the users provided feedback. Although many conflicting points of view emanated from these sessions, the result was a prototype that grew to reflect a corporate mental model of decision-making needs.

The business content of the system evolved as all information users or providers described what they would need in a tool set for informed decision making. The initial set of interviews focused on the needs of upper-level management to make daily and weekly decisions regarding key customer attention, sales objectives, and product margins. As more executives were interviewed, concepts of total quality management and Motorola's six-sigma software quality program became a dominant theme.

The decision analysis yields groups of questions known as decision classes. These decision classes can be satisfied, for the most part, from a core set of data. For example, data on domestic sales of a product line can satisfy such questions as, "What are total revenues, year to date, in the northeast region?" or "What are the revenues by product, month to date, for each region?"

The CIW menu structure is composed of the decision classes shown in Figure 1. These decision classes allow for efficient information retrieval that is based on the user's stated require-

Figure 1 CIW Menu Structure

ment for decision support. One of the primary functions of the Codex Information Window is to provide on-demand access to many levels of detail of corporate information. This drill-down structure of each decision class then provides successively more detailed information, high-lighted by executive reporting capability. This drill-down menu hierarchy is displayed in Figure 2, which shows that revenue and booking information can be accessed in successive slices of detail, arranged to directly correspond to the Codex financial reporting structure.

For example, if an exception threshold is set for ±15% variance and a variance in product bookings at the product family level exists, the screen highlights this variance. The user may then select, with a click of a mouse, that line item, and the interface will then descend to the next layer of detail on that line item: the product series.

System Architecture

As the prototyping effort matured, the architecture for the system crystalized. The critical success factors set for the architecture were designed to:

- Use the existing AppleTalk network and installed base of Macintosh workstations.
- Extract reports from IBM mainframe and VAX minicomputers.
- Consolidate the reports into a multidimensional data model.

Figure 2 CIW Menu Hierarchy

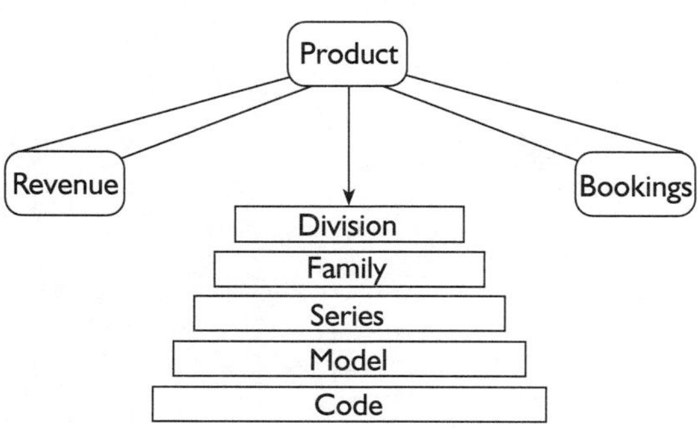

The objective of the architecture is to develop an information base on the minicomputer or server and to extract information updates from the various data sources as appropriate. The result was the three-tier system architecture shown in Figure 3. Report updates are generated weekly and monthly. The data extraction controller coordinates the efforts of the individual extraction programs and is written in IBM JCL.

IMPLEMENTATION ISSUES

Many issues surfaced during the design, prototyping, development, and rollout phases of this project. The following recommendations are critical to the success of the project.

ENGAGE THE USER COMMUNITY EARLY IN THE SYSTEMS LIFE CYCLE

The user community and those individuals who ultimately must support such a system must be on the project team from the start. Their insight into the initial focus of the decision support system is absolutely essential. As is the case with most executive and decision support systems, the use of these tools will always be optional. Therefore, a feeling of ownership and pride in the accomplishment of seeing the system come to fruition is critical to the long-term use—and success—of the system.

MONITOR SYSTEM USE BY KEY CHAMPIONS

It was critical to the project members to understand who was using the system from the beginning and how often they were using it. This monitoring can be done in many different ways, although automating the monitor function is most desirable because of its inherent objectivity. For CIW, the implementation team checks the number of log-on days per user. By checking this number during the sensitive period of a system rollout, the team can spot trouble in the user's training or interest level or in the basic design of the system. In addition, a brief questionnaire or interview is also needed to understand specifically which

Figure 3 CIW System Architecture

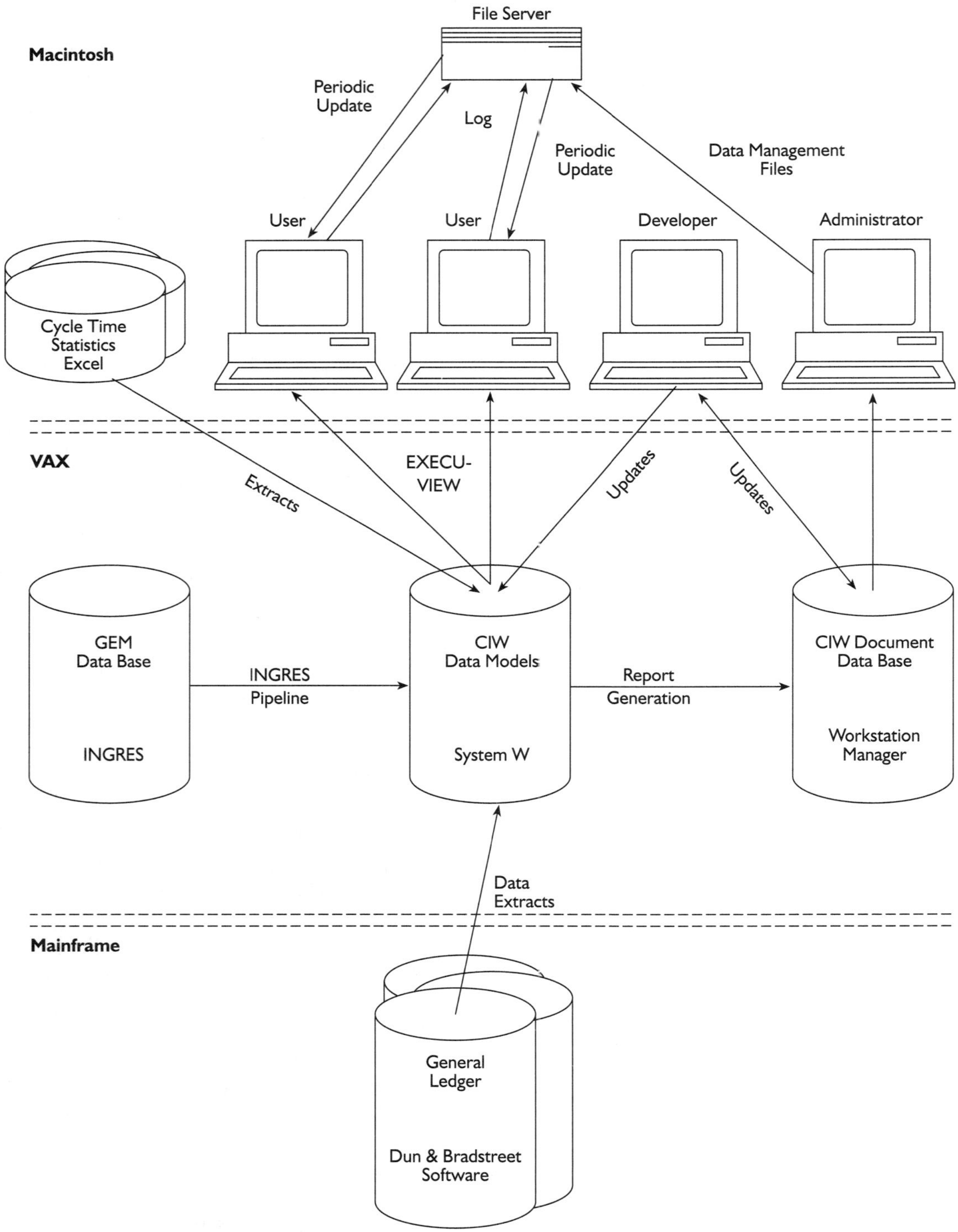

parts of the system are being used. In this way the rollout and maintenance team will continually focus and mold the system according to evolving user needs.

DEVELOP MOMENTUM FOR ORGANIZATIONAL SUPPORT OF THE DATA

During the design phase of CIW, it became clear that many territorial battles were being fought over whose data is right and whose data is wrong. Many senior management meetings resulted in data validity debates instead of productive problem-solving sessions. It was also clear that for CIW to succeed, the data sources and models should be as correct as any other figures used, at the same time providing the organization a focal point for continuous improvement of the data. Without support for CIW as the definitive data source for organizational decision making, the CIW system would be subject to constant question of its value.

After the organization recognized the data integrity qualities in the system, a few key champions—including the company's president—announced support for the data. This turned the tide of organizational support for CIW.

CONCLUSION

The system is up and running and is being used by senior executives and mid-level managers to analyze data and formulate business decisions. A plan is in place to gradually make the system available to a greater number of executives. The primary obstacles to implementation are the non-Macintosh-like interface and the lack of historical data with which to compare current and past performance.

With CIW, management better understands the business and is better prepared to make decisions, and as each month passes the system incorporates one more month of historical data. Comparative analyses become possible and the information therefore becomes more valuable to decision making.

Learning issues are being addressed, and personalized training is being conducted. Use by executives is monitored by the systems operator; managers who aren't using the system are polled, and actions are taken to make the system attractive to them, either by providing them with customized training to make the system easier to use or by incorporating their suggestions for enhancements into future system design.

While it is too early in the systems life cycle to definitively determine the system's success, it does appear that the system has been accepted, is being used productively, and has become a useful decision support tool. The information generated is helping managers drive the business transition from a communications product company to a networking systems company, and the EIS is being used to make the choices necessary for the company to achieve its business objectives.

QUESTIONS

1. How do executive information systems differ from decision support systems and conventional management information systems?
2. What are the major quantitative and qualitative benefits of executive information systems?
3. Do you think executive information systems primarily improve executive productivity or organizational productivity? Give reasons to justify your position.

3.5 Electronic Data Interchange: The Elements of Implementation

JAMES A. SENN

The nature of transactions among firms has undergone a major change in the last several years as the concept of Electronic Data Interchange, or EDI, took hold. EDI involves the exchange of data among trading partners using an agreed upon format. It is used for such purposes as ordering parts from vendors, sending bills, and receiving payment.

At one level, EDI is very scary. You have access to other people's computers and they have access to your computer. A large amount of faith must exist between you that you will not corrupt or steal each other's data. Furthermore, small firms (who act as contractors) must adopt computing rules set up by the large firms that give them business. A contractor working with several large clients may have to adapt to different standards for each client. Although small firms are often computer naive, they know that they must adopt their client's EDI if they are to hold the account.

Yet, the payoffs from EDI overwhelm the difficulties. Business is done much more quickly and accurately. Information does not have to be keyed in over and over. Just-in-time manufacturing (Section 4.1) becomes possible. Data is available for running computer-based analyses. The speed of transactions is enhanced. The benefits go on and on.

Senn's article is an overview of EDI. It describes not only the technology of EDI, but also the organizational and strategic implications. Senn argues that firms must have an EDI policy if they are to be successful.

ELECTRONIC data interchange (EDI) is one type of interorganization system; others include electronic mail, electronic funds transfer, and videotext. All types of interorganization systems are increasing in number as business processes are modified so that corporations can respond to new opportunities as well as to the constant pressures for greater responsiveness to the needs of customers and trading partners.

Source: Information Systems Management (Winter 1992): 45–53.

DRIVERS OF INTERORGANIZATION SYSTEMS

Although there are numerous information technology alternatives available to facilitate the communication of business documents between organizations, information technology by itself is not the principal force behind the spread of these systems. Depending on the industry, the following factors are the driving forces.

CUSTOMERS DETERMINE BUSINESS ACTIONS

Across virtually all industries, customers (not dealers or suppliers) determine which products and services will be successful. Whether at the individual or corporate level, buyers are stating what they will accept, when they will accept it, and under what conditions. Mass-produced goods of the past are being replaced by customized products.

QUICK RESPONSE IS BECOMING THE RETAILING STANDARD

Time is increasingly being recognized as a barrier to business. Delays are caused in addressing market opportunities if the elapsed time between the need for a product and its fulfillment is excessive. Time requires decisions to be made far in advance of the point at which their impact will be noticed. Quick response methods are redefining the way business is transacted by shortening the time periods around which commercial activities are arranged.

THE JUST-IN-TIME MANUFACTURING TECHNIQUE HAS GAINED INCREASED POPULARITY

The barrier of time also forces businesses to accumulate inventories and stockpile goods to ensure that demand can be accommodated when it occurs. Manufacturers are revising their strategies so that plenty of materials are on hand and that the materials arrive in time for manufacturing, but not so far in advance that they accumulate as inventory.

ESSENTIAL INFORMATION TECHNOLOGY IS AVAILABLE

Hardware, software, and communications networks provide the ability to link organizations and transmit business documents electronically. The level of information technology needed for interorganization systems is generally available and component reliability and cost levels are acceptable.

STANDARDS HAVE BEEN FORMULATED

For customers, suppliers, and other business partners to be linked, each party must use an agreed-on format for communications. Standards for such transmission have been established and are used in many industries. They serve as a foundation on which to build interorganizational relationships.

THE GLOBAL NATURE OF BUSINESS

At one time, the notion of a global business community was of concern only to international or multinational companies. Today, there is a truly global market for sales and sources of materials.

Even if an organization does not directly market its products in another part of the world, it may interact with a distributor or representative who does. Material and service sources may also be internationally based. The capability to transmit and receive business communications over long distances and between distinct time zones is usually the reason interorganization systems are created. Electronic data interchange systems in particular are of increasing importance because they link trading partners and redefine the way business is transacted.

FEATURES OF EDI

Electronic data interchange is a form of electronic communication that allows trading partners in two or more organizations to exchange business transaction data in structured formats that can be processed by computer applications software. EDI is fundamentally a data processing concept that spans multiple businesses. It does this by relying on data communications methods and standard formats for the transmission, acceptance, and understanding of business data (see Figure 1).

The basic features of EDI include the following:

- Intercompany communications. EDI links trading partners (e.g., organizations and their customers and suppliers) in a cooperative relationship that offers the potential for improved levels of profitability, service, and productivity.
- Computer-to-computer linkage. Data is exchanged electronically, in machine readable form, and without the need for manual intervention in the data entry or manipulation process. Networks link computers within the departments of participating organizations.
- Transmission of business documents. Business documents, rather than messages or narrative reports, are the substance of EDI transmissions. Typical business documents include purchase orders, purchase order acknowledgments, request for quotations, quotes, invoices, bills of lading, and shipping notices.
- Formatted data. To ensure that the data is received and properly understood, carefully defined formats specify the order of data items in a transmission and the characteristics of each item.
- The capability to overcome technological differences. Distinctions in the computer technology used by communicating parties are overcome through translation. Thus, hardware, protocol, and data format differences do not limit participation in EDI activities.

Figure 2 describes the flow of information between the EDI trading partners. The EDI network interconnects the partners, either directly

Figure I Intercorporate Transactions Using EDI

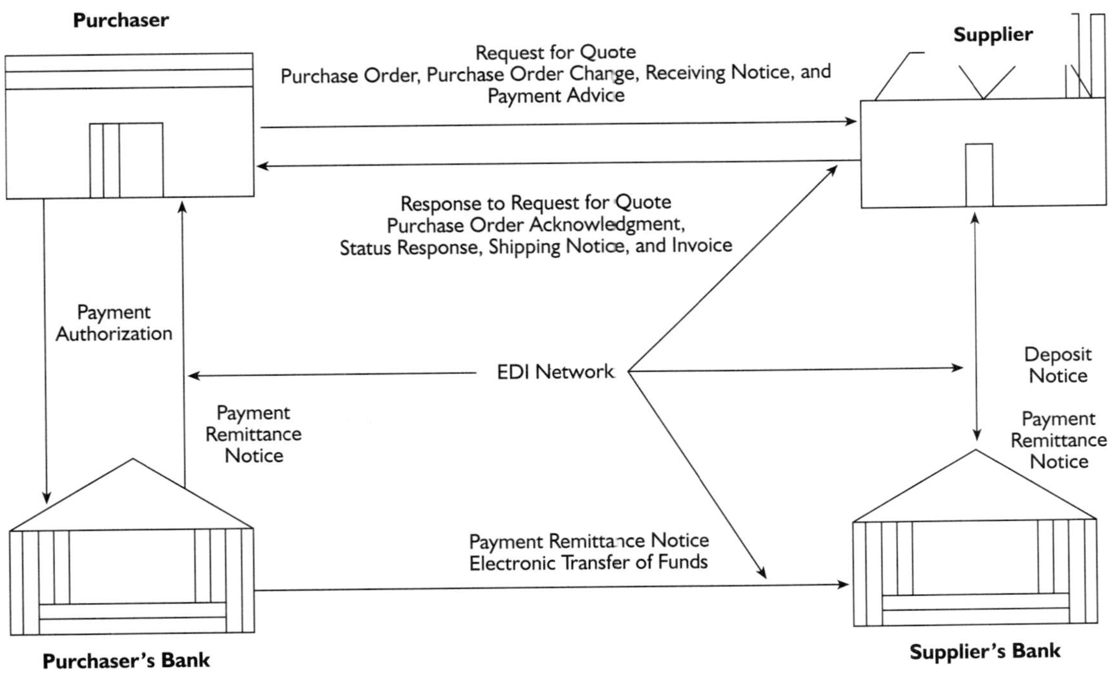

Figure 2 EDI Links Trading Partners

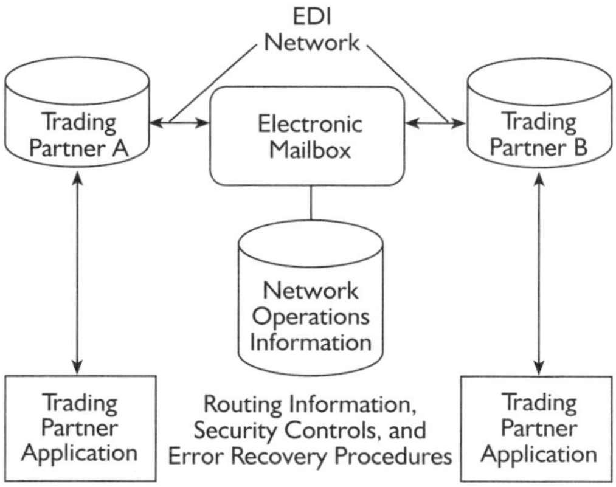

(i.e., computer to computer) or indirectly through the use of electronic mailboxes. Located within the network, the mailbox holds documents transmitted from a source authorized to add to its contents. Documents remain in the mailbox until they are checked and retrieved by the intended recipient.

The network capabilities include more than just transport capability. Routing and security is an essential feature that protects against unauthorized access to potentially sensitive data. Built-in mechanisms for recovery protect against loss of data if the network fails. As these characteristics suggest, EDI consists of much more than electronic transmission of orders.

REASONS TO ADOPT EDI

Organizations migrating to EDI are driven mostly by business needs. Improved productivity for the entire organization and the need to reduce the number of paper documents used are the broadest objectives.

An organization should consider itself a candidate for EDI when the following conditions exist:

- Loss of time due to paper flow is substantial.
- Management of paper flow consumes excessive personnel or financial resources.
- Tracking of specific sales or manufacturing orders is essential.
- The cost of an out-of-stock condition is high or unacceptable.
- Trading partners are concentrated.
- Customer service expectation is high.
- Current internal systems are effective and adaptable to EDI.

The business case is typically made using one or more of the five strategic and operational reasons discussed in the following sections.

Strategic Reasons

The drivers of EDI point to two strategic reasons (i.e., strategic in that they change the way an organization competes) to consider adoption.

COMPRESSES BUSINESS CYCLE

Electronic transmission of business data compresses time. Details of transactions are sent, received, evaluated, and processed in a fraction of the time usually associated with business processes and with enormous benefits. Evidence is growing: One manufacturer, Lithonia Inc., Conyers, Georgia, has become a leader in the lighting industry by developing applications that incorporate EDI to cut in half the time it takes to receive and fill orders from its customers. The company is now able to receive, process, package, and ship orders in the same amount of time it takes competitors just to receive orders from their field locations. This industry leader also releases its invoices as the orders are shipped, providing an additional cash flow advantage over its competitors.

An IBM industry consultant study of the textile industry, which is notorious for long lead times in its order-to-manufacture-to-ship cycle, has demonstrated lucrative benefits from speed-of-business improvements. The cycle has been reduced by 25% and overall industry savings are projected to exceed $12 billion annually. In addition, both buyer and manufacturer share the benefits.

IMPLEMENTS JUST-IN-TIME PRACTICES

Just-in-time inventory management, which involves shipping parts to factories and assembly facilities as they are needed and thereby eliminating the need for large inventories of safety stock, is dependent on rapid and accurate transmission of shipping notices. Ordinary mailing of documents is virtually out of the question. Delay in receiving the shipping notice, for any reason, or an error in specifications could lead to the shutdown of a production line. It is even conceivable that the parts could be on the receiving dock, even though the organization's records would not indicate that.

More businesses are developing systems to send shipping notices through EDI and route them directly into the recipient's computer for evaluation. The supplier's computer remains online to the recipient while the specifications are verified. This promising method is an alternative to leaving notices in an electronic mailbox, which recipients may not check for several hours. It compresses the time interval because the document is both transmitted and accepted in a matter of moments. Chrysler Corp., Highland Park, Michigan, is continually evaluating the method for enhancing its just-in-time inventory practices. Chrysler management anticipates that all transactions with its more than 2,000 suppliers are or will soon be handled through EDI.

Operational Reasons

Potential strategic benefits often drive many organizations to use EDI as a basis for formulating external linkages with traditional partners. They are important because they change the way the organization competes. At the same time, there are significant internal benefits which improve operational performance. Three operational benefits include the elimination of errors, the rapid detection of document errors when they do occur, and the reduction of business document costs.

ELIMINATES ERRORS

The Electronic Data Interchange Association reports that 70% of computer output becomes input to another computer somewhere else and that 25% of the cost of a business transaction is in data entry and reentry. Each time data is manually entered into a system, there is a chance of error. Computer-to-computer transmission of business documents not only eliminates costs, it also eliminates errors.

DETECTS ERRORS QUICKLY

Errors may not be due to data entry, but a result of differing business records, for the price charged versus the price quoted for a product. The following example illustrates how EDI can help managers detect errors of this nature and correct them immediately.

A select group of grocery suppliers and their customers are testing a direct-to-store EDI system. For example, Frito-Lay Inc., the Dallas-based snack food manufacturer, worked with Ralph's Grocery Company of San Francisco to develop such a system. The route salesperson in the Frito-Lay truck uses a hand-held computer which contains daily delivery information for the driver. By simply plugging into the in-store computer, the driver is able to transfer the delivery information from the hand-held device into the store's computer (product and quantity details are coded in a previously agreed upon data format).

In only a few moments, the receiving manager electronically compares the delivery information with the records that describe the expected shipment, including prices. If they agree, the driver's entry becomes a document of record. Disagreements are detected and resolved on the spot. Having the store computer accessed directly eliminates the need for checking every delivery transaction and eliminates the need for paper. The company anticipates that the prototype will prove itself, and the process will become common throughout the industry within a five-year period.

REDUCES DOCUMENT COSTS

The levels of costs savings resulting from EDI use can be substantial. A widely publicized study performed by Arthur D. Little, Inc. for the grocery industry concluded that if only one-half of the industry's businesses implemented EDI, the industry would save in excess of $300 million per year. This level of cost saving is particularly significant because of the low profit margins that characterize the grocery industry.

RCA recently estimated that each purchase order it processes costs approximately $50. EDI, it estimates, reduces the cost to approximately $4 for the same purchase order ($1 of which is communications costs).

Most significantly, the cost savings are not one-time. They continue for every transaction that follows, growing as business volume increases. A bottom-line impact is likely.

Responding to Strategic Imperative

A decision within an organization to develop its EDI capability may not always be a deliberate choice. Instead, it may be the result of pressure placed on it by other events in the industry in which the organization competes. Such pressures create strategic imperative, and, therefore, are a strong driving force.

In these settings, EDI is not an alternative but a requirement for remaining competitive. In some industries, such as pharmaceuticals, transportation, and automobiles, EDI is already the standard conduit for business transactions. In others, the message is rapidly spreading to link up or lose out. The threat of losing business is indeed a very convincing argument.

Some organizations have told their suppliers in no uncertain terms what is expected: "Not developing EDI capabilities will undoubtedly lead to fewer orders from us." Others have emphasized the effect on revenue: "If you begin transmitting business data to our firm electronically, you may take an additional 1% cash discount on all orders." Or to suppliers: "Your participation in our EDI project should noticeably increase the amount of business the company does with you annually."

TECHNOLOGY REQUIREMENTS

EDI relies on widely used computer networking and information systems technology. However, several key activities and decisions are essential to ensure that its advantages are obtained.

Industry Standard for Product Identification

Although fundamental in concept, the establishment of a standard method for identifying and marking products is an EDI requirement. Proprietary identification systems limit the number of organizations with which a corporation may interact for electronic trading. Partners may be unwilling to adopt the particular system requested if they are unwilling to use product identification or information-handling procedures dictated by another organization. Or they may be unwilling to use multiple systems, one for each major partner with which they interact.

Some industries currently struggling with the development of an identification system are finding that the process is by no means quick or without controversy. It is clear that the universal product code (UPC) in the grocery industry has been a major enabler of EDI. Because the standard bar code is printed on more than 95% of all products handled by the industry, all participants can use a common code (often in ways not previously anticipated). There is little question that industry standard coding systems, such as the UPC code, will be a fundamental element in the successful development of EDI at both the individual organization and the industry level.

Standard Format for Exchange of Data

Business documents transmitted by EDI methods are included in an envelope (Figure 3) that contains both the data itself as well as identification and routing details. The communications transport protocol indicates the method by which the document is sent (e.g., SDLC, ASC, or BISYNC), which enables the coordination of transmission between sending and receiving computers. Interchange control segments delineate a set of documents transmitted between organizations at one time. Within each group, headers and trailers separate the transaction sets; that is, the actual business documents themselves.

Standard data formats, which prescribe the

Figure 3 EDI Data Transmission Format

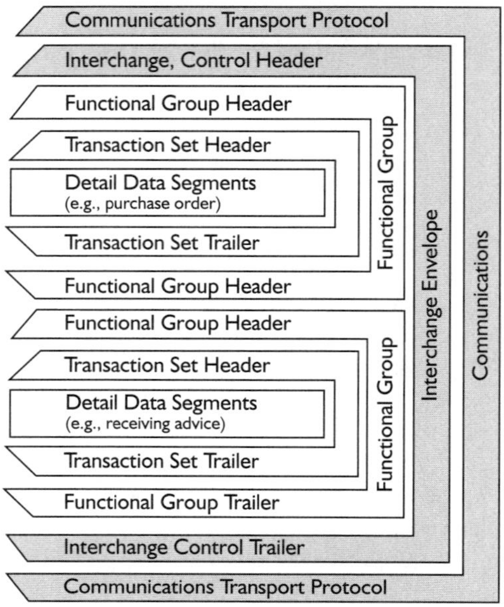

contents of a business transaction (e.g., order number, quantity required, and shipping dates), tell trading partners what to expect in the contents, order, and form of data transmitted through EDI. A family of standard formats, known as the X.12 format, is widely promoted.

Among its attractions is the existence of a variety of transaction subsets that specify the content of electronic purchase orders, packing slips, invoices, and payments. Each subset identifies the contents of the transactions (i.e., the data items to be included), the order in which each item is presented, and the specifications for the data. Both sender and receiver must expect and use a particular format containing data in a specific order or an exchange will not occur.

The following is a list of current and proposed industry transaction subsets and data formats.

- Purchase order
- Purchase order acknowledgment
- Purchase order change request
- Purchase order change acknowledgment
- Invoice
- Payment and remittance advice
- Request for quote
- Reply to request for quote
- Shipping notice
- Receiving notice
- Receiving advice
- Planning schedule with release capability
- Price and sales catalog
- Data element dictionary

Table 1 identifies several industries that have developed unique transaction formats.

TABLE 1
Status of Industry EDI Formats

Industry Segments Using EDI Formats
Using Standard (X.12) Transaction Formats
Electrical Equipment
Telecommunications Equipment
Chemicals
Textiles
Textile and Apparel Linkage
Fabric and Suppliers Linkage
Automotive Industry Action Group
Retail Merchants
Paper
Other Industry Formats (Not Using X.12 Standard)
Medical and Surgical Suppliers
Oil and Gas
Retail Hardware
Health Insurance Claims
Insurance
Wholesale Pharmaceuticals
Trucking
Optometrists
Retail Grocers
Warehousing
Industry Segments Planning to Use EDI Formats
Metals
Electronics
Petroleum
Aerospace
Hospital Supplies
Electric Utilities
Federal Government Procurement (Department of Defense and General Services Administration)
Federal Government taxation (Internal Revenue Service, Treasury Department, and Customs Bureau)

EDI Software

Several levels of software are used in EDI networks (Figure 4). The communications software connects the sender and receiver systems to the transmission network. Working in conjunction with the carrier source (e.g., a value-added carrier), the software provides for the proper protocols and speeds.

The existence of the communications software is usually taken for granted and thus reference to EDI software includes the components responsible for translation and application activities. Translation software is any computer code that transforms data from one form to another as a way of facilitating the exchange of data. The sender, for example, may use an internal proprietary format whereas the receiver may rely on an X.12 specification. Translation software would handle the conversion between the formats.

Application software performs the typical transaction-processing activities (e.g., preparing purchase orders, acknowledging receipts). However, in an EDI environment it also includes the capability to interact with the translation software and to use agreed-upon data formatting guidelines.

Figure 4 EDI Software

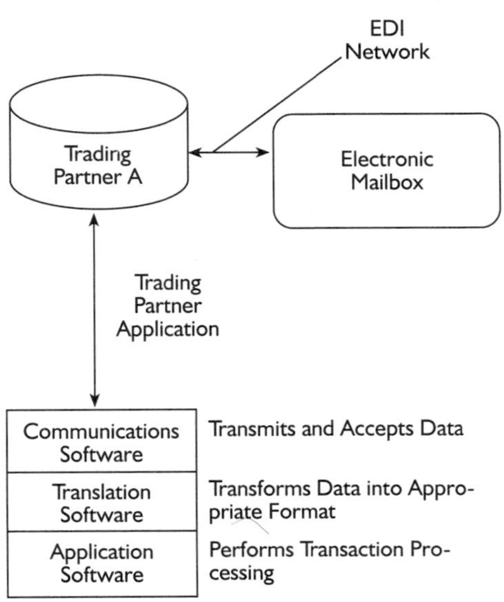

Common Data Dictionary

Standard transmission formats must agree with internal data standards in an organization. The responsibility for setting and maintaining these standards falls within the IS department.

A data dictionary records the number of characters or digits allowed for each item. This is an essential element for ensuring that data transmissions will be understood. Therefore, any organization or industry must not only maintain a common data dictionary of data elements, but must also ensure that the dictionary is constantly brought up to date, usually on a quarterly basis. Each partner must always know or be able to determine the makeup of each electronic messaging format it sends or receives.

Network Transportation Source

The transmission network may be proprietary, consisting of transmission lines established for the sole use of the trading partners, or for general use, involving public data networks. Although the trend is moving away from proprietary systems, this does not mean that organizations will use only a single network. Many corporations, particularly those that are involved in several industries, are finding that they must seriously consider using multiple commercial telecommunications networks. Each trading partner may use a different network for data transmission.

The following list identifies the most widely used EDI networks and their providers in order of market share.

- GEISCO (GE Information Services Co.). Industry focus includes apparel, automotive, petrochemical, electronics, steel, and international trade.
- British Telecom/Tymnet. Industry focus includes grocery, transportation, chemicals, and electronics.
- Sterling Software/Ordernet. Industry focus includes pharmaceutical, warehouse, medical, transportation, and grocery.
- IBM Information Network. Industry focus

includes insurance, retail, apparel, and electronics.

- Kleinschmidt. Industry focus includes transportation, grocery, warehouse, electronics, and chemicals.
- Control Data Corp. Industry focus includes automotive, manufacturing, and chemicals.

Other specialty providers include:

- CompuServe. Electronic network provider, seeking to provide niche EDI services.
- Harbinger. Providing selected services for banking, utilities, and electronics industries.
- Sears Technologies. Commercial service is an outgrowth of its internal EDI network; it provides new insurance interface traffic for the IVANS insurance network.
- Regional Bell Operating Companies (RBOCS). Seeking to develop services that will leverage their telephone customer network.
- Western Union. Seeking to provide services in the distribution, electronics, and petrochemicals industries.

Financial Partner

The electronic funds transfer component of EDI involves a banking partner. The originator bank concept has caused financial institutions to increase their consideration of EDI. General Motors initiated this practice when it said it would make electronic payments to any of eight specified banks (the originator banks) in the United States. Trading partners are paid only if they make arrangements to receive payment from one of the eight banks. Facing the possibility of being left out of other corporate EDI systems if not included as an originator bank, many financial institutions began to take EDI seriously.

Yet while some have established services, most others are still creating an EDI strategy, often very slowly. To date, it appears that many institutions are taking a wait-and-see stance, not having established an EDI program.

FORMULATING AN EDI STRATEGY

Organizations evaluating EDI alternatives face several key concerns. The strategy itself will be most successful if it results from an evolutionary, rather than a revolutionary, process.

Key Concerns

When general management and the IS department in an organization formulate their strategy for moving ahead on electronic data interchange, the following concerns must be addressed.

HOW TO ENCOURAGE TRADING PARTNERS TO PARTICIPATE?

A decision by management to invest in EDI does not mean that trading partners will go along. For example, Hewlett-Packard Co., Palo Alto, California, recognized as a pioneer in EDI within the electronics industry, found that even after seven years of EDI experience only nine of its trading partners were participants.

IN WHAT WAYS WILL THE ORGANIZATION BE REQUIRED TO CHANGE THE WAY IT DOES BUSINESS?

The question of whether or not to adopt EDI is essentially a business issue. The technology for EDI is in place and well-proved, and any remaining technical issues have been identified and are correctly being addressed. Standards are available to support EDI. The essential question is whether an organization wants to change the way it does business (or whether there is even a choice). The other question to consider is should it change the way it handles business transactions, at least during the initial phases of implementing EDI?

CAN THE BANK PROVIDE SERVICES?

Can the organization's bank handle the volume of its remittance data? Many banks do not yet have the capacity to handle the high volume of activities that full-time EDI use will require. The banking industry recognizes the problem; an individual corporation must decide whether it can afford to wait for its bank to solve the problem and ensure adequate capacity.

In some instances, the banks are still deciding how to manage EDI as a revenue source and at the right profit level. They are also caught up in debates about whether to handle non-bank data fundamental to their clients' activities.

Some organizations may decide they can no longer wait and will take steps to initiate the service. For example, the National Wholesale Drug Association recently decided it could no longer wait for the banks. Consequently, it established its own task force, which created a record structure that a third-party clearinghouse could handle and circulated details of the format for evaluation.

CAN THE ORGANIZATION'S ACCOUNTING SOFTWARE ACCOMMODATE EDI?

Accounting software is an essential component in electronic data interchange, although most discussions of EDI do not address this element. Among the concerns are the ability of the software to match orders and requisitions against invoices and orders received, and the capability for line-by-line reconciliation of quantity received and quantity billed. These concerns transcend data coding and go to the heart of procedural control.

SHOULD STANDARD OR PROPRIETARY METHODS BE USED?

Some organizations have developed their own proprietary, corporate EDI formats. Among the largest are J.C. Penney, K-Mart Corp., McKesson Corp., Wal-Mart Stores, Inc., General Motors Corp., and Ford Motor Co. However, many organizations using proprietary formats are rethinking their strategy as they increasingly find that partners are refusing to use a proprietary standard. (In some cases, organizations are adopting a second format that meets the X.12 standard, while also maintaining their proprietary format.)

Other Implications for the IS Department

The information systems department in an organization is a key player in the development of an effective EDI strategy, as the preceding discussion indicates. In addition, unexpected challenges can arise, including inadvertent overselling to management, restructuring of business elements, managing maintenance requirements, and underwriting costs.

OVERSELLING TO MANAGEMENT

Electronic data interchange often has an immediate appeal to business managers when it is presented to them, particularly when it is emphasized that the underlying technology already exists and has been perfected. Unfortunately, the attraction can lead to expectations that will not be met in the time frame most business managers anticipate. Once the enthusiasm is generated, it is difficult to explain that almost a year may pass before full implementation occurs. The peculiarities of testing data formats, communications links, and interconnection of multiple vendor systems are lost on most business managers.

When the system is conceptually understandable, it is a challenge to explain why the service cannot be delivered immediately. Initial applications should be selected to demonstrate the impact of EDI while testing procedures at both ends of the communications link. The most successful organizations will be those in which business procedures are not modified right away, but evolve through experience. Starting with current transactions is the safest route, with enhancements to follow.

RESTRUCTURING BUSINESS ELEMENTS

EDI is not another layer of technology that is added to existing applications. Instead it often involves a restructuring of key elements of business systems. Applications systems become increasingly interdependent through the sharing of product and process information. Accounting, shipping and receiving, inventory, manufacturing, and sales support (even compensation) systems become intertwined.

Data standards take on a new level of importance because all applications must expect the same details. Ownership of information and the authorization to change details becomes a particularly sensitive subject. All parties must have a clear understanding of their responsibilities for providing and maintaining operating information.

MANAGING MAINTENANCE REQUIREMENTS

The development of interorganization systems is a continuing process. EDI activities cannot be installed and then left to operate unattended. Rather, like information systems in general, they require ongoing support, either because processing requirements change or because new business partners must be accommodated. In addition, ongoing changes in transaction systems linked to the EDI process continue their typical evolution. On both the customer and the supplier end of the link, the implications are the same: Changes must be accommodated.

Organizations experienced in EDI typically find that they have underestimated the level of maintenance and support required. Many of these organizations today have as many staff members focused on EDI maintenance as they do on new systems development.

Underwriting Costs

The decision to proceed with interorganization systems is also a decision to invest additional resources in information processing. The internal IS development costs are often recognized and dealt with according to typical resource allocation guidelines (e.g., chargeback to the functional unit). Other costs, however, may be questioned (e.g., how is the cost of the value-added carrier providing the EDI network charged?).

Frequently, organizations must underwrite the costs associated with bringing their trading partners into their EDI environment. It is quite common for an organization to provide the software the partner needs to get started; in some instances, personnel are sent to the trading partner's site to help with startup. Although these costs are not insurmountable, they are internal issues to be addressed. Often they do not fit within existing IS operating policies in most organizations.

CONCLUSION

The growing emphasis on compressing business time, enhancing the linkage between customers and suppliers, and improving the effectiveness with which businesses operate, for both strategic and tactical gain, has brought fundamental change to business as usual. The potential of EDI as a vehicle for interorganizational business is substantial. It can enhance both local and international commerce.

However, the full perspective of the business change must not be overlooked. EDI alters the way organizations do business and therefore management must be willing to adjust fundamental activities, such as when and how orders are processed or funds transferred, to participate successfully. Above all, EDI is not a technology change but a restructuring of business practice.

QUESTIONS

1. What are the advantages of EDI?
2. What are the principal forces for adopting EDI in business firms?
3. What major issues need to be considered in implementing EDI in a firm? To answer this question, pick a medium-sized organization in a particular manufacturing industry.

3.6 Achieving World-Class End-User Computing: Making IT Work and Using IT Effectively

JOHN P. HALLORAN

The coming of the personal computer created the age of end-user computing. That is, people in offices and factories used the computer on their own, rather than relying on a central MIS staff. The trend toward using computers personally moved from nerds up the company ladder to the executive suite, and in the home the PC moved down to the preschooler. When totaling up the business cost of the hours spent using the computer and adding to that the average investment of thousands of dollars in hardware, software, and local area networks, the amounts involved are truly staggering.

It is Halloran's conclusion that end-user computing is badly managed in American firms. Halloran's is an important voice, since he serves as managing partner of U.S. operations for Nolan, Norton & Co., a prestigious global management consulting firm that specializes in information systems. In the argument presented in Chapter 1 of this book about productivity gains from computing, he comes down on the side of small gains from computing.

The data for the article is based on a statistically large sample of interviews. To interpret the data, Halloran uses the concept of stages of computer adoption, originally developed by Nolan and Gibson in 1974 when the world was a centralized mainframe environment. Be warned that this concept is controversial. Many people feel it is too simplistic a representation and may not be borne out by the data, particularly for end-user computing. Nonetheless, the stages model has great face validity in that it seems to correspond to a lot of people's experience.

Source: Information Systems Management (Fall 1993): 7–12.

A Nolan, Norton & Co. (NNC) study revealed that 55% of a company's employees use microcomputers, and of that group, 70% are now connected by a local area network (LAN). The study also established that while the number of microcomputers used by companies has doubled since 1989, the growth in LANs has quadrupled.

This rapid influx of end-user technologies is creating a major challenge for both IS and business management—and they're losing the battle. The cost of supporting the end-user computing (EUC) environment is high (as much as $20,000 annually per workstation), while the benefit (measured in worker productivity) is low (1.3% annual improvement from 1982 to 1992, according to the U.S. Bureau of Labor Statistics). What does management need to do to reverse this trend and balance the value equation?

THE EUC ENIGMA

For its study, NNC interviewed hundreds of users and IS personnel to determine how they use and support the EUC environment. The study found confusion, a lack of role definition, poorly defined management practices, and weak organization structures for dealing with the issues that face both constituencies daily. This lack of definition and management infrastructure results in the high "peer support" costs that were identified in the study.

Peer support is defined as the amount of time non-IS professionals devote to supporting the EUC environment in their areas or departments. These people are usually hired as engineers, cost analysts, accountants, or for some other functional specialty, but if they show a talent for mastering the complexities of departmental hardware and software, they are sought out by their peers to help in problem solving, training, or otherwise rendering assistance to colleagues. Given that these extracurricular activities were not in the job descriptions of these technological good Samaritans, three things are found to occur.

- The true costs of supporting the EUC environment are not being identified.
- A department's measured productivity suffers because people are actually performing tasks not included in their job descriptions.
- The organization does not receive the necessary feedback to support the case for building an enabling infrastructure to support the EUC environment.

From the IS professional's perspective, the EUC environment may be an enigma. In some companies, IS professionals reported feeling less competent than the superusers providing the peer support. Cost-conscious IS organizations frequently look at their EUC support function as the most expendable area when budget cuts are required. As a result, in tracking the growth of EUC support personnel from 1989 to 1993 and relating it to the growth of microcomputers and LANs, NCC found that IS departments have been losing serious ground in their efforts to keep up with the surge of distributed technology (see Figure 1).

DOMAINS OF RESPONSIBILITY

A clear first step toward remedying this situation and moving to a high-performing and benefit-rich

Figure 1 Comparison of PC, LAN, and IS Support Growth

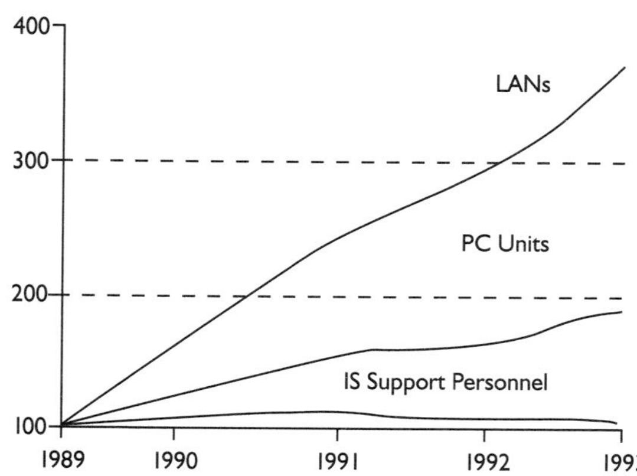

Source: Nolan, Norton & Co.

EUC environment is defining the domains of responsibility for the IS and user communities. On the basis of the in-depth knowledge gleaned from its study and from various consulting engagements on EUC, NNC has developed a domain of responsibility framework. This framework assists companies in defining who has responsibility for what function in the area of EUC support. The framework identifies specific activities that are the responsibility of each party as well as shared responsibilities.

As Figure 2 shows, the domain for IS is to make it work. The domain for business management is to use the network effectively. The activities identified in this framework are not meant to be comprehensive, but they serve as examples of how the framework can provide a basis for developing service level agreements, job descriptions, organization charters, planning processes, and other requirements for building a world-class EUC environment.

MAKE IT WORK

The responsibility for managing the network has evolved slowly in most organizations. Initially, companies were concerned only with telephony or the voice networking capabilities of their networks. The once stable and slow-moving communications industry has, with the advent of new technologies, been transformed into an innovative and dynamic area, and a fiercely competitive battle is being fought for domination of the airwaves.

In many companies, the response to the changing communications environment has been to beef up the communications function within the enterprise and make voice and data communications the responsibility of the IS department. This has resulted in sophisticated and technically talented communications organizations sprouting up to lead the effort of connecting far-flung and not so far-flung operations. These largely centralized communications departments have been charged with making the network work, in both efficiency and effectiveness. For the most part, they have done their job as well as it relates to the wide area network. However, with the emergence of LANs, the cycle of technology assimilation began to play out in companies.

THE STAGES MODEL

This technology assimilation process was described by Richard L. Nolan and Cyrus F. Gibson

Figure 2 Domains of IS and User Responsibility

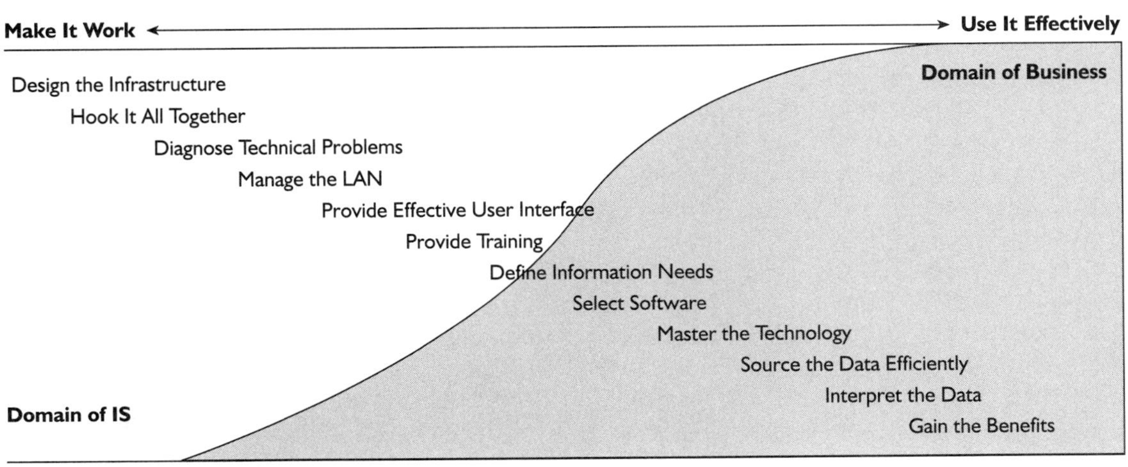

Source: Nolan, Norton & Co.

in their landmark 1974 *Harvard Business Review* article "Managing the Four Stages of EDP Growth." Nolan has embellished and applied the Stages theory in hundreds of companies in the 20 years since that article was published. The Stages model (Figure 3) identifies a series of stages that an organization goes through as it successfully assimilates a new technology. Adapted to the end-user computing environment, the Stages theory reveals how organizations assimilate and employ the hardware and software associated with this brand of computing. Based on NCC's experience and research over the past six years, a Stages model for EUC has emerged that describes the evolution of this unique technology within organizations (Figure 4).

The importance of the model is that it describes how individuals and, ultimately, organizations evolve in their utilization of end-user technologies. This model supports the data collected on the growth of LANs over the past four years. This growth will not abate any time soon and poses a fundamental challenge to the large, centralized communications organizations entrusted to manage the enterprisewide network.

The challenge lies in the ability of communications organizations to deal with the consumer attributes of the EUC environment. In company after company, network managers complain about their lack of involvement in LAN activity in the organization. Often, LANs are linked to the backbone network without proper planning or even the knowledge of communications management. Consequently, capacity spikes and performance

Figure 3 Stages Model

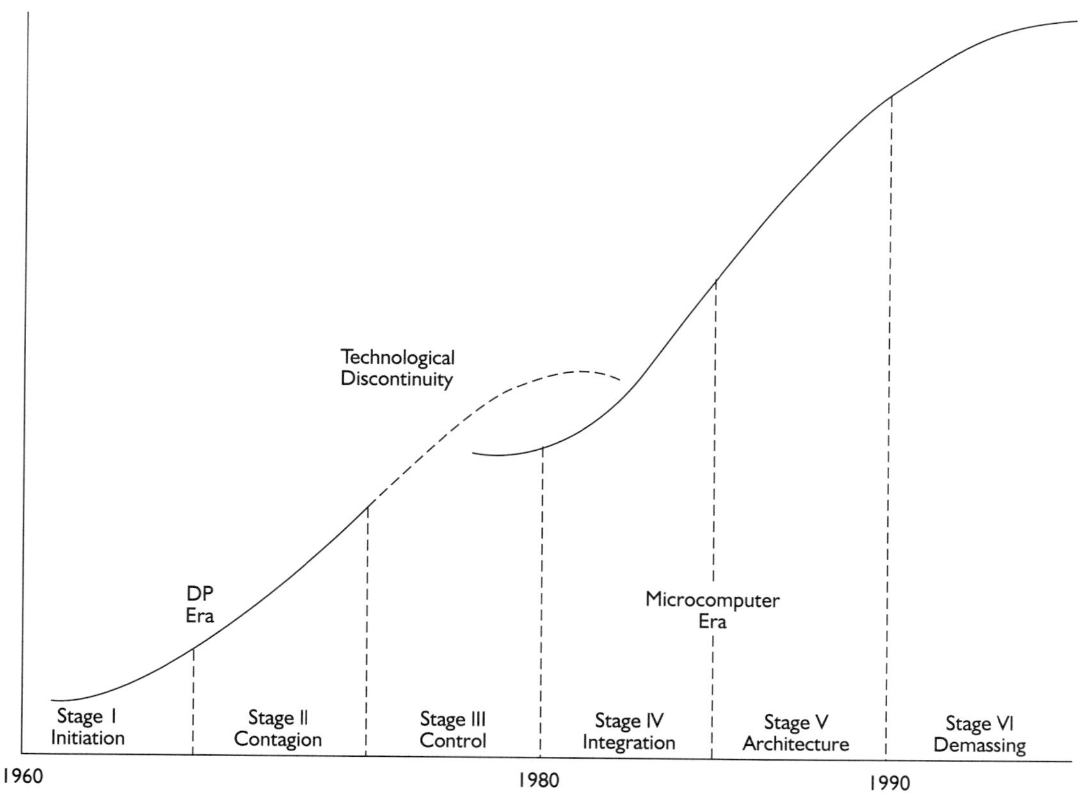

Source: Nolan, Norton & Co.

Figure 4 Stages Model for End-User Computing

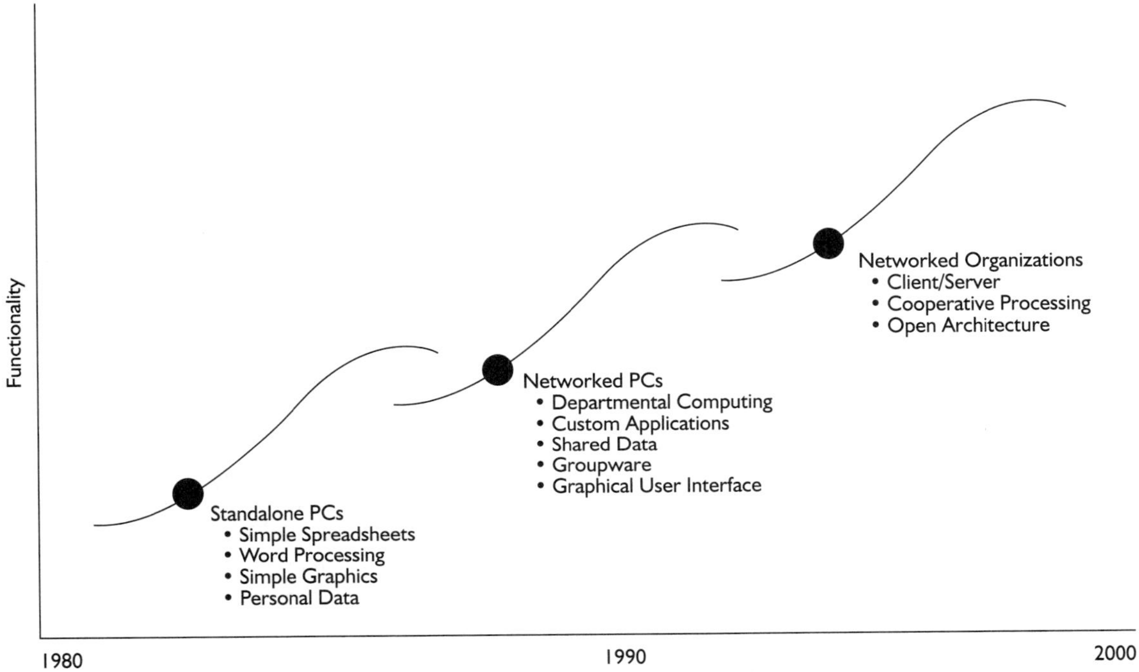

Source: Nolan, Norton & Co.

degradation are common. In many cases, companies lack the process and procedures to effectively integrate planning for wide area networks, mainframes, terminals, backbones, LANs, PCs, and workstations. This technical infrastructure design process for an organization is one of the most important activities IS departments can undertake in providing the enterprise with an enabling infrastructure. Using the domains of responsibility as a template, an organization can define what is expected of IS and the end-user department. This action will help to clarify the specific roles and responsibilities ascribed to each party as they develop service level agreements and other charters to guide decisions and behaviors in the organization.

Designing the infrastructure is a complex task that must be assigned as a top priority to the IS organization. Although IS professionals are best qualified to undertake such an endeavor, they must engage the organization's most senior business executives in the process. In architecture, "form follows function." This is why it is imperative to understand the enterprise's strategy, key performance measures, critical processes, and how the broad array of information technologies can be designed into a cohesive architecture. Most organizations today are not following a prescribed blueprint for how the spectrum of technologies—from mainframe to desktop to laptop—need to be linked together to best support the business needs of the organization. If companies are to reap the promised benefits for their IT investments, the first step is to establish domains of responsibility.

THE CHANGING OF THE INFORMATION SYSTEMS DOMAIN

Traditionally, the IS department has been responsible for the physical installation of IT. This responsibility was consistent with the IS charter to plan for, purchase, and install the information technology required to support the business. The IS department was the central focus for hardware and software vendors alike. The close relationship of the vendor's sales representative and the IS manager became the cornerstone of growth for such companies as IBM Corp., Digital Equipment Corp., and Hewlett-Packard Co., for example. As personal computers became more broadly accepted in business, however, a retail market mentality developed. In this new paradigm, the IS manager became the middleman, viewed by the end user and the vendor alike as an unnecessary intermediary. Because microcomputers are relatively easy to install as standalone units, the once unique charter of the IS organization was further compromised.

The result is that companies experience the proliferation of myriad and often incompatible technologies. Some organizations have acted quickly to establish standards and institute central purchasing policies, but most are living with the consequence—an inefficient hodgepodge of technology that does not support the business objectives of the enterprise. Sophisticated users have assumed direct responsibility for hooking up departmental LANs, undermining the ability of the IS organization to implement its still-unchanged charter to connect all the technology of the enterprise. When problems develop around issues of incompatibility, data security, data access, and backup and recovery, however, the IS organization is looked to as the responsible party. If, indeed, organizations are going to receive the full value from their IT investment, they need to understand not only the importance of the technology infrastructure design but the importance of the physical implementation of that design.

USE IT EFFECTIVELY

Just as it is critical to have the enabling infrastructure to support business strategy, it is even more important to use that infrastructure effectively. This is primarily the end users' domain of responsibility, since it is the user who ultimately defines the value gained from available tools. In a collective sense, this value is translated into how IT is benefiting the organization as a whole. This is always the central question asked by business executives.

Unfortunately, most of these executives look to the IS organization for the answer. They confuse the domains of responsibility themselves. They feel that the technologists in the organization are solely responsible for the company's reaping the rewards of its IT investment. This is tantamount to holding the tool makers responsible for the consumers' effective use of a tool. They may have designed the most user-friendly and technically advanced tool on the market, but if consumers do not have the basic knowledge required to use the tool effectively, they will not reap the value from the purchase. The same is true for end users. If they do not understand how to use IT to more effectively perform their job functions, they will not gain any value from IT investment.

During the past few years, organizations have experienced enormous pressure to change the hierarchical and function-oriented structures that are a carryover from the industrial era. No matter what this phenomenon is called—business reengineering, transformation, business process redesign, or innovation—one thing is clear: organizations are fundamentally changing the way they conduct business, and IT is a critical component of the change. End users must understand their company's strategy, critical processes, performance measures, and how they can add value to achieve enterprisewide goals and objectives. Only by doing so can they use IT to improve business performance.

Successful companies recognize the importance of having the domains of responsibility clearly defined, especially if meaningful performance measures are to be developed and linked to the organization's overall success. Once the domains of responsibility of IS and users are made clear, then a true partnership can be forged to build a world-class EUC environment. The result can be superior performance throughout the enterprise and an environment that provides for the flexibility and adaptability so important to competitive positioning.

QUESTIONS

1. End-user computing can either increase or decrease organizational productivity. Under what conditions does end-user computing enhance productivity? Under what conditions does it decrease productivity?
2. What should IS do to make end-user computing more effective?
3. Some people think that end-user computing will eventually merge with traditional information systems. Do you think this viewpoint can be justified? Explain your reason for agreeing or disagreeing.

3.7 Sales Force Automation

CLARE GILLAN

Of the people in a corporation, the ones who are perhaps least tuned to the computer are the sales force. Selling is inherently a verbal operation whereas computing is oriented toward the written word. Yet, sales operations in organizations require a large amount of paperwork. The marketing managers in the home office want to keep track of which clients were contacted, the orders received, the prospects for future purchase, the state of the market, and more. The sales force are the market manager's eyes and ears. The sales force itself needs to place orders, receive leads and information about clients, and keep track of commissions. Despite these needs, in organization after organization there is a culture clash between the go-go marketeer and the computer professional. As a result, the sales force is one of the least automated areas in business. The situation is changing as a difficult economy forces companies to reduce costs and optimize resources.

The realities and potential of automating paperwork for the sales force are described in Gillan's paper. This material first appeared as a commercially written "white paper" bound into the weekly trade newspaper *Computerworld*. It is the result of studies done by a consulting firm, International Data Corporation. Unlike many of the papers in this volume, it represents a commercial rather than an academic view.

THE convergence of economic issues and technological advancements is moving the sales function into the IS limelight. In the wake of this convergence, automated sales support and the integration of sales data with other corporate information—Sales Force Automation (SFA)—have become necessary tools for many companies.

Other tools—lightweight portable computers, intuitive graphical user interfaces, open access to heterogeneous data, and advances in communication technology—are all stimulating broad demand for SFA applications considered too complicated for previous technologies.

Despite this demand, however, sales has been one of the least automated areas in business. That is finally changing as tough economic conditions force companies to reduce costs, optimize resources, and improve quality and service. Now companies are turning their attention not only to

Source: International Data Corporation White Paper, *Computerworld* (December 7, 1992).

sales, but also to related functions such as customer service.

The benefits of a properly implemented sales system go well beyond increasing a sales person's productivity. Sales data is critical to the decision-making process at all organizational levels, from the executive making corporate decisions to the sales representative canvassing a territory.

To date, few companies have successfully implemented comprehensive SFA systems. International Data Corp. (IDC) attributes this low activity to a lack of resources, awareness, and know-how among many prospective users. Over the next two to three years that will change as radical advancements are made in technology aimed at improving sales productivity and the quality of customer relationships.

KEY FEATURES OF SFA

SFA encompasses technologies tailored to the sales function, in particular, providing field sales people the necessary automated tools to increase their productivity. Attributes of SFA include territory management, for example, a customer database, electronic calendar/tickler files, integrated word processing with mail-merge capability, and reporting. More advanced systems may support sales forecasting, two-way communication between the field and group offices, links to external and corporate data, and a high level of reporting and analysis.

The most valuable SFA systems do more than increase the productivity of the individual sales person. The truly influential SFA systems integrate with corporate information systems and facilitate the flow of information across sales and related functions such as marketing or customer service.

ENTERPRISE INTEGRATION

The currently emerging generation of business applications is best referred to as "enterprise integration." The dominant trend of this generation is the act of bringing information and people together. Applications facilitate communication among people and the integration of disparate information located in corporate, departmental, and personal databases. Key technologies enabling this movement include:

- Open systems. Open systems promote the deployment of SQL-based applications, data access middleware for access to heterogeneous databases and hardware platforms, and Unix for platform portability.
- Group collaboration. Group collaboration is characterized by integrated office systems—in particular electronic mail—for communicating and coordinating with others; groupware applications like Lotus Notes, which allows people to share hard and soft information; and workflow applications that optimize processes involving multiple people.
- Distributed information. Critical characteristics of this category include client-server technologies, LANs, WANs, and portable computers. Wireless communications, allowing access to remote computers without a modem, will also become important as its costs drop.

Sales organizations stand to benefit from these technologies, particularly in companies that embrace the concept of team selling. A recent IDC study indicates that 73% of IS managers at large organizations consider sales to be an integrated part of their corporate IT user community.

THE CURRENT STATUS OF SFA

There are three primary stumbling blocks to SFA: poor market awareness, apathy, and a lack of sophisticated technology. Poor market awareness is a problem despite the fact that nearly 100 U.S. vendors provide SFA packages. Unfortunately, the majority of these vendors are small companies that lack marketing muscle. Since active market demand is often a function of marketing dollars

Figure 1 Which of the Following Best Describes Your Current Sales System?

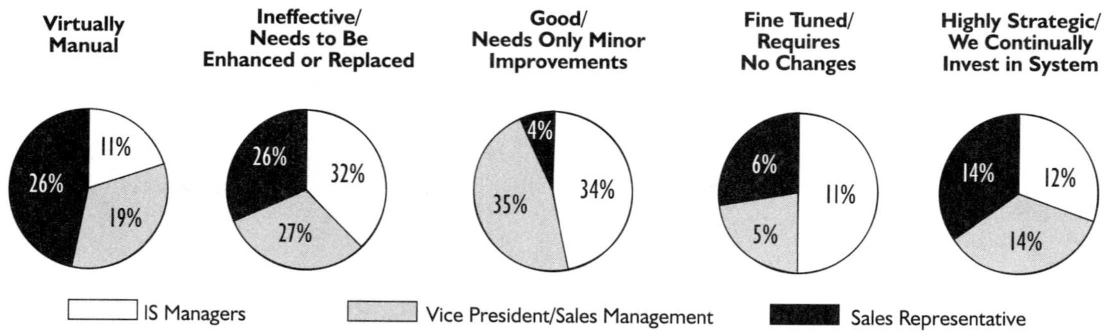

Note: Based on 100 responses from each of the three respondent categories.
Source: International Data Corp.

spent, corporate focus on SFA has been relatively low.

Lack of interest in this application came partly from the fact that the 1980s was a good decade for business in general. So long as the sale function met its quotas, increasing sales productivity was not a broad market issue. Other business issues, such as controlling inventory overhead and more quickly informing management on its previous month's performance, took precedence.

Perhaps the biggest deterrent to SFA was the lack of sophisticated technology. Only recently have the necessary tools successful SFA implementations require become available. These technologies include lightweight and affordable portable computers, such as Apple's Macintosh Powerbook, intuitive graphical user interfaces, and flexible integrated applications. More companies are using them as they attempt to utilize every possible weapon in their competitive arsenals.

SFA ACTIVITY AMONG U.S. COMPANIES

IDC recently interviewed 100 large U.S. companies (5,000 or more employees) regarding their use of sales technology. The participants were senior IS people, vice presidents and senior managers of sales operations, and sales people—the three groups generally involved in successful SFA projects.

More than half of the companies consider themselves to be somewhere between "efficient" and "bleeding edge" when asked how they apply IT toward increasing performance and competitiveness. IDC estimates that less than 15% of sales-support systems are highly strategic, that is,

Figure 2 How Familiar are You With the Available Range of Hardware and Software Sales Force Automation Solutions?

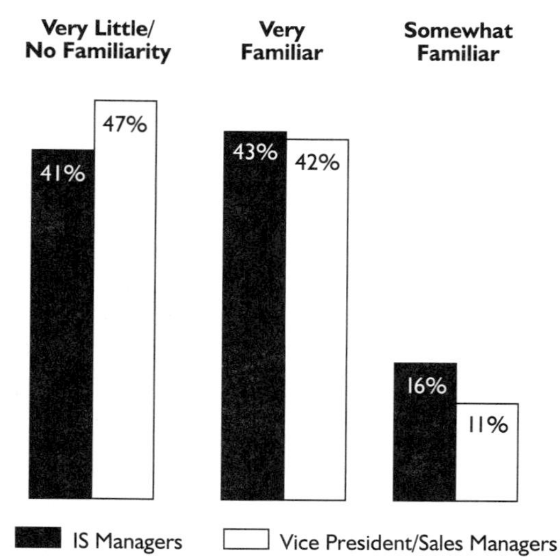

Source: International Data Corp.

instrumental in helping the sales organization and the overall corporation meet or set objectives.

Seventy-five percent of the surveyed companies are in manufacturing; the balance are spread across other non-government, non-education industries. Sixty-six percent of the sales forces are centrally organized, while the others operate on a distributed structure. Fifty percent of the companies rely completely on direct sales people. Another 35% claims more than 40% of their staff focuses on direct sales. Twenty-seven percent have less than 10 sales people, while 38% have more than 10 but less than 50. Only 12 sites have more than 500.

Most large companies support their sales effort with some level of computerized technology. These sales support systems vary in sophistication. Some are composed of a single terminal located in the sales area. This terminal is used to send sales data to a corporate database or a PC in the sales manager's office. It may run a spreadsheet, calculate commission, adjust quotas, or forecast sales activity. Although these systems are used in the sales area, their back-office focus and limited integration with a sales strategy strongly reduce their effectiveness.

Investment in SFA is picking up. Forty-eight percent of the IS managers interviewed report that they are actively improving automation in the sales area; another 34% are in the planning stages. IDC attributes this increased SFA activity to a reversal of the three previously negative factors holding SFA in check: poor market awareness, apathy, and the lack of sophisticated technology.

PORTABLE COMPUTING

Portable computing is one of the strongest drivers of current SFA activity. Clearly, the bulk and weight of first-generation portables deterred their use in SFA. Even now, despite the advances made in portable computer technology, about 35% of sales organizations in large companies still do not have any portable computers. Of those that do, portables are only used by about one-third of the sales representatives and managers.

This is changing as the allure of portables grows while prices and sizes shrink. Consider Apple's Mac Portable: In August 1991, it sold at an average street price of $3,170 and included 2M bytes of RAM, a 40M byte hard drive and 16 MHz clock speed. The Mac Portables have since been succeeded by Apple's Powerbook line of notebook computers. Today the top-of-the-line Powerbook 180 weighs just 6.8 pounds, comes with 4M bytes of RAM, up to a 120 byte hard drive, a 33 MHz processor, and sells for an average street price of $3,595.

Although the price gap between desktop computers and portable computers has not disappeared, it is shrinking to the point where it is no longer an automatic impediment for companies that are serious about automating field sales.

THE VIRTUAL OFFICE

Enthusiasm will spread as laggard companies come up against competitors armed with a "virtual office," which is essentially a portable computer equipped with all the information and function a sales representative needs inside or outside the office. For example, portables in the SFA environment hold all customer account information, order forms, pricing information, electronic sales presentations, and electronic mail access. A modem allows the representative to download or interact with information from the home office. A modem and facsimile software enable the representative to use any available facsimile machine as a substitute for the printer. However, portable printers are available from a number of vendors.

This virtual office concept allows sales reps to function on a sales call as efficiently as if all their prospects had come to their office. Ultimately, such a properly planned and implemented system will generate more and higher quality sales calls. The virtual office implies corporate integration, but corporate integration requires a large investment of hard and soft (time and resources) dollars. There are less expensive, unintegrated approaches, but they may be insufficient when applied to a highly competitive situation.

IDC asked sales reps about their views of portable computers as a sales support tool. Forty-five percent indicate that portables only enhance sales performance if they are part of an automated and integrated sales automation effort. Integration becomes particularly important in a team selling environment. Notably, 40% agree that even using portables as personal tools supporting daily sales-related activities would contribute to increased performance.

Many companies choose to start at this personal level. However, it is important that they do not get caught in a dead-end strategy. As the system grows, back-end data servers and integration with other corporate information become more important. Portable computer users must adapt to such changes.

REMOTE ACCESS TO CORPORATE INFORMATION

Not surprisingly, most reps still use the telephone to access sales-related information located back at the office. This practice will decrease dramatically over the next few years. Seventy percent of the sales managers report to IDC that their reps access this information by telephone today. The second most common data access method is downloading via on-line portable computer communication. Within three years, this method will be number one. With technology such as AppleTalk Remote Access, users can act as end nodes on the network no matter what their location, whether at home or via cellular connection. Printouts are now the third most popular information access method.

Figure 3 How Do Your Field Sales Representatives Access Office Information?

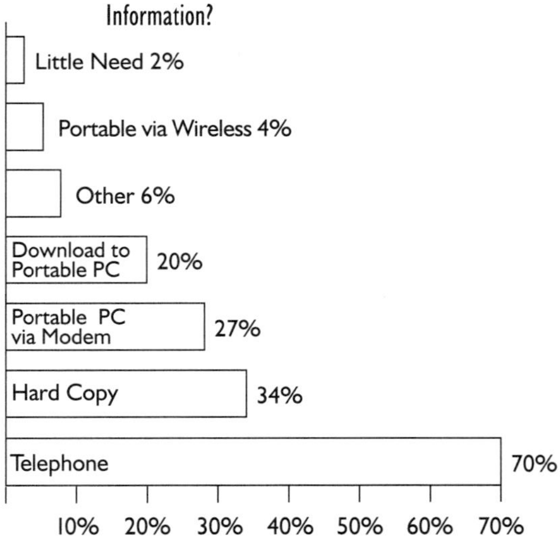

Note: Graph based on data published with original article from responses of 100 Vice Presidents/Sales Managers.

While SFA vendors frequently compete against internal development, they also run up against more general packaged business applications that offer sales-automation capabilities. Examples of non-sales-specific solutions include spreadsheets, word processors, and Personal Information Managers (PIMs).

SFA vendors have begun to realize that the opportunity lies not in competing with generalized solutions, but in leveraging them. For instance, more mature SFA solutions initially included a proprietary word processor and a proprietary electronic mail capability, largely because de facto application standards had not yet emerged.

Since that time, many of these SFA packages have been altered, allowing users to integrate popular word processing and electronic mail products such as Microsoft Word, WordPerfect, and Lotus's cc:Mail. Leveraging third-party technology allows SFA vendors to be more focused on the value-added features, such as optimizing profit margins and coordinating workflow in team selling environments.

SFA software vendors continue to evangelize the market. About 40% of IS and sales managers report that they are being consistently contacted by vendors about SFA solutions. Another 20% report that they have been contacted by a small number of vendors, but that these vendors called infrequently. Despite the various contacts between users and vendors, more than 40% of IS and sales managers indicate they have little or no familiarity with the range of available SFA solutions.

Figure 4 How Would You Describe the Impact of the New Generation of Smaller, Lighter PCs on Your SFA Strategy?

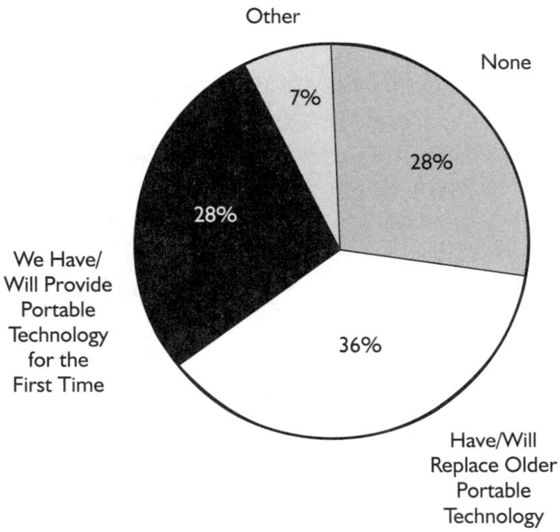

Note: Based on Responses from 100 Vice Presidents/Sales Managers.
Source: International Data Corp.

APPLICATION WISH LIST

IDC asked each of three groups (IS, sales management, and sales representatives) to identify the three areas of sales they would most like to automate if their management gave them a blank check. IS says it would focus on sales analysis and remote access to corporate data. Sales management and sales reps agree that account management would be their number one investment. When asked what area they would automate last, all three groups indicate telemarketing. Customer presentations and external database access also rank low. The importance of customer presentations could increase now with the advent of multimedia capability on notebooks such as the Apple Powerbook.

INTEGRATION WITH OTHER INFORMATION

The newfound emphasis on team selling, particularly among companies selling into large accounts, is boosting SFA. Team compositions vary. For instance, the team might be composed of multiple sales people, or a sales executive and a technical engineer. It may also involve people from various parts of the company, for example, customer service, marketing, engineering, and sales. Whether or not team selling has been formalized in a given company, integrating the sales information system with other information systems not only benefits the overall organization, but also enables individual reps to simulate the shared-knowledge environment offered by team selling.

Today, more than 75% of SFA systems in large companies are integrated with order processing and general accounting systems. Seventy percent of these large-company SFA systems are integrated with customer service. IS, sales management, and sales reps all agree that order processing, customer service, and price/specification data are the three most important areas for SFA integration.

Frequently, companies do not consider what areas besides sales can benefit from sales data. For instance, executive information systems provide senior management easy access to critical business information from a variety of data sources. Sales information should be included in the mix. Although financial data is the most common source of information, sales and marketing data are also vitally important.

It is interesting to compare views on how IS management, sales management, and sales reps view the automation level of their current sales system. IDC interviewed representatives from all three groups across 30 companies to find out. Ten percent of IS management feels their sales support system is manual compared with 23% of sales management and 30% of sales reps. By comparison, 17% of IS report that the sales support system is highly strategic, compared with 13% of sales management and 7% of sales reps. Most people, regardless of group, report that their system requires some level of improvement. Only a handful of people report full satisfaction.

STORAGE TECHNOLOGY TIGHTENS UP WITH SFA

■ ■ ■ ■ ■

THE COMPANY

Storage Technology is a $1.5 billion manufacturer and supplier of storage systems to computer manufacturers. The firm describes its selling style as "relationship selling." Sales cycles often take up to one year, requiring an average of 50 on-site sales calls. It has about 200 field sales people in North America and more than 85% of its sales volume is direct.

THE MISSION

"Apollo" was the name given to the Sales Force Automation (SFA) system developed by John Williams and Tom Campbell. Williams, vice president of North American field operations, conceived the system, and Campbell was made project manager in February 1991. He was given eight months to have the system up and running. This aggressive deadline was driven by Storage Technology's anticipated launch of a major new product line.

HARDWARE AND SOFTWARE SELECTION

Prior to this project, the sales force had limited exposure to computer automation. Campbell involved the regional sales support managers and one sales rep from each of six regions in evaluating alternatives. They agreed that portable computers—notebook computers in particular—were a necessary component.

Before committing to a specific notebook configuration, Campbell and his team investigated a variety of off-the-shelf SFA software solutions. They concluded that, considering the cost of packaged software and the specific functions they desired, it would be better to design and build their own single-user system. They also saw this as a way to increase their strategic advantage, because it would make their system more difficult to duplicate.

The company evaluated a number of notebook computers during a period preceding the introduction of Apple Computer's Powerbook notebook model. Under terms of non-disclosure, Apple representatives described Powerbook to Storage Technology's SFA project team. Based on that presentation, Campbell and his team selected the Powerbook. They were particularly attracted to the quick prototyping software facilities that would enable them to deliver a highly functional and easy-to-use application within four months.

The team selected Fourth Dimension, a

INFLUENCERS AND DECISION MAKERS

Just who drives an SFA project? Vendors report that the initiators range from an individual sales person to IS vice presidents, noting that the sales manager is typically the most influential. IS management, sales management, and field sales people agree that IS, marketing management, and sales management are the most active SFA decision makers.

Senior corporate management generally observes SFA projects without intervening. The IDC-interviewed groups all agree that the actual users of the SFA system are more involved in the selection, design, and implementation processes than the CEO. Despite this, CEOs generally sign off on these expensive projects.

In smaller companies, or even in smaller-scale implementations, the decision-making roles may

Macintosh-compatible database application, as the software development platform for their custom application. Microsoft Office was chosen to provide integrated word processing, spreadsheet, and presentation graphics. The storage systems company selected an external messaging service as an interim solution for remote communications and electronic mail.

CONSULTING ASSISTANCE

Storage Technology commissioned Andersen Consulting to develop the functional design specifications and data conversion plans, handle project management and training materials, and the actual training. The objective was to have each rep's portable equipped with all necessary information upon delivery. Metropolis Software Inc., an experienced Fourth Dimension developer, provided most of the technical leadership and programming for the custom application.

PRODUCTION SYSTEM LAUNCH

The system went live in November 1991. Each sales rep received two full days of training and a notebook equipped with a broad suite of applications. Applications included territory management, calendaring, account information, pricing configuration, communications, and a proposal generator. All data could be stored locally or in corporate data bases. Product descriptions and competitive comparisons were also scanned and downloaded to the portables.

Campbell estimates that about 95% of the sales force currently uses Apollo. Some use it casually, others heavily. The casual users tend to be older sales reps who have established their accounts and territory, as well as a refined style of account management. The more aggressive users tend to be the less experienced reps who are toiling to develop their territory.

ESTIMATED RETURN ON INVESTMENT

Storage Technology has done some preliminary analysis of how the system has affected its business. The cost of implementing the system, including development, software, hardware, and training was about $1.5 million. Spread over the 200-person sales force, the cost per rep is about $7,500. Campbell estimates the system will pay for itself within eighteen months.

He also points out that, prior to implementing the SFA system, the average number of sales calls per sale was 50. As a result of the system, that number has dropped to 40. He adds that as the system is increasingly used, the number of calls will drop even further.

change. For instance, in a smaller company, a sales manager or senior sales rep may be able to drive an SFA project from beginning to end. It should be noted that these projects are relatively simple, often using off-the-shelf single user software.

CONCLUSIONS

The need for SFA varies by company. The answer may be as simple as hooking sales into a corporate electronic messaging system, or as complicated as building a highly strategic sales system that integrates various departments across multinational companies.

When making the decision whether or not to invest in sales automation, be aware of the many available alternatives. If your company has decided, or is even thinking about moving down the SFA path, consider the following "dos" and "don'ts."

TABLE 1
SFA Grows from Three Generations of Computing

Glass House	*Personal Productivity*	*Enterprise Integration*
Central Batch Processing or Host Systems	Local Processing on Departmental Desktop Computers	Client/Server Distributed Processing
		Open Systems/Networked PCs
Character-based Interfaces	Graphical Interfaces (Macintosh, etc.)	Paradigm-Based GUIs
Routine Reporting	Interactive Analysis/Reporting	Mobile Computing
Central Databases	Personal Databases	Group Collaboration/Communication
1970s	Mid-to-Late 1980s	1990s

- Don't underestimate the importance of ease of use. The weakest link in the system can be the one between the user and the computer.
- Begin with an effective sales strategy. Throwing technology at a dysfunctional process will only set your company back. Also, consider the sales process as it involves other departmental functions.
- Meet with a broad range of SFA software vendors. Many have gone through great pain to overcome potential SFA pitfalls. Their references will provide you with the opportunity to compare notes on SFA strategies and expectations.
- Investigate new technologies. An effective SFA system does not necessarily require bleeding edge technology. However, as when making any IT investment, beware of dead-end technologies—such as those that may lock you into a single-user approach—as well as the obsolete (transportable computers). Select products that receive healthy and regular investments by the manufacturer or provider.
- Leave room to grow. In addition to considering current and future needs in the sales organization, consider how the use of the system may change. Take into account central IS issues even if rolling out independent systems, thus allowing potential integration down the road. Avoid the trap of outgrowing your first SFA investment.
- Don't alienate potential users or supporters. Despite the potential benefits of SFA, some may perceive it as a prying Big Brother. Interview users extensively not only to find out what will excite them but also to make them feel like contributors, not victims, of the process.
- Don't encourage abandoning unique selling styles in favor of generic system prescriptions. Customization should be considered at the sales organization and at the personal level. The system should complement rather than inhibit a rep's style of selling.
- Stay in touch. View the SFA implementation as an evolution, not a revolution. Let the users know you care. Check back with them after several months of usage.
- Invest in the system continually. Sales people will pay more attention to a system that receives regular attention than one that is established and forgotten. Sending out regular notices regarding system enhancements and training classes can be a passive, yet effective, way to encourage system usage. Furthermore, as the system is used, leave open the possibility of incorporating new needs or suggestions.

The SFA evolution is underway. Becoming SFA-aware is an inexpensive first step for companies that have not fully extended computer automation to the sales function. This simply requires understanding the potential benefits of SFA and how the various alternatives might apply to your sales organization. If nothing else, becoming SFA-aware will help to prevent competitors from taking your company by surprise.

QUESTIONS

1. What are the major components of a sales automation system? Which features are most likely to provide the largest benefits?
2. What are the major barriers to implementing sales automation systems? How can these barriers be reduced or eliminated?
3. How can present-day sales automation systems, such as those described in this section, be made more effective?

4 Factory Productivity

FACTORY automation reduced the touch-labor content in most manufacturing operations in the United States and other advanced economies so that touch labor is now only of a small part of total production cost. The challenge of improving manufacturing productivity is no longer automation of individual manufacturing steps, but improving the ability to move materials, parts, and products efficiently and being able to adapt rapidly to changing customer demands and preferences. Just-in-time delivery, flexible manufacturing, integrated design and manufacturing through CAD/CAM, and superior logistics operations provide competitive advantage in today's business environment.

This chapter introduces the major applications of information systems into manufacturing. The first four sections describe significant changes enabled by advances in information systems that link design and production. The next section shows how information systems speed the flow of goods in and out of the plant through improved logistics. These gains do not come for free. Seraph and Guimaraes discuss the requirements they impose on the information systems group. The chapter concludes with an integrated case study in a manufacturing environment.

4.1 Systems Support of Just-In-Time Manufacturing

JOHN W. ROBERTS

Just-in-time manufacturing differs from conventional manufacturing in the way materials are handled. In the traditional system, assemblies are built in lots, whereas in just-in-time manufacturing, each item is built from beginning to end. The traditional system involves large amounts of in-process inventory to make sure that materials and subassemblies are available when needed. Such in-process inventory results in large costs. With the just-in-time system, the in-process inventory is largely eliminated. However, in a just-in-time system, not having materials or parts available at the time you need them stops the entire production line. Therefore, to make just-in-time work, computers are used extensively—to keep track of where a material is and to make sure that necessary pieces arrive at the place where they are needed at the time they are needed.

Just-in-time manufacturing proved extremely popular in Japan and was credited as one of the reasons Japanese manufacturers were outperforming those in the United States. To remain competitive, many U.S. operations adopted Japan's just-in-time procedures. As a result, some of the terminology, such as kanban, involves Japanese words that have migrated to English without translation.

ONE of the most significant differences between just-in-time and traditional manufacturing is the flow of material. In traditional manufacturing, a work order for a certain quantity of an assembly is scheduled by production planners. The material required is then picked by stockroom personnel and placed in a staging area on the production floor. After each assembly is completed and inspected, it is placed back in the stockroom until it is picked again as part of another assembly. The process is repeated until a finished product is created. Although this method appears efficient, the waste inherent in the system becomes apparent on close examination.

Because assemblies are built in lots rather than one at a time, the entire work order quantity may be at risk as a result of a problem with materials, design, or quality of work. The problem may not be caught until the assemblies enter final test. The material review board then decides whether

Source: *Information Systems Management* (Winter 1990): 22–29.

the assemblies should be scrapped or sent back to the production floor for rework. This possibility for waste is repeated all through the manufacturing process until the finished product passes through inspection and is shipped. As a safeguard, material is stored in quantities much larger than required to prevent a delay in the manufacturing process due to faulty components. There is even additional waste in the process. The production path zigzags across the production floor and the stockroom until the final product is produced. The movement of the material takes time and effort, places parts at risk more often, and adds to the manufacturing lead time of the product.

In just-in-time manufacturing, the production lines are set up to build each product from beginning to end. There are no defined subassemblies, and JIT bills of materials (BOMs) are flat, consisting of only the pile of parts required to build each product. Although the process appears to flow in a straight line, it is not completely linear. Therefore, feeder production lines are used to perform processes that are not part of the critical path of the product. The synchronization of the feeder lines and main line is extremely important to ensure that one line is not waiting for the other to complete its operations. This is accomplished by balancing lines. A line is balanced when each operation takes about the same amount of time to complete.

Often, the output from a feeder line must also be considered a product for the purpose of repairing or upgrading products already shipped to a customer. This output, a field replaceable unit (FRU), is treated like any other product. FRUs are master scheduled and have their own bill of materials. The FRU part number, however, is not listed on the BOM of another product. Instead, the components of the FRU are listed separately, keeping each product on a single level. This is extremely important because it keeps the focus on materials rather than on subassemblies. In some environments, it may be advantageous to produce FRUs on one production line and the final products on another. Treating each product as a separate pile of parts allows for this flexibility.

There are no work orders in just-in-time. Products are made on the basis of a daily rate determined from bookings or the master schedule and are pulled through the manufacturing process on a demand flow basis. Bins (i.e., kanbans) are used to control how many parts are located at each operation. Each kanban is assigned a unique part and quantity. When the kanban is empty, it is sent back to the stockroom or supplier for replenishment.

In-process kanbans are used to signal demand from one production step to another. Units flow through the JIT production line one at a time. As the worker from the last step completes a unit in the in-process kanban, a unit from the previous step is pulled. The demand ripples back all the way to the first step. As final products are completed, inventory balances are backflushed by processing them through each product's BOM to determine the materials used.

The goal of JIT is to achieve the highest quality product at the lowest possible cost. To help accomplish this, work is checked for quality throughout the production process at Total Quality Control (TQC) points. If a problem is encountered, the few assemblies affected are reworked at the steps that take place after the previous TQC point. This method has a significant advantage over batch manufacturing, in which the entire work order quantity is at risk if a problem is detected. TQC checks are performed just like any other step in the production process. All steps are illustrated on method drawings, which visually describe the production steps that take place at each operation.

JIT is much more than a manufacturing philosophy. When it is implemented successfully in the production area, the efforts must turn to improving the quality of parts purchased from suppliers. The goal of the purchasing group is to make suppliers an extension of the manufacturing process. This means developing long-term relationships with suppliers and working with them in improving the quality of the incoming material until it no longer needs to pass through receiving inspection.

UNDERSTANDING THE JIT SYSTEM

The raw material and in-process inventory (RIP) stored in kanbans on the production floor is the equivalent of work in process (WIP) in traditional manufacturing. The JIT system must be able to keep track of the levels of such inventory, just as the batch manufacturing system keeps track of WIP. Anytime material is moved into or out of raw and in-process inventory, the appropriate transaction must be entered into the JIT system. If materials are transferred from stores, the stores' on-hand balance is decreased and the in-process inventory on-hand balance is increased. If materials are received into in-process inventory directly from the supplier, a receipt transaction is created and the on-hand balance is increased.

For raw and in-process inventory balances to remain accurate on the computer, use and scrap must also be recorded in detail. Transactions that reduce raw material and in-process inventory include material issues, component scrap, and miscellaneous use. Materials issues are created by the computer backflush transaction when products are completed and placed in finished goods. This transaction reduces the on-hand balances of the in-process component parts and increases the finished goods balance by the quantity of the product that is completed.

Scrap transactions are also entered into the JIT system whenever components are scrapped. These transactions reduce the inventory on-hand balances and charge the scrap account numbers. In JIT, scrap is not planned for with a yield or scrap factor assigned to BOM. Miscellaneous use takes place when a department takes a part out of in-process inventory. The on-hand balance is reduced and the appropriate department is charged.

Some computer systems actually record the movement of materials within the production floor (e.g., from in-process kanban to in-process kanban). This is usually unnecessary because in-process balances in individual kanbans are small and easily managed manually. Detailed reporting on kanban balances is often required not because it is needed but because management is accustomed to the level of detail provided in work order–based systems.

RIP Cycle Counting

The accuracy of the computer on-hand balances is extremely important in JIT. Because the goal is to keep raw material and in-process inventory balances as low as possible, the chance of running short on parts is greater. An effective cycle counting system must be in place to ensure that counts are performed accurately and regularly. Cycle counting of raw material and in-process inventory differs from cycle counting of stores because parts in in-process inventory can be located anywhere between the start of a manufacturing process and the backflush point. This includes the materials in kanbans as well as material that has been used in assemblies that have not yet been completed.

The A-B-C method of cycle counting can still be used in JIT; however, the classification system should consider not only dollar value but also use activity. High-use parts should be monitored more frequently because of the potential impact that a shortage may cause, even though it may have been considered a B or C item in traditional manufacturing.

The computer is an essential tool in the A-B-C classification of raw materials and in-process inventory and in producing cycle count reports. Cycle count lists should be printed in order of the kanban locations to make the counting process as easy as possible. If a part is used in more than one kanban, it appears in multiple places on the list. The cycle count system then consolidates the individual counts and provides overall count variances for each part.

Cycle counting of inventory is more difficult than cycle counting of stores because raw material and in-process inventory is often a moving target. The computer system must facilitate this process as much as possible. The fact that cycle counting is difficult, however, does have one advantage: manufacturing has an incentive to keep raw material and in-process inventory balances as low as possible.

Kanban System

Although the demand flow system is primarily manual, the computer can be used effectively to identify and size the kanban containers. Each kanban must have a card or label providing information on how to replenish the material or where to pull the partially completed assembly from. The information on the card includes:

- The materials part number for materials kanbans or the product part number for in-process kanbans.
- A description of the part.
- The quantity to be replenished when the kanban is empty.
- The location from which the kanban is replenished. This can be stores, another location in in-process inventory, or a supplier.
- The location at which the filled kanban resides.

Each kanban can also be bar-coded with a unique ID (e.g., a sequential number) to facilitate the materials movement transactions. This is an effective way to make transaction entries as accurate as possible. Instead of the part number and quantity being keyed into the computer each time a materials kanban is replenished from the stockroom, the bar-code label can be scanned and the kanban information can be automatically processed. If the kanban quantity cannot be refilled because of a parts shortage, the stockroom person simply overrides the quantity displayed on the terminal. The computer system can then produce a list of all parts that are short. This list is provided to purchasing for expediting purposes.

Bill of Materials

The JIT BOM system is similar to the traditional BOM system in that each BOM consists of a list of parts required to build an assembly. The JIT BOM, however, has one level, consisting of a list of purchase parts required to build the final product. Subassemblies don't usually exist on the JIT BOM because they are not recognized in the manufacturing process.

The JIT BOM is also used somewhat differently than the traditional BOM. A traditional BOM is used to produce a pick list for a specific quantity of an assembly when a work order is released; the JIT BOM is used to relieve component inventory when the product is completed. Although the JIT BOM is simpler in structure, it includes additional information used in the computer backflush transaction. A part number may be listed multiple times on the JIT BOM, representing the various locations at which it is used on the production line. Each line on the BOM contains the following information.

- Part number: The component part number required to build the product.
- Quantity: The quantity of the component that is required to make one unit of the product or FRU. This quantity does not include an allowance for planned scrap or yield.
- Backflush location: The location at which the materials should be deducted during the backflush transaction. This is usually raw materials or in-process inventory. On advanced JIT implementations, however, a purchase may be relieved directly.
- Deduct point: The location on the production line where components are consumed or scrapped as a group. This is used for scrap or intermediate backflushes.

The JIT BOM system must be able to support multiple revisions of a BOM at the same time. In JIT manufacturing, a product revision can happen at any time. The computer system must allow the units completed before the revision to be backflushed at the old revision and the units completed after the change to be backflushed at the new revision. If all units were backflushed under the new revision or if the components on the BOM are wrong, the resulting inventory balances would be incorrect. In JIT, it is critical that product BOMs are kept accurate.

Backflushing

When the product is completed, a backflush transaction is performed to decrease in-process inventory for the materials used and increase finished goods inventory for the completed unit. The backflushing process makes the assumption that if a unit is completed, all the components in the unit's BOM must have been used. If the BOM is not accurate, all the materials will not be relieved correctly.

The backflushing program takes the quantity of each component and extends that by the quantity completed of the product or FRU. This does not include an allowance for scrap or yield because scrap is reported as it is detected. For each component, an issue transaction is created that reduces the inventory on-hand balance by the computed quantity. In addition, a receipt transaction is created for the final assembly. For example, if one chair has been produced, the finished goods on hand for chairs is increased by one, the in-process on-hand inventory for legs is decreased by four, and the on-hand quantity for seats is decreased by one. This is assuming, of course, that legs and seats are purchased from a supplier. If they were produced in-house, the on-hand balance of their components would be reduced.

For some products, the total production time extends beyond one work shift. In this situation, it is useful to perform an intermediate backflush to relieve the material up to a certain step in the manufacturing process. Certain component parts may be no longer identifiable or no longer physically in the production area (e.g., in a test or fabrication process) and cannot be cycle counted. These components are assigned an intermediate deduct point on the BOM. The intermediate deduct location is entered in the backflushing program, which backflushes only the components related to the deduct point. The inventory balance of affected material is reduced, and an in-process balance for the material is increased. When the backflush transaction is performed for the completed unit, the in-process balance is relieved for material that has had an intermediate backflush and the in-process inventory balance is relieved for additional material. Some products have more than one intermediate deduct point, especially if the production time extends over several days.

The backflush transaction can also be used to facilitate scrap reporting. Often, it is not individual components but partially completed units that are scrapped. Scrap commonly exists at TQC points, at which a defined set of components has been combined into a partial assembly. Rather than scrapping each component individually, a scrap deduct point can be assigned to the associated components in the BOM of the unit being assembled. The part number of the product and the scrap deduct point is entered into the backflushing routine. Scrap transactions are then created automatically for each component that has been related to the scrap deduct point.

A backflushing program would require the following information.

- The part number of the unit: The part number must be for a final product or an FRU.
- The revision level of the completed unit: If a revision to a product is in the middle of the day, the part number is backflushed twice—once under the old revision and once under the new one. Otherwise, the wrong components would be backflushed.
- The quantity of the backflush transaction: The quantity of each component on the BOM will be extended by this quantity to determine the quantity to be relieved. In addition, if the units have been completed, finished goods inventory will be increased by the quantity entered here.
- The intermediate deduct point or the scrap deduct point, depending on the type of backflush: If no deduct point is entered, the program would assume the units have been completed.

Because of the number of part balances updated by the backflushing routine, it is important to allow the backflushing transaction to be reversed. One way of doing this is to allow input of a negative quantity. The reversing backflush transaction would do the opposite of a regular one.

Process Steps and Method Drawings

In JIT, the manufacturing process is broken down into a sequence of activities, or steps, that must be followed to create a product. This is similar to the routing in traditional manufacturing; however, in JIT the sequence of steps describes the entire manufacturing process rather than the steps required to build each subassembly. Each step must be described in detail and be specific enough to allow an accurate time standard to be determined. After the steps are defined, they are accumulated into JIT operations, with each operation taking the same amount of time to complete. The time of each operation is called the cycle time and is based on the daily rate of the product.

For example, if the daily rate for a product is 32, the cycle time of each operation must be 15 minutes to complete 32 units in an eight-hour day (four an hour). This means that one unit passes through the final operation every 15 minutes. If the daily rate increases to 48, the process steps must be regrouped so the cycle time is 10 minutes.

Sometimes a process limitation requires the cycle time for one operation to be much longer than for the others. If this is the case, the in-process quantity of the operation must be sufficient to satisfy the demand while the operation is taking place. In the foregoing example, if a certain operation requires an hour to complete, an in-process kanban would be set up for four. This way, the subsequent operations can still pull at a rate of one every 15 minutes. When the four units in the kanban have been consumed, four more will come out of the operation.

The computer can be an extremely useful tool for developing operations that have similar cycle times. The line-balancing program would take the list of process steps with time standards and the desired cycle time as input and automatically produce a list of operations. If the desired cycle time changes because of a change in the daily rate, the computer could rapidly recompute the new list of operations. In addition, the computer could also compute the in-process kanban size of those operations that must take longer to complete than the standard cycle time.

When the cycle times of the various operations are balanced, method drawings are developed to visually describe the process steps required at each operation. In addition to showing each production step, they also illustrate the TQC checks that must be performed at the particular workstation. The drawings are much more effective than written instructions and act as a quick reference of the steps that are required. Computer-aided design (CAD) packages are extremely useful for creating and maintaining the method sheets.

A powerful CAD system makes revising method drawings relatively simple. In some JIT manufacturing companies, terminals rather than method books are placed at each workstation to provide the method diagram of the operation to be performed. This is extremely useful when one workstation is used to perform a variety of operations on a variety of products or product configurations. Most JIT implementations, however, do not require this.

MATERIAL REQUIREMENTS PLANNING

The material requirements planning (MRP) system in JIT is similar to MRP in batch manufacturing, except that requirements are generated for purchase parts only. With the demand flow system implemented on the production floor, work orders are no longer required. The complicated scheduling problems associated with work orders are eliminated. The primary function of the MRP system in JIT is to provide the purchasing department with visibility of the purchase parts required to support the master schedule.

Few changes have to be made to the MRP system to support JIT. Without manufactured parts, the system will not attempt to schedule work orders. The MRP process is simpler because the bills are on a single level. The MRP netting process must be modified, however, to include raw material and in-process inventory as well as stores inventory. Otherwise, material will be ordered when it is not required.

Potential materials shortages may not be as visible in a JIT environment as they are in a batch manufacturing environment. Work orders are not scheduled to be picked, so shortages cannot be foreseen in between MRP runs. One solution is to run MRP more frequently. This is more feasible in JIT; because the BOMs consist of purchase parts only, MRP often takes less time to run.

Purchasing

The relationship between the manufacturer and the supplier is much closer in JIT than in traditional manufacturing. The goal of JIT is to have a single source for a part whenever possible rather than multiple suppliers. Single sourcing makes sense only if the supplier is financially secure, has a stable manufacturing process that ensures consistent deliveries of quality material, and is willing to work in a JIT environment.

Single sourcing provides several benefits. The relationship with one source can be developed much further than with multiple sources. The purchases are of greater volume, allowing lower prices to be negotiated. The quality of parts can be monitored more closely. When the parts received from the supplier are of perfect quality, the material no longer needs to pass through receiving inspection.

The JIT computer system is useful in evaluating which suppliers should be considered for single sourcing. As material is received, information is collected on the material's price and quality and timeliness of delivery. Reports can then be produced to identify suppliers who are consistently delivering quality products on time and at a reasonable price. Suppliers that meet the requirements for single sourcing are awarded JIT supplier contracts.

The JIT contract specifies the duration and terms of the contract and a forecast of the material to be purchased. The duration is usually for 12 months but can extend for many years if the supplier proves to be a stable source. Included in the contract are delivery terms, packaging requirements, and inspection criteria. The contract allows for multiple release dates, with the actual delivery quantities not varying more than a predetermined percentage from the forecast ones. The JIT supplier is usually responsible for ensuring the quality of the parts so that the buyer need not perform receiving inspections.

Single sourcing opens up a variety of possibilities for automating the purchasing process. In fact, the increased deliveries that are the result of JIT often necessitate automation. Purchase orders can be produced automatically based on the requirements generated from the MRP system and the JIT contract information on the computer. The computer system can communicate the purchase order information to the supplier electronically, so little paperwork is required: The backflushing process can then be extended to backflush material directly against the JIT contract, with the assumption that if the material is used, it must have been received. This allows the supplier to deliver material directly to kanbans on the production floor rather than going through a receiving process.

Statistical Process Control

Statistical process control (SPC) is extremely useful in the JIT environment to gain control over the processes and to eliminate problems that result in poor-quality products. The SPC system is used to collect data on the problems that are encountered at the TQC points. A Pareto analysis or histogram of the more frequent problems is then produced. Working from the top of the list, the TQC teams attack the problems until their causes are eliminated or minimalized. The teams then move on to the next set of problems.

Because parts from certified JIT suppliers are usually not inspected at receiving time, the SPC system is a useful tool for monitoring suppliers. If a purchased part is continually shown to have defects, the supplier can be contacted and the inspection efforts can be increased. If the problems continue, supplier contracts may be terminated.

An SPC system not only highlights problems in the manufacturing process but indicates when problems are resolved. When certain defects no longer appear on the SPC reports, the TQC teams know their efforts are worthwhile.

Cost Accounting

Cost accounting in a JIT environment is significantly different from cost accounting in a batch environment. JIT cost accounting is closer to process accounting than to job cost accounting because there are no work orders and direct labor is not tracked discreetly. This is because the percentage of the total product cost that can be attributed to direct labor is relatively small. Labor variance reports have little meaning because the manufacturing process can be managed visually by the production supervisor. Instead, labor becomes part of the manufacturing overhead pool and is allocated to the products on the basis of the total manufacturing time each product takes. The overhead rate (including labor) per hour is multiplied by the total manufacturing time of each product to determine the total overhead per product.

The materials cost of producing each product is determined during the backflushing transaction. When this is added to the overhead cost, the total product cost is known. The actual cost can then be compared with the standard cost to arrive at a variance.

The value of work in process is determined by taking the total value of materials in process and adding the overhead cost that can be attributed to the products in process. Assuming the production lines are balanced, the in-process overhead cost is computed by taking 50% of the standard overhead cost for the units in process. This means that on the average the units are half completed.

Because of the absence of work orders, cost accounting is actually simpler in the JIT environment. The major difficulty is getting the manufacturing and cost accounting groups to understand that the detail they are used to in traditional manufacturing is not necessary in JIT. The production process is primarily managed visually, so by the time the computer system produces variance reports, the problems are already known.

ACQUIRING A JIT SYSTEM

JIT systems can be developed in-house or purchased from software vendors. Which avenue to take depends on the type of system, the availability of software, the availability and background of the MIS personnel, and the budget. The system must be able to work in tandem with the work order system because work orders are still an effective method of monitoring costs on new products until the product cost stabilizes. In addition, the switch to JIT does not take place all at once, but on a product-by-product basis. Either way, MIS personnel must have a thorough understanding of JIT concepts for the implementation to be successful.

There are advantages and disadvantages to both choices. Developing the JIT system in-house ensures that the system interfaces with other applications and provides the users with exactly what they want. Because JIT is relatively new, however, the users may not know what they want, and projects may pull MIS, manufacturing, and cost accounting staff from other important projects. In addition, MIS departments are usually understaffed. MIS should probably not be the pacing item on the JIT implementation.

Purchasing a JIT system from a vendor can save a lot of time. Introducing an entirely new computer system while implementing the JIT program in the organization, however, should be avoided. This may strain resources and make both projects unsuccessful. MIS managers will find that the best alternative is a software package that can plug into the existing computer system and applications with minimal modifications. This depends, however, on whether such an application is available.

QUESTIONS

1. What are the differences between JIT and traditional manufacturing?
2. What is the role of information technology in JIT?
3. How does JIT improve manufacturing productivity?
4. Japan found that JIT manufacturing led to continuous traffic jams because trucks are on the highways all day and all night to deliver just in time. Is the situation in the United States different than it is in Japan? Should business or the public pay for the costs of the congestion created?

4.2 Fulfilling the Promises of CAD

JEFFREY K. LIKER, MITCHELL FLEISCHER, AND DAVID ARNSDORF

American industry made huge investments in both Computer Aided Design (CAD) and Computer Aided Manufacturing (CAM). These investments are still growing. The basic ideas of these technologies is that many engineering design and many manufacturing tasks can be automated to increase the productivity of people. As Liker and his co-authors point out, the promise of productivity, of being able to design in three dimensions, the rapid transfer of data, and the promise of less paperwork are reasons for purchasing CAD. Unfortunately, industry surveys show the reality of CAD differs from the promise.

That is not to say that CAD offers few benefits. Rather, the experience is that in many firms the gains actually achieved are far less than those promised. As in much of the computer business, there is often a gap between promise and performance.

Why the gap? This study of CAD emphasizes that merely computerizing existing jobs is not enough. In many organizations, CAD is simply a computerized drawing board. To gain the benefits available, the process of design has to be thought through and changed. In other words, to make full use of CAD, management must re-engineer the design process.

AN aerospace engineer is designing a new plane wing. He puts the preliminary sketch on the computer-aided design (CAD) surface modeler, which generates the mesh for a complete structural analysis. After making a few modifications, he feels confident the wing will withstand the stresses specified by government regulations. No prototypes are necessary; the computer analysis tells him everything he needs to know and far more than physical tests would. The final database goes to a computer-aided manufacturing (CAM) system that will cut the molds for the internal wing supports.

The CAD family of technologies has experienced phenomenal growth over the past few years. The sidebar, "Glossary of CAD Features," describes some of these technologies' features. Although CAD has been around in various forms since the late 1960s, the industry has seen a nearly eightfold increase in revenues in recent years (from

Source: Sloan Management Review (Spring 1992): 74–86.

GLOSSARY OF CAD FEATURES

■ ■ ■ ■ ■

We use "CAD" (computer-aided design) broadly to include four levels of geometric complexity, ranging from 2-D drafting to solid models; use of a database; interference checking; mechanical and electrical design; electronic communication of CAD to other CAD systems and to CAM (computer-aided manufacturing); and CAE (computer-aided engineering) methods like finite element analysis. There is no universal vocabulary for CAD; below are our definitions:

2-D Drafting. This is an electronic emulation of basic drafting. The only geometric entities stored are points, lines, and curves in a two-dimensional plane. 2½-D simply adds an extruded depth to the basic 2-D information and is considered to be the same level.

3-D Wireframe. As with 2-D drafting, the only geometric entities are points, lines, and curves, but they can be defined in 3-D space. This allows the model to be rotated and viewed from any angle.

Surface Models. Surface models define corners, edges, and surfaces, and add surfaces to the geometric entities of points, lines, and curves that are in wireframe systems. Surfaces are required to describe certain classes of parts, such as a front fender on a car, and to do certain CAD operations, such as cutting a section through a part.

Solid Models. A solid model is a mathematically complete and unambiguous representation of a part. In addition to defining the corners, edges, and surfaces, it also defines the volume that is inside of the part and outside of the part.

Standard Parts Database. As noted elsewhere in this article, much design work consists mainly of using standard parts in varying configurations. In order to do this, a "database" of parts must be created. As used here, the standard parts database is the collection of standard parts.

Interference Checking. Use of the CAD system to check the fit between different parts.

Printed Circuit Board Design. This includes several specialized design programs including circuit board layout, which is mostly manual, and routing design, which allows the designer to specify the connections between components and then automatically connect them.

CAD Customer/Supplier Transmission. The stated goal of many large companies is to link their suppliers into their design systems so the suppliers can directly access their CAD designs and save redrawing time.

CAD-CAM Transmission. Often the CAM system is different from the CAD system. This means that the design must first be created in the CAD system and then transmitted to the CAM system for definition of the CNC machine code.

Finite Element Analysis (FEA). This is a CAE program originally created to allow the analysis of stresses in complex structures. It is now being used for other areas as well, such as heat transfer and electric field analysis.

Circuit Analysis and Simulation. These types of programs allow a designer to simulate a circuit design in software and test it without having to build it.

$0.6 billion in 1980 to $4.5 billion in 1988). Market penetration is also high for such a relatively new technology. A 1987 Industrial Technology Institute survey of Midwestern manufacturing firms showed that 49.5% of all manufacturing establishments with 250 or more employees use CAD. Despite this growth, the scenario above remains a dream for most companies with CAD.

Companies typically buy CAD on the basis of five promised benefits.

1. Automation of routine design tasks will increase productivity of designers and design engineers.
2. The ability to design in three dimensions will increase the designer's conceptual capacity and creativity, improve design quality, and create a geometric database useful for downstream analysis and production operations.
3. Design databases will be electronically transferred to manufacturing or CAD/CAM, improving quality and productivity and reducing lead time.
4. Design databases will be electronically transferred to engineering analysis and simulation programs, which will alleviate costly testing of prototypes.
5. The design process will become paperless as databases, rather than drawings, are sent to customers and suppliers. This change will reduce the costs of producing, storing, and managing blueprints and will speed communications with customers and suppliers.

While in theory CAD can do all of these things, in practice the technology has not lived up to its billing in any of these areas. While productivity gains of up to 5,000% were once predicted (and vendor claims of 1,000% are still common), 100% to 200% gains are the best that most firms can realistically attain.[1] Many firms gain no productivity at all, and some claim that their staffs are less productive using CAD. Success in the electronic transfer areas is even less evident. True CAD/CAM is still quite rare and requires a degree of cooperation across organizational boundaries for which most companies are ill equipped.[2] Design transfer between different CAD systems is highly problematic, even using the latest transfer protocols. And, if anything, CAD has produced more paper; minor design changes are so easy that drawings are now reissued where once changes would be marked up directly on old blueprints. The gap between the promise and reality of CAD is wide.

If CAD were cheap none of this would matter. However, a typical investment in the hardware alone can cost $50,000 a seat, and the costs associated with training, user support, and reduced productivity during the inevitable learning curve can be much higher. We recently visited a division of a Fortune 100 company that makes large mechanical assemblies and has over 3,000 engineers. It is developing an integrated engineering design workstation that will tie together most engineering analysis and design functions. The company expects to deploy more than 800 of these workstations over the next few years at a cost that we estimate to be over $40 million for hardware and base software alone. The additional costs of software, support, training time, and lost productivity can best be left to the reader's imagination.

While hardware and software costs are now declining dramatically as PC-based CAD systems become more sophisticated, they will remain high. Failure to achieve benefits will be extremely costly. Thus, in this paper we ask, "Why is CAD so underutilized and its promised benefits so unachieved, and what can be done about it?"

To understand CAD's use, we assembled representatives from ten sites in six Fortune 100 companies that are heavy CAD users. These firms included two aerospace companies, one automotive firm, a large consumer products company, and two automotive parts suppliers that were divisions of large conglomerates. Table 1 lists the companies (names disguised) and shows the extent to which they use specific CAD features, as reported by designers using CAD. In 1988 we interviewed CAD system managers, designers, design engineers, sales

TABLE 1
Average Percentage of Designers' Projects that Used Various CAD Features

Study Sites	2-D Drafting	3-D Wire-frame	3-D Surface	3-D Solid	Use Old Designs in Database[1]	Interfer-ence Checks	Circuit Board Layout	CAD to CAM	CAD to Customer/ Supplier	Finite Element Analysis	Circuit Analysis/ Simulation
Consumer Prod	45	53	6	0	21	5	na	7	2	18	na
AutoCo	25	73	20	3	15	15	na	15	2	47	na
Machined Parts	92	4	4	0	23	18	na	0	2	0	na
Stamped Parts	0	95	50	0	33	25	na	25	50	0	na
Spare Parts	78	0	0	0	14	31	na	0	0	30	na
Small Parts	91	2	1	1	21	9	na	0	2	20	na
Airplane Structures	80	4	6	1	13	15	na	8	1	16	na
Airplane Electronics	na	na	na	na	17	42	100	20	7	na	9
Weapon System Structures	62	13	3	0	14	8	na	0	0	0	na
Weapon System Electronics	na	na	na	na	15	13	100	35	12	na	81
Total[2]	62	26	7	<1	18	14	100	11	4	25	52

Designers were asked what percentage of the projects they worked on in the last year used each of these CAD features. Shown is the average percentage across designers.

1. The average percentage of the design project that was done using pieces, or modified pieces, of old designs stored in the computer database.
2. Based only on sites where the CAD feature was applicable.

engineers, drafters, and others at each site, for a total of 77 interviews, and we surveyed 243 designers and design engineers, both CAD users and those who work closely with CAD during the design process. In addition, our team (two social scientists and a CAD-experienced engineer) observed each CAD system in use, observed the manufacturing processes where possible, and examined typical products of each firm. In addition to our formal research, we have spent the past several years working with a large number of firms to assess their design processes and CAD needs. When we began the study, we expected significant variation in how different sites, and even different departments within sites, used CAD depending on their organization and management. We were very surprised to find that, despite all of the research on the relationship between technology implementation and organization, none of the companies had seriously changed their organizational structure when they implemented CAD.[3] They simply replaced old tools with new ones—"move out the drawing boards and give the drafters a CAD system." These were not fledgling companies living in the dark ages; they were highly respected market leaders, on the forefront of adopting and utilizing CAD.

We had planned to statistically correlate variations in organization to outcomes of CAD use.

We made the site visits primarily to help us understand the general organizational context in order to interpret the statistics. However, there was surprisingly little variation in organization of the design process. Moreover, the main story was not how differently the organizations used CAD, but in their almost uniform underutilization of high-level features.

We define "high-level use" as the use of features beyond two-dimensional drafting. Many design tasks require nothing more than 2-D, but using such a sophisticated technology exclusively for this purpose underutilizes many of its most powerful features. Zuboff uses the term "automating" to describe the process of replacing manual functions with a computer. She argues that the real power of computer technology comes from "informating," that is, using the computer to add new information to the work process.[4] Applied to CAD, this means developing the design in 3-D so that the designer can visualize it in new ways as it is being created, checking for interferences between different components before the product is built, and calling up related designs as other designers develop them in order to improve the fit of interrelated components. Although each of these functions can be done manually, CAD can perform them quickly and without physical mockups or drawings, thus giving much more information to the designer in real time—as the design is being done.

During our interviews and site observations, we were struck by some of the reasons that these high-level features were underutilized, notably the extensive division of labor and the high degree of fragmentation and segmentation of knowledge and skills in the design process. In these companies, the design process was divided into many little pieces and delegated to different job specialties. Stand-alone features of CAD systems were used quite well within these job specialties. For instance, engineering analysis specialists used analysis software, and former drafters used CAD to make drawings. However, CAD features that require integration across boundaries—integrating CAD and CAM, integrating the work of designers of separate components, integrating results from engineering analysis specialists with design—were not well utilized. Based on our observations, we believe that the high degree of specialization and fragmentation of the design process is at the heart of the difficulties U.S. companies are having exploiting the potential of CAD.

We are not the first to observe that CAD is being underutilized. A 1989 two-part special issue of *IEEE Transactions on Engineering Management* includes a broad array of studies of CAD around the world (Asian companies are not represented).[5] The guest editors, Majchrzak and Salzman, conclude that most companies in the studies, regardless of how they vary in size, base country, and product, do not achieve CAD's full benefits. However, these papers focus mainly on the social impact of CAD, and only a few address the reasons why CAD is not being fully utilized.

Several papers do address barriers to one of our five promises—the integration of CAD and CAM.[6] The authors observe that managers often assume that CAD/CAM technology will, by itself, automatically integrate design and manufacturing. In fact, lack of social integration is a major cause of the failure to utilize CAD/CAM's potential. The "brick wall" between design and manufacturing does not automatically come down, nor is it traversed, when computer technology is introduced. Forslin and associates sum up the issues:

> The crossing of departmental borders that an integrated system presupposes, brings, at least initially, conflicts concerning power and authority, territory, and status. In hierarchic organizations competition between functions and defending of territories are inherent. This creates barriers for the necessary cooperation in technical changes that affect more than one part of the organization.[7]

In short, although the promises of CAD are technically feasible and although they are being realized in isolated instances (e.g., special pilot projects), in most cases CAD is merely being used as an expensive replacement for the drawing board. The design process has changed little. Given the common belief that the five promises of CAD can be achieved through principally technical means, it is worth taking a closer look at each promise to describe the reality as we have seen it. We will show how organizational fragmentation contributes to the failure of CAD. We will also discuss the implications of our observations for integrating technology and organization. Again, what follows is based on our interviews, user surveys, site visits, and consulting experiences in a small number of firms, and not on a more systematic survey of a large number of firms.

THE UNFULFILLED PROMISES OF CAD

CAD Will Automate All of Your Routine Design Tasks

CAD's productivity potential cannot be realized when entering a new design from scratch into the system. The real power comes when the design is modified or when it is built from pieces of old designs already on the system or by using parametric design programs that automatically create a particular object based on a few parameters.

In reality, designers typically function as islands, developing their own designs from scratch and borrowing little from others or from their own previous designs. They seem to feel that they do not have the time to store and retrieve primitives for future designs.

Our survey asked designers what percentage of their designs in the past year were done by revising existing designs or using design primitives already on the CAD system. The answers ranged from a low of 13% (at Airplane Structures) to a high of 33% (at Stamped Parts, see Table 1). Yet the main products of all of these companies were mature, and most designs were modifications of a dominant design (e.g., a 15-year-old design at Airplane).

At Machined Parts, a design engineer estimated that "97% of all new work requires going back to the records and looking at what we have that meets the requirements." Yet only 23% of their design work actually used elements of old designs, according to our survey. They were slowly developing a set of design primitives, but in the meantime CAD users designed the same thing in slightly different ways depending on their personal preferences.

At Weapon Systems Structures, one designer explained that the company had libraries of standard parts, but they were in a book, not on the CAD system. Entering the standard parts into the computer for each design defeated the purpose of the standards. This company was organized around project teams working on different contracts. To computerize the standard parts library, the project managers would have to pool their money and invest in a long-term effort, a degree of cooperation that was not typical in the company's hierarchical culture.

Your CAD System Will Use a Complete Three-Dimensional Picture

Advertisements for CAD typically display a complete 3-D rendering of a part that can be rotated, viewed from a variety of vantage points, sliced to reveal detailed cross sections, sent for analysis to computer-aided engineering (CAE) packages, and used to develop CAM data sets. From this 3-D data set, an infinite number of 2-D images can be quickly and easily produced. However, the predominant use of CAD at our study sites was for 2-D drawings that were essentially electronic renderings of blueprints. When 3-D was used, it was mainly to construct wireframe models, rarely surface models, and almost never solid models. Experienced design engineers rarely used CAD themselves, even for simple 2-D sketches. Rather, they delegated CAD use to drafters, designers, and new engineers.

One problem with using 3-D was its complexity. At Consumer Prod, CAD users were required to develop all of their designs in 3-D, and they mainly used wireframe models. One designer explained that on his system you could only be in 2-D or 3-D at a given time. He said that in 3-D wireframe the multiple lines were difficult to sort out, but by blowing up a section you lost the big picture. He said, "You see thirty to forty lines in a small area and have no idea what they are—there is nothing to relate them to." The implication is that only very experienced designers can work in 3-D wireframe if the product is at all complex, and even they experience difficulty from time to time. In some cases, large screens are needed to see the detail and larger context simultaneously. Moreover, not all designs require 3-D CAD.

Another problem is the politics of getting engineers to develop conceptual models in 3-D. At Small Parts (small stamped and assembled auto parts), a Canadian operation where designers are unionized, the CAD systems group working with management determined that solid modeling, as a complex tool for conceptual design, should only be used by engineers. Designers should then work off these models to develop 2-D drawings. However, the union, fearing designer layoffs if engineers began to develop their own 2-D drawings as well, objected to this new division of labor. In the meantime, the engineers we talked to had no desire to learn CAD at all, arguing, for example, "It is a total waste of time and talent. I have more important things to do than make a drawing. You get trapped into trying to make the drawing look neat. Pretty soon you're not doing engineering."

You Can Automatically Download CAD Designs to CAM

A number of the design departments we visited were called CAD/CAM engineering, propagating the myth that the systems were being used for integrated design and manufacturing. As many before us have observed, in reality the traditional brick wall between design and manufacturing in U.S. industry is as sturdy as ever. In the companies that did any CAD/CAM, CAD and CAM were separate processes done by different people who had little interaction. Often data for CAM was entered into a computer system by hand based on blueprints.

At one extreme, Machined Parts used no CAD/CAM at all, even though its potential for the company's parts was clear. A design engineer explained:

> Manufacturing process design is totally separate from product design. There is very little that goes right into production. Usually we release designs several years before production. We design the part, build a prototype, and wait a few years before we make it [in order to fit into the design cycle of their automotive customers]. Manufacturing engineering designs the process at a different plant, and there is very little communication.

At Stamped Parts, product and process design were also separated, even though they were in the same building. While the product designers were downstairs using CAD, the manufacturing engineers sketched out the dies and fixtures for stamping the parts on drawing boards upstairs. One manufacturing engineer explained that "we engineer the dies, but they are designed by outside design houses using CAD." He went on to say that "we do have CNC [computer numerical control] lathes at our die model shop, and die models are typically CAM manufactured, but the die makers work from drawings."

Even companies that used CAD/CAM only did so on a limited number of designs, generally prototypes. Actual part production was handled by a myriad of suppliers who rarely had CAD/CAM systems compatible with those of their customers. Consumer Prod only used CAD/CAM for development of prototype dies. Of the hundreds of vendors who made tooling, only 14 used numerical control (NC) tapes generated by the CAD system, and even that was on an experimental basis with only varying degrees of success. In two cases, vendors were sent CAD-generated NC tapes that were used successfully and "those two were among the best tools we have ever bought." Similarly, we were told at Airplane Structures that "suppliers are not equipped for using CAD databases. We send electronic data to only three dozen of over 2,000 suppliers." At Weapon System Structures, "we don't have CAD links with our vendors. We buy small production runs in the thousands and can't dictate to our suppliers what CAD/CAM systems to buy."

A separate study examined the problems that arose when 53 small tool and die firms tried to use their customers' CAD data for CAM.[8] The CAD data they received had a variety of problems. In almost half the jobs, lines did not meet at corners, or lines crossed at corners. The problems were caused by both technical translation problems and the errors of designers who apparently did not understand their impact. In 51% of the jobs, the supplier had to fix the data, and in 25%, the supplier *completely recreated the dataset.* Users in the shops estimated that fixing dataset problems increased the lead time to produce tooling by 23% to 27%.

There Will Be an Iterative and Seamless Link Between CAD and CAE Analysis

This promise assumes that the CAD database can be directly uploaded to an analysis module and that CAD and CAE are iterative processes in which analysis results inform design decisions and design changes are fed back to CAE for further analysis. It also assumes that computer models reduce or even eliminate the need for building and testing prototypes, thereby cutting costs and reducing the length of the design cycle.

The reality is that engineering analysis is done on a relatively small number of designs and that it is done parallel to CAD, by different people, on different systems, generally from blueprints rather than the CAD database. Typically, analysis occurs *after* most design decisions have been made. CAE is generally a small department that serves design engineers, or in the case of small companies, it may be subcontracted to outside firms. As such, it has little power to influence the design process. In fact, analysis is often an after-the-fact confirmation of decisions that have already been made based on physical test results. As the head of the CAE department at Consumer Prod put it: "Analysis is used as secondary supporting evidence. No one is waiting for the results. Decisions precede analysis results. If the decisions are confirmed, fine. If not, they may ignore or hide the analysis results."

Since engineering analysis is not considered an integral part of the design process, the department is generally understaffed. In the body division of AutoCo, one of the biggest users of finite element analysis (FEA) in our study, there were over 600 CAD users but only 12 CAE analysts. The latter spent much of their time training design engineers, as design engineers were expected to do most of their own analysis. Analysts generally thought this was unrealistic. As one analysis specialist explained, "Even with a four-month training course, design engineers forget FEA in about one week. They have too much to do visiting vendors to sit down and relearn FEA." Design engineers agreed. An engineer at Airplane Structures started, "I have more than enough to do than to extract my own FEA data. I am talking to manufacturing, tooling, vendors, fatigue and fractures, the stress group—I simply don't have time." As a result of lean staffing, analysis often takes too long to be of value to design engineers. A Consumer Prod design engineer said, "The senior analyst is so busy, he has seven to eight months lead time. I can build a prototype and test it faster than CAE and feel confident it is accurate. Mostly junior people are assigned to CAE and lack job experience."

In theory, engineering analysis should greatly reduce the need for costly physical prototypes. But according to the analysts, managers and design engineers are often hands-on types who believe their senses, not an abstract computer model—if they can't touch it, it isn't real. For instance:

> Cars have historically been designed by building them and running them on the road. Managers are reluctant to risk betting the company on the skills of engineers running CAE (CAE analyst, AutoCo).

> The question is always asked, does the analysis confirm the test results? Design engineers be-

lieve the test results, not the analysis. There have been enough wrong analyses that I can see their point. But they have developed negative attitudes and believe good analysis can't be done (CAE analyst, Consumer Prod).

With CAD, the Design Process Will Become Paperless

When we first discussed our questionnaire with CAD system managers, some strongly resisted including the word "drawing." They claimed that drawings are obsolete: "We release databases, not drawings." By now it should be clear that, in reality, paper is still the dominant mode of communication. This is encouraged by both the input and output side of the design process. Rarely do parts suppliers and customers have compatible CAD systems. Moreover, designers like to look at the whole picture rather than a small window and to physically mark up the drawings. And meetings seem to require prints. A designer at Weapon Systems Structures admitted that "drawings are easier to huddle around. Supposedly we don't make any drawings. But we fall down; we make drawings all the time. Today, 99.9% is on paper." AutoCo invested in a computerized conference room that can display drawings from the CAD system on a large screen. But a designer claimed, "We are still running drawings for meetings with design engineers. You can only see one drawing at a time on the screen and you can't see it from some places in the room. You also get a distorted view of the edges." The result was that the expensive large screen was plastered over with full-size automobile drawings.

WHY IS CAD UNDERUTILIZED?

In each of the above examples, a major source of the problem is lack of integration, both technical and organizational. Table 2 summarizes the barriers to each CAD promise that involve organizational fragmentation, along with relevant solutions.

Lack of Organizational Integration

Design is typically organized as a linear, sequential process, with people organized by function to perform one step in the process. Yet advanced engineering workstations are designed and advertised as though an engineer sits at the workstation, designs and analyzes a complete part or component system from start to finish, and perhaps even prepares a manufacturing database that will ultimately drive manufacture.

Figure 1 illustrates the many interfaces that need integrating. There is lack of integration between people who perform different job functions (e.g., engineering analysts and design engineers), people within the same function working on different parts of the overall design (e.g., designers working on different components), and people working on different phases in the process (e.g., styling design and detailed design). Even this figure is a simplification. For example, other functions that might need to be linked to CAD include purchasing, industrial design, materials engineering, process planning, and reliability engineering. In addition, the design process is generally far more complicated than this set of discrete, linear steps suggests.[9]

Many of those interviewed expressed frustration with the fragmented process:

> It is important that designers and design engineers work closely together, but it just doesn't happen. Design engineers travel so much they are hard to get hold of (Mechanical designer).

> An inexperienced circuit designer simply puts symbols on paper and doesn't even consider implications for packaging (Electrical packaging designer).

> I remember working with the analysis group on one part, and it was nearly useless. Finite element analysis was handled by an inexperienced program engineer who made bad assumptions about boundary conditions. So I had to pull the stresses out of my pocket (Mechanical design engineer).

TABLE 2
Overview of Organizational Fragmentation Barriers to CAD Promises and Solutions

CAD Promise	*Organizational Fragmentation Barrier*	*Integration Solution*
Automation of routine tasks	Coordination needed to develop design standards and parametric design programs	Integrate across CAD users
Complete 3-D picture	Design engineers rarely use CAD	Better integrate jobs of design engineers and designers
	Designers threatened by engineers using 3-D	Better integrate jobs of design engineers and designers
	3-D design often overkill for the design task	Integrate design task and CAD use
Integrated CAD/CAM	Brick wall between design and manufacturing	Integrate design and manufacturing
	Incompatibility of CAD with vendors who NC machine parts	Integrate vendors into process
	CAD designers do not understand the requirements of CAM	Integrate CAD and CAM users
Integrated CAD/CAE	Analysts use stand-alone CAE tools but don't use CAD	Better integrate jobs of CAE specialists and designers
	Design engineers trust physical models over abstract analysis results	Better integrate design engineers and CAE specialists
	Automatic mesh generation programs are inadequate	Better integrate CAD and CAE technologically
	CAE departments are understaffed and too slow	Increase staff capable of using CAE, and integrate it into the design process
Paperless design process	Suppliers and customers rarely have compatible CAD systems	Integrate customers and suppliers
	Protocols such as IGES are not very good	Develop better technology for integration of disparate CAD systems
	CAD systems do not effectively support group work	Develop technology suited to groups at work

The elaborate division of labor and poor integration across functions help explain why many of the higher-level features of CAD, CAE, and CAD/CAM are rarely utilized to best effect. Yet social integration is rarely considered when purchasing expensive CAD systems.

Difficulties in Technical Integration

Integration difficulties also present very real and thorny technical problems. The problems of developing effective communication protocols between different CAD systems, for example, are even greater than previously realized. While the Initial Graphics Exchange Standard (IGES) communication protocol at one time seemed like a quick fix, it has failed to develop into a true standard, and different CAD vendors "comply" with IGES in different ways. Added to this are the data transmission problems of ordinary telephone lines. Automatic CAD/CAM data transfer is also far more complex than originally thought. The process of preparing a database for operating automated equipment is typically a separate design process, and geometry is not always directly downloadable. Downloading data from CAD to analysis programs is also complex, as the CAD output is often in a different form than that needed by the analysis programs.

Work is proceeding on all of these problems,

Figure 1 Interfaces in the Design Process

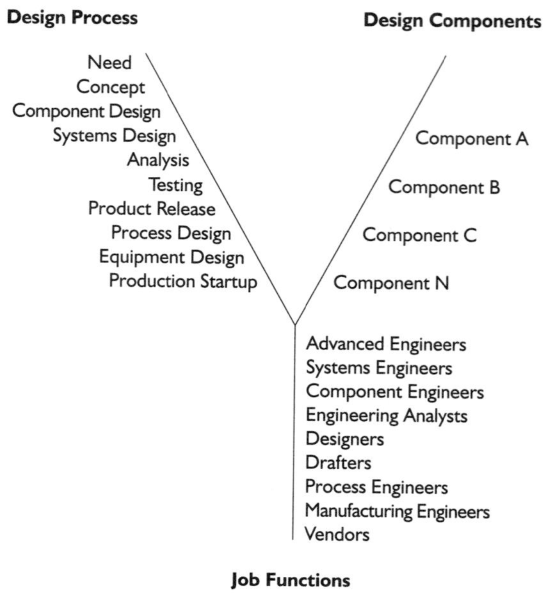

but fully satisfactory solutions may be years away. For example, the replacement for IGES, Product Data Exchange using STEP (PDES), is generally thought to be 10 years away from full usability.

FULFILLING THE PROMISES OF CAD

How can we remove the gap between expectations and accomplishments? We offer a solution that involves understanding the fit between three areas—design tasks, CAD integration, and organizational integration mechanisms.[10] CAD integration and organizational integration go hand in hand, and the appropriate levels of CAD integration and organizational integration depend on the design task. In particular, they depend on the design task's *information processing requirements*.

This concept is derived from Galbraith's seminal information processing perspective on organizational design.[11] Galbraith's theory is a contingency theory—it assumes that there is no one best way to organize but rather that the best organization depends on the nature of the task. Galbraith focuses on task uncertainty as the key characteristic of a task that conditions organization design. By task *uncertainty*, he means whether the process and content of the task can be reliably anticipated or whether they evolve in novel and unpredictable ways. This has also been referred to as the degree to which tasks are "programmable," that is, the degree to which all the steps can be anticipated and programmed in advance.[12] A task is "unprogrammable" if it varies unpredictably each time it appears, it has many alternative solutions, and the information and skills needed for the task are widely dispersed in a group.

Galbraith argues that high task uncertainty increases the organization's information processing requirements. When information processing requirements are low, organizations can use rules, extensive division of labor, and a hierarchical organization structure to produce a design. For example, if the task is to design a product that consists of a few components and that is a minor variant of an existing product, it is probably possible to delegate the design of the components to different designers. Rules can probably be devised so the designers can simply follow them and produce a manufacturable design. It may even be possible to build an expert system to make many design decisions. These designers could in theory work with minimal communication and still produce components that form a functioning product.

For new, nonroutine designs with high uncertainty, the specifications are likely to evolve with each component. In this case, the only way the parts will fit properly is if the designers communicate their decisions to each other as they make them. This task has high information processing requirements. Our fragmented design organizations in the United States provide numerous barriers to this communication. The problem then is to increase the organization's information processing capacity where high task uncertainty requires it. Galbraith says this can be handled technically by investing in information systems, such as CAD integration, or organizationally by creating lateral relations that cut across barriers. We consider each in turn.

Match the Level of CAD Integration to the Design Task

An overly elaborate CAD system not only will be expensive, it will be difficult and frustrating to use. The CAD equipment that gets sold is often either overkill or simply inappropriate for the application. For example, some design tasks do not require 3-D images, and it is not worth the time or investment when 2-D is perfectly adequate. In many other cases, simple PC-based CAD systems are quite suitable. We saw many companies that bought sophisticated systems without examining the actual design tasks, and then expected everyone to use high-end features.

For example, one CAD user at Consumer Prod complained that his manager insisted that he should use 3-D CAD for a particularly complex plastic part and go directly from the database to the model. The designer argued it was infeasible because of the part's complexity. He could not draw it accurately enough and get all of the angles and intersecting lines geometrically correct. But according to the designer, "the engineer's boss's boss told him to do it. He has no idea what the part looks like or what the manufacturing process is."

In general, it costs more to integrate different CAD systems (e.g., CAD-CAD communication, CAD-CAE, CAD-CAM) than to buy stand-alone systems. Robertson found that use of CAD for communication is much more likely when 3-D as opposed to 2-D capabilities are used.[13] One interpretation of this finding is that there is information in a 3-D design that can only be communicated CAD to CAD—mailing 2-D blueprints will not exploit the power of a 3-D database. Thus a company investing in 3-D design may also need to invest in mechanisms for CAD to CAD communication to get the full benefit of the technology.

We are suggesting that the most highly integrated CAD systems are warranted only when there are high design task coordination requirements. This will be the case when task uncertainty is high and design decisions are interdependent (see Figure 2). We define levels of CAD integration in terms of the communication's speed and accuracy. The slowest and in many cases least accurate method is to mail blueprints, which must be reentered into the next system (CAD, CAE, or CAM). The next level is to mail data files on tape or disk, which can take considerable time if the mail system is not working properly. Transmitting files that require translation may not be fast and may not be accurate. The situation improves when files are transmitted between compatible systems and translation is unnecessary. Asynchronous use of a common database can be highly accurate although there are time lags between use by separate parties. The most highly integrated level is working concurrently in the same database, thus communicating in real time. In general, new and complex design projects have greater task uncertainty and thus warrant higher levels of CAD integration. Also, we believe that more complex designs, for example, with more complex curved shapes, will have a greater degree of task uncertainty. That is, their

Figure 2 Matching CAD Integration to the Design Task

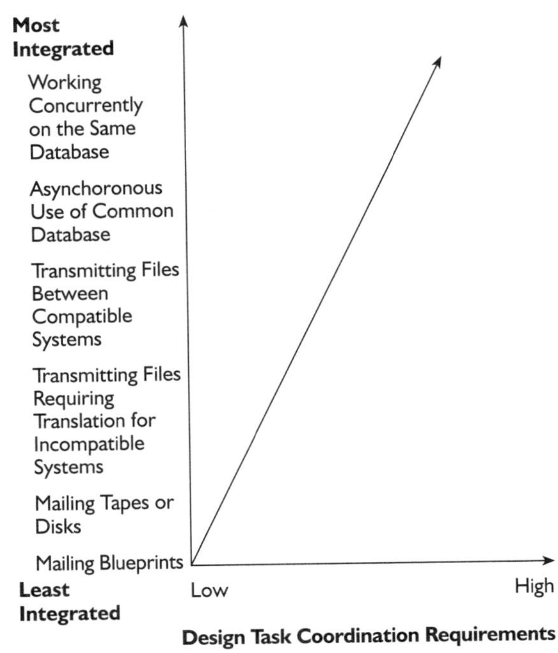

complexity will defy simple design rules and require more real-time communication between persons working on different aspects of the design.

Firms work on a variety of design tasks, and they may need several CAD solutions. At one firm, which had no CAD system when we visited, parts came in a range of complexities, but they had been made for decades and fit together in easily understood ways. We recommended that the company purchase PC-based 2-D CAD systems for most of their designs and a single high-end 3-D system, with the ability to do sculpted surfaces, for the small proportion of designs that required it. The savings over purchasing several high-end compatible systems and making all designers use the more cumbersome interface was considerable.

Toyota is envied for the speed and quality of its design process. It is generally regarded as having an impressively well-integrated CAD/CAM system that it developed in-house. This is true for body design.[14] Toyota focused on body design because it is a bottleneck that increases product lead time, and the company realized great benefits from integrating styling, structural design, and CAD/CAM, all driven from a central database. However, the drive train required a lower level of integration and could be developed after body design. Toyota decided to purchase an off-the-shelf Computervision system that had strong 3-D capabilities appropriate for the task but that was incompatible with its own CAD/CAM system.[15] It knew that more integration is not necessarily better. Thus the level of technical integration should fit the design task.

Match the Level of Organizational Integration to the Design Task

Some of CAD's most powerful productivity and quality-enhancing features require integration across internal and external organization boundaries. For example, CAD/CAM will only be effective if design and manufacturing coordinate and communicate closely. This may be difficult to achieve if an outside supplier does the manufacturing. One approach is to coordinate CAD selection and purchase with your suppliers and customers and build close, long-term ties to a small set of dedicated suppliers.

We saw several U.S. companies that had only developed successful CAD/CAM links with a tiny fraction of their suppliers. In most of these cases they could not even transfer electronic data files for drafting; their systems were not compatible. In Japan, large companies cultivate relationships with dedicated suppliers, thus it is common for suppliers to have the same CAD system as their major customer. We recently visited Hiroshima Press, which does most of its business with Mazda and has the same CAD system. Moreover, several of its designers work at Mazda's design center, also in Hiroshima, and actually use Mazda's CAD system, directly accessing Mazda's engineering database. Toyota developed its own workstations with proprietary CAD/CAM software to sell to its suppliers and has now commercialized its CAELUM system. The workstations are compatible with its mainframe-based system but are also available to suppliers at low cost.

A variety of linking mechanisms can be used to integrate organizational units. These vary in obtrusiveness, as shown in Figure 3. By "obtrusiveness" we mean the degree of change required for a fragmented design organization to implement the mechanism. At the lowest level are design rules. In this case the organization remains highly specialized, and each function has a set of rules to guide its decisions. For example, the Boothroyd/Dewhurst Design for Assembly (DFA) system enables product designers to anticipate manufacturing costs associated with design decisions without ever talking to manufacturing engineers.

At the next level, ad hoc communications and cross-functional teams do not require fundamental change in organizational structure. They are simply communication mechanisms. With cross-functional teams, individuals remain in their functions but go to cross-functional meetings to make key design decisions. Job rotation is obtrusive to the individuals who change functions, but the basic organizational structure remains intact.

Figure 3 Matching Organizational Integration Mechanisms to the Design Task

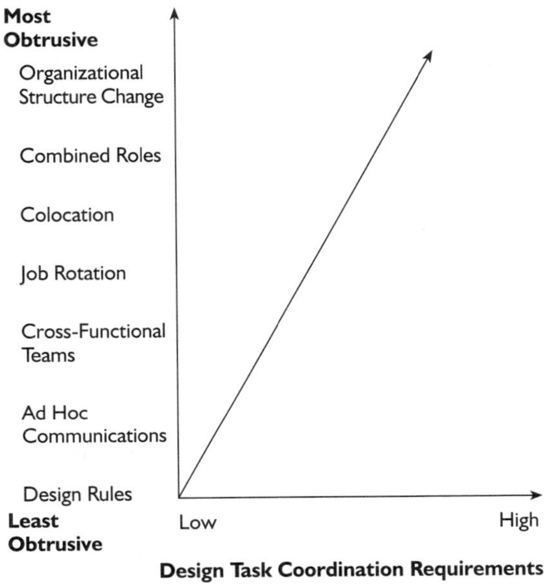

The hope is that people trained in design will bring their perspective to manufacturing and then will return to design with a manufacturing perspective. Several of the companies we studied rotated engineers fresh out of school to orient them to various design and manufacturing aspects. However, it is hard to imagine that they can develop any real depth in these short stays in different functions. Colocation (e.g., design and manufacturing engineers working in the same area) is a more powerful way to ensure ongoing, real-time communication about design decisions. It is being practiced by some companies that are aggressively pursuing concurrent engineering.[16]

The most obtrusive mechanisms require an actual change in job design and organizational structure. Several of the companies we visited planned on combining roles. One company intended to combine the roles of design engineer, engineering analyst, and designer into one person's job. This strategy reduces information processing requirements; Galbraith calls it the creation of self-contained tasks. Another division of this company planned to combine the roles of product design engineer and manufacturing engineer. At the highest level, organizations change their structure; they become matrix organizations or product organizations, and they get all the functions designing a single project to work as a team and report to a common boss, perhaps in the same location.

Companies should proceed cautiously into high levels of organizational integration. Companies that are apt to mandate one technology solution for all designs are also apt to mandate one organization solution for all designs. Figure 3 suggests that design tasks with high design task coordination requirements will benefit from higher levels of integration, whereas most routine tasks or those that are self-contained will not.

TOWARD REALIZING CAD'S PROMISE

As with many technologies, the learning curve for utilizing CAD's potential has been slow. U.S. companies, under competitive pressure, have been challenging the premises under which they initially invested in CAD. It was viewed as a solution before there was an understanding of the problem. What are we trying to accomplish in our design organization? Where does CAD fit in? How can we plan and implement an integrated social and technical system to achieve our goals?

Perhaps we must experience disappointments before we can challenge basic assumptions about implementing technology. Perhaps engineers are naturally enamored by technology and cannot see the big picture of how CAD fits into their design processes. One thing is clear—the companies in our study gave little consideration to organizational and human resource issues before implementing CAD. They took the path of least resistance by replacing manual drawing boards with electronic ones. In the meantime, more advanced CAD capabilities have been greatly underutilized.

We recommend that companies examine their design tasks and select appropriate CAD system features and the appropriate level of social integration. This will not necessarily lead to complex organizational changes and major investments in high-level technologies. In fact, you may discover that relatively inexpensive 2-D systems are perfectly adequate. Opportunities for organizational improvement may reduce the need for technological fixes. The important point is that managers must understand the process they are attempting to infuse with new technology, understand why they are investing in the technology, plan for appropriate organizational changes, and start with realistic expectations about what the new CAD systems can contribute.

NOTES

1. See for example J. Krouse, R. Mills, B. Beckert, and C. Potter, "CAD/CAM Planning Guide." *Industry Week* 17 (July 1989).
2. Paul Adler reports that CAD/CAM was greatly underutilized by companies that are supposedly leading-edge technology users. He suggests that major organizational changes are needed to utilize this technology. See: P. S. Adler, "CAD/CAM: Managerial Challenges and Research Issues." *IEEE Transactions on Engineering Management* 3 (1989): 202–216. For a good discussion of the barriers to utilizing CAD/CAM, see: C. A. Beatty and J. R. M. Gordon, "Barriers to the Implementation of CAD/CAM Systems." *Sloan Management Review* (Summer 1988): 25–33.
3. A. Majchrzak, T. C. Chang, W. Barfield, R. Eberts, and G. Salvendy, *Human Aspects of Computer-Aided Design*. London: Taylor & Francis, 1987.
4. S. Zuboff, *In the Age of the Smart Machine: The Future of Work and Power*. New York: Basic Books, 1984.
5. A. Majchrzak and H. Salzman, eds., *IEEE Transactions on Engineering Management* 36 (1989).
6. Adler (1989); Beatty and Gordon (1988); and J. Forslin, B. M. Thulestedt, and S. Andersson, "Computer-Aided Design: A Case of Strategy in Implementing a New Technology," *IEEE Transactions on Engineering Management* 36 (1989): 191–201.
7. Forslin et al. (1989): 201.
8. M. Fleischer, T. Phelps, D. Arnsdorf, and M. Ensing, *CAD/CAM Data Problems and Costs in the Tool and Die Industry: Final Report on a Study Conducted for the Michigan Modernization Service*. Ann Arbor: Industrial Technology Institute, 1991.
9. L. Tornatzky and M. Fleischer, *The Process of Technological Innovation*. Lexington, Massachusetts: Lexington Books, 1990.
10. This formulation is based on Leavitt's systems model of organizational effectiveness. See: H. J. Leavitt, "Applied Organizational Change in Industry: Structural, Technological, and Humanistic Approaches." In *Handbook of Organizations*, ed. J. G. March. Chicago: Rand McNally, 1965.
11. J. R. Galbraith, "Organization Design: An Information Processing View." *Interfaces* 4 (1974): 28–36.
12. D. M. Herold, "The Effectiveness of Work Groups." In *Organizational Behavior*, edited by S. Kerr. Columbus, Ohio: Grid, 1979.
13. D. Robertson, "CAD Systems and Communications in the Design Engineering Process: An Information Processing View." Ph.D. diss., MIT, Cambridge, MA, August 1990.
14. M. Ohara, "CAD/CAM at Toyota Motor Company." In *Computer Science and Technologies*, edited by T. Kitagawa. Amsterdam: North Holland Publishing, 1988.
15. H. Takarada, T. Yokoi, and T. Goto, "Application of Computer Graphics to Drive Train Engineering at Toyota Motor," International Congress and Exposition, Society of Automotive Engineers, SAE Technical Paper Series, 1986.
16. J. K. Liker and M. Fleischer, "Organizational Context Barriers to DFM." In *Managing Design for Manufacturability*, edited by G. I. Susman. Oxford: Oxford University Press, 1992.

QUESTIONS

1. What are the potential benefits of CAD?
2. What are the barriers to achieving the benefits of CAD?
3. How can CAD systems be introduced and used for maximum productivity gains?
4. Is it possible for an engineering design firm to compete without CAD? Explain your answer.

4.3 Flexible Manufacturing Systems: Implementing an Automated Factory

MOHSEN ATTARAN

Flexibility is the ability to produce many products quickly using the same equipment. Its main advantages are that it makes it possible to provide customized products, reduce setup time (the bugaboo of manufacturing) and respond more quickly to market demand changes.

The systems that perform flexible manufacturing need not be at the technological forefront. For example, creating clusters of "cells" of machines that are not fully automated to perform specific tasks has proven advantageous. Flexibility is gained by the computer performing three functions:

1. Running supervisory programs that provide an overview and coordinate the activities of all or part of a plant. These programs are known as "cell control systems" in discrete manufacturing where parts are assembled and "supervisory control systems" in process manufacturing where ingredients are blended. The outputs of these programs go to both plant managers and to the human or machine controllers.
2. Translating the outputs of the supervisory programs into instructions that operate machines.
3. Creating a factory floor network that transmits the supervisory instructions to the machines in real time.

Flexible manufacturing systems offer high payback, but require large initial investments. Total costs can range from $10,000 to several million dollars, depending on the complexity of the process and the number of machines involved.

Attaran presents a series of examples of flexible manufacturing in both the United States and Japan. He also discusses the critical success factors for flexible manufacturing, and the technical planning required. Planning is particularly important to ensure successful implementation.

Source: Information Systems Management (Spring 1992): 44–47.

FLEXIBLE manufacturing technology enables production systems that are capable of creating a wide product variety at prices that approach those of mass production. American manufacturers that employ flexible manufacturing systems report that machine setup times are getting shorter, responses to changes in demand are becoming faster, inventories are dropping, and machine use is increasing. Quality, consistency, and performance are improving as well.

Significantly improved flexibility in the factory has many strategic advantages including the production of low volumes at competitive costs. Within their envelope of capability, FMSs can produce varied products with almost the same speed and efficiency as in the manufacture of identical products. This enhanced flexibility permits manufacturers to make a far broader product line, simplifies style changes, and may stimulate new product design. Faster throughput from the better use of all machines, shorter machine setups, and automatic material handling all allow FMS users to respond rapidly to changes in market demand. In addition, an FMS enables manufacturers to proliferate product designs to enter into market segments that are not economically feasible for competitors using traditional technologies.

This article discusses the economic benefits that a manufacturer can expect from a flexible manufacturing system, the prerequisites for successful implementation of such a system, and the FMS applications of some American companies that are capitalizing on the advantages of flexible manufacturing technology.

APPLICATIONS IN THE U.S.

Although pioneered in the United States, flexible manufacturing systems were not implemented in American companies until fairly recently. By 1986, American manufacturers, following the influence of the Japanese, had installed more than 50 FMSs at a cost of about $200 million. The following sections describe some of the American manufacturers that have capitalized on the advantages of flexible manufacturing to achieve required standards of flexibility, quality, cost, and delivery.

Deere & Co.

At its $1.5 billion complex in Waterloo, Iowa, Deere's FMS is capable of producing 5,000 variations on 10 basic tractors. Customers can select the parts they want on their tractors and the FMS produces them to order. For almost a decade, Deere has been hailed as a pioneer of this manufacturing revolution. Its flexible manufacturing system has realized all of the cost reductions and productivity improvements that the company had aimed for since its decision to implement the system.

Allen-Bradley Co.

At its 45,000-square-foot facility in Milwaukee, Allen-Bradley is using an FMS to build electric motor starters and contact devices for the world market. The company implemented the new technology when it switched its product line from electromechanical to solid-state technology in 1979. The FMS has reduced direct labor costs to 1.5% of total sales dollars, down from 15% 10 years earlier.

General Electric Co.

GE has installed FMS technology in several of its plants. At GE's Erie, Pennsylvania, locomotive plant, an FMS has reduced the time required to make locomotive motor frames from 16 days to just 15 hours. In GE's Somersworth, New Hampshire, plant, an FMS can be programmed to make up to 2,000 variations on 40 basic models of electric meters.

Cincinnati Milacron, Inc.

In 1982, Cincinnati Milacron decided to install an FMS in its plastics machinery division (PMD), which produces plastic injection-molding equipment. PMD is recognized as one of the most important segments of the corporation, yet its lead time for producing the injection-molding machines was approximately twice the established norm. By 1985, however, the flexible manufacturing system had resulted in PMD sales that constituted 26% of the corporation's total, up from 15% in a three-year period.

Increased machine use, inventory reductions, shorter lead time, and shorter machine setup time have resulted in higher profitability. Cincinnati Milacron is saving 25% in both direct and indirect labor costs. The biggest contribution of the FMS has been as a catalyst for change within the entire company. The installation of the FMS not only forced better management practices, but increased the visibility of those areas that needed the attention of management.

Vought Aero Products Co.

A superb example of how an FMS can outperform conventional systems is provided by Vought Aero Products Co., a subcontractor to Rockwell International Corp. The plant's 19 employees including maintenance staff are running a $10.1 million, eight-machine FMS through three shifts to build 568 parts for the B-1B bomber. This is compared with the estimated 72 workers and 24 machines that would be required to produce the same output using a conventional system. The plant is operating in 40,000 square feet of floor space, in contrast with the 120,000 square feet that a conventional system would need. The implementation of the FMS has allowed the company to go from being a subcontractor on government projects to being nearly a prime contractor. Vought's strategy is to have the manufacturing technology in place that will allow the company to bid on any piece of subcontract work and win, no matter what the work requires. Vought's management views flexible automation as a strategic weapon for enhancing the company's competitive position.

INTELLIGENT BODY ASSEMBLY: JAPANESE AUTOMAKERS

Japanese auto manufacturers are increasing the flexibility of production lines to handle a mix of models and options and to permit easy changeover. Efficiency is achieved through variety rather than repetition. This enhanced flexibility lets the car manufacturer create a broader product line; it also stimulates new product design, accommodates style change, and makes efficient processing on small production runs feasible.

Nissan Motor Co.'s Zama No. 2 assembly line is one of the most highly automated in Japan, with 97% of the 3,000 to 4,000 welds per car done by robot. Thirty-five robot arms hold the main parts of a Nissan Sentra body in virtually perfect alignment, and 16 others weld the spots together in 62 places. Within 45 seconds, the four-door sedan body moves along the assembly line and a computer automatically readjusts the bank of robots to assemble the next item on the line, a hatchback. After the computer does its work again, the robots assemble the body of a station wagon. Down the line, each body will be painted a different color and receive different parts, all determined by computer command.

The new intelligent body assembly at Nissan's Zama plant is the first in the world to use computer-programmable robots instead of fixed jigs for aligning a car's main body parts, including floor, roof, and side panels. As a result, one assembly line can handle up to four models and eight body types at once. Lengthy and expensive retooling of jigs and production machines for model changes typically results in extensive downtimes for production lines.

With a flexible body-assembly system, many changes can be made by modifying the computer program. The system greatly reduces the time and cost of model changes—for example, the retooling costs for a new model are cut in half. The computer program can also be taken to another plant and production can begin in three months; previously it took 10 to 11 months to transfer production. The system now used in Nissan's plants at Zama and Tochigi will be introduced at two other Japanese plants and at Nissan's U.S. factory in Smyrna, Tennessee.

Other Japanese car manufacturers have also introduced a high degree of flexibility in their production facilities. Toyota makes Supra, Lexus, and Soarer models on the same line in its Tahara plant. Honda produces Accords and Integras on a single line at its Suzuka plant.

THE EVOLUTION OF FMS

Automation is not new in the manufacturing environment; numerical control machines have been around since the 1950s. A computer program guided the production of a tape containing the instructions for numerical control machines, which automatically cut, bore, milled, and performed other operations as instructed. More sophisticated technology developments have permitted the application of robotics in manufacturing. Robots are programmable, multifunctional machines capable of moving materials, often with many degrees of freedom, as well as performing repetitive tasks. Robots are flexible; they eliminate the need for operators, and they provide consistent quality.

Various types of materials-handling systems have been used in manufacturing since the 1920s. New developments in data processing technology have led to the application of automated, guided vehicles to large, complex facilities. These trackless, self-propelled vehicles are automatically routed and provide a flexible means of moving parts over random routes within a factory. Automatic storage and retrieval systems are used to locate and retrieve needed components. Automatic storage and retrieval systems can be integrated with the new automated factory without altering the basic mechanism by upgrading the FMS computer controller.

The introduction of robotics and computer-controlled materials-handling equipment into the manufacturing process has permitted the linkage of these standalone machines into a complex FMS. An FMS uses information technology to integrate computer-controlled machine tools (which consist of numerical control machines), robotics, materials-handling systems, and a system-level controller to form a highly automated and flexible system.

KEYS TO SUCCESSFUL IMPLEMENTATION

The central core of an FMS is the manufacturing information system that contains all product- and process-related data. This system captures and stores relevant information necessary for the production of different parts. The integration of the different components of an FMS, computer-aided design and manufacturing (CAM/CAM) systems, robotics, and materials-handling systems depends on a carefully designed system of information flow. A properly designed manufacturing information system provides the FMS with the necessary information structure for the efficient conduct of the manufacturing activities, including the development, evaluation, testing, and production of a variety of products.

The adoption of an FMS depends upon the initial implementation of a manufacturing information system. Active participation by the IS manager greatly facilitates the integration of such an information system.

In addition to reviewing current manufacturing information systems, IS management must work with senior management to investigate changes in plant layout, equipment requirements, accounting and costing functions, as well as workers' attitudes. The following sections list the prerequisites to the successful implementation of an FMS.

Review the Current Manufacturing Information System

It is the responsibility of the IS manager to facilitate this function. The IS manager needs to evaluate the manufacturing information system to identify and rank opportunities and to develop the specific objectives that guide the implementation of the FMS.

Evaluate the Company's Current Position

Senior management and IS management must determine and evaluate current practices in terms of the company's business goals and the competitive environment. A careful analysis of the needs and priorities of the company is part of this evaluation process.

Establish a Desired Outcome

The company must identify what it hopes to achieve from the implementation of an FMS. The benefits might range from increased productivity to improved quality of the workplace. The list of improvements expected from automation should be consistent with overall corporate goals and strategies.

Conduct a Cost-Benefit Analysis

New technology is highly capital intensive. Capital budgeting procedures should be enhanced to incorporate the intangible benefits generated from flexibility, improved efficiency, and higher productivity to justify the high capital investment in FMS technologies.

Formulate Long-Run Objectives

The implementation of new technologies is a long-term decision. Instant results cannot be expected. Long-term plans should provide room for technological change and expansion.

Encourage Worker Involvement

Senior and IS management must explain the reason for changes to those employees whose jobs will be affected. It is important that employees be informed and made a part of the planning process to improve their perception of the changes.

The Integration Challenge

Linking the different IT components of an FMS poses a serious technological problem. Although the computers, robots, and sophisticated machines exist, the necessary systems software and common data bases to control and link the equipment are not yet sufficiently developed.

Another serious problem is the lack of expert knowledge in the field—both managerial and technical. Senior management must ensure that its suppliers will provide extensive assistance with the design and modification of hardware and software. In addition, senior management should develop a comprehensible master plan and implement it one step at a time. Improvements should be made in steps, rather than in one quantum leap.

TECHNICAL PLANNING

The technical problems of implementing an FMS are often underestimated. Proper planning enables senior and IS management to take the steps necessary to minimize the chances of failure. To ensure a successful implementation of the FMS, IS managers can take the following actions:

- Obtain a clear understanding of the new technologies involved. One method is to talk to others who have implemented FMS technologies. IS managers can get accurate cost and operating information through suppliers, competitors, and professional organizations that will make it possible to determine how the new technologies can satisfy the existing needs of the organization.
- Form a multidisciplined project team to bring all the necessary technical skills and experience to the FMS project. Outside assistance may be needed to ensure objectivity.
- Select qualified vendors who will provide the essential hardware and software support.
- Formulate a training program to ascertain what level of training is required by employees who will be involved in the supervision and operation of the new technology.
- Monitor the initial performance of the FMS and maintain a basis for evaluating feedback.
- Assign adequate resources to support and maintain the technology after installation.

Because implementation of flexible manufacturing systems has not yet been reduced to a simple set of textbook rules and procedures, there are no detailed guidelines for manufacturers to follow. Yet meticulous planning enables a company's management to take the steps necessary to minimize the impact of possible pitfalls.

CONCLUSION

Increased national and international competition in manufacturing has stimulated the adoption of new technologies such as flexible manufacturing systems in the United States. Those companies that fail to focus on the demands and opportunities presented by technological improvement are risking their continued profitability and long-term survival. The many challenges facing American manufacturers today indicate that technology management is a critical strategic issue. Successful design and implementation of new technology requires the close integration and coordination of variables ranging from technical issues to human and managerial ones.

Senior management must recognize the internal and external benefits and costs of new technology in the automated factory. Management must assess the technology's compatibility with the company's goals, as well as determine the company's financial ability to support an FMS project. Automation must be viewed as a manufacturing project with long-term objectives that encompass all aspects of the production facility. The involvement of the IS department and other employees in the planning as well as the installation process is critical to reduce resistance and skepticism.

QUESTIONS

1. What are flexible manufacturing systems? How are they used?
2. How are flexible manufacturing systems making American manufacturing firms more productive?
3. What are some of the challenges in implementing flexible manufacturing systems?

4.4

Robotics Applications in Manufacturing

MOHSEN ATTARAN

The popular notion of robots as humanoids (as in classic films like *Star Wars*) is far off the mark in industry. Robots are used when tasks are:

- Highly repetitive
- Dangerous to health (e.g., spray painting)
- Manipulating large weights

In other words, when the job is not really suited to humans. Although robots require maintenance and tending, they do not get benefit packages or take sick leave.

The reality of robots is that they have been used more outside the United States than within. They require a high initial capital investment, which often rules them out when short payback periods are desired. Nonetheless, they are part of the automation that is coming to the factory.

Attaran's article classifies robots and shows when they are used. It also explores their economic benefits and the factors that point to their increasing utilization.

DURING the past 10 years, many U.S. manufacturers have accepted and implemented robotics technologies into their manufacturing processes. Although the United States has not yet met its projections for robotics installations (250,000 predicted by 1990), 25,000 robots were installed by 1987.[1]

The three basic characteristics of a robot are that it is reprogrammable and multifunctional and can manipulate objects for the performance of its tasks.

ROBOT COMPONENTS

The typical industrial robot currently in use has three basic components: a manipulator, a computer controller, and a power source.[2]

The manipulator is a mechanical system of grippers, robotic hands, or other adapted, special-purpose devices, which moves tools or material to complete the task.

The controller is the robot's brain. It contains the specific programming tasks and sequences

Source: Information Systems Management (Winter 1990): 14–21.

that direct the manipulator. It also provides for interface to sensors (e.g., switches or cameras) to determine the motion the robot performs.

The power source for the robot motion is hydraulic, electric, or pneumatic. A hydraulic power source, the oldest type of power for robots, is generally used for lifting heavy weights or for performing potentially harmful tasks (e.g., spray painting). Production of this type of robot is being phased out by the major robot manufacturers. Electric robots have a much higher degree of accuracy than hydraulic ones and are being used to carry light to medium payloads; lower maintenance costs are increasing their use. Pneumatic robots are very light duty and are often used for simple, pick-and-place tasks. Pneumatically powered robots are used in factory environments in which hydraulic or electric robots cannot be used for safety reasons. They also have an industrial advantage because these robots can be installed directly to a compressed-air supply usually available in factories, without a separate power source.

CLASSIFICATION OF INDUSTRIAL ROBOTS

One way of classifying industrial robots is based on their control system, which can be servo or nonservo. A servo-controlled robot uses software to control the manipulator's motion. An advantage of servo-controlled robots is that they can be programmed by walking them through a task to teach them its required sequence, location, and range of motion. A combination of offline programming and walkthrough programming provides greater flexibility and controllability of the motion sequence. Due to their complexity, these robots are more expensive than nonservo-controlled ones; however, their flexibility and versatility allow them to accomplish a greater variety of tasks.

A nonservo-controlled robot uses mechanical stop devices or limit switches to control its motion. These robots are ideal for such tasks as materials handling or point-to-point transfers, die casting, injection molding, and machine loading. They are simple to operate and inexpensive and offer high reliability.

There are three levels of sophistication and performance ability for industrial robots: simple, or low technology; medium technology; and sophisticated.[3]

- Simple, or low-technology, robots. The capabilities of these robots are limited to such rudimentary, repetitive operations as pick-and-place tasks. These robots are relatively inexpensive, easy to maintain, and highly reliable.
- Medium-technology robots. These robots have larger storage capacities and perform more complicated but still repetitive tasks. They are designed to be used in single-machine load or unload situations and are not capable of performing such continuous tasks as spray painting or welding.
- High-technology robots. These robots are very flexible and easy to program and can perform sophisticated jobs. They have the capability of storing and changing a large number of programs. They can be equipped to sense heat, objects, pressure, and electrical impulses. Equipped with sensory devices, they can monitor the work they are performing, learn and remember tasks, respond to the work environment, and operate other machines.

ROBOTICS APPLICATIONS IN MANUFACTURING

Robots are more advanced than traditional automated process systems because they can be designed to perform a variety of tasks and can outperform human workers in repetitive tasks. This nonstop ability is a robot's best qualification for displacing or assisting a human worker.

Faced with worldwide competition, many manufacturers are installing robots as part of a national drive toward reinventing factories. Purchases of factory automation systems increased from $7 billion in 1980 to $18 billion in 1985, an increase of more than 150%. The market is expected to double again by the early 1990s (about $40 billion), with industrial robots accounting for almost 9% of this total expenditure. The general price categories for industrial robots range from less than $20,000 for material handling robots to more than $100,000 for multifunctional ones.

There are several principal uses for industrial robots, as shown in Figure 1.[4] They are:

- Welding
- Finishing operations
- Assembly operations
- Materials handling

These and other applications are described in the following sections.

Welding

Welding was the number one application of industrial robots as of 1980. Both spot and arc welding are hazardous jobs, and workers dislike the protective equipment that must be worn. In addition, the consistent quality and product uniformity that these machines attain makes them popular for both types of welding. The majority of applications of spot welding are in such assembly-line industries as automobile manufacturing, appliance industries, and sheet-metal fabrication.

Figure 1 Robot Applications

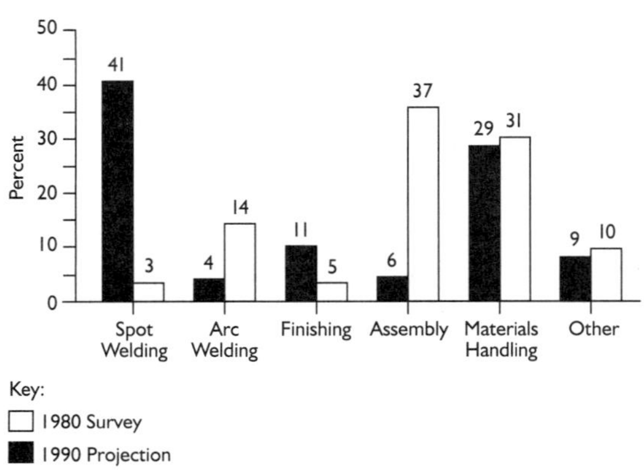

Source: Modern Materials Handling (July 1986): 9.

For example, robots perform more than 98% of the spot welding on Ford's Taurus and Sable cars. At Chrysler's Newark, Delaware, plant, 98% of the welding is done by robots.[5]

Arc-welding robots applications include the automobile industry, shipbuilding, and various assembly lines. At Litton Industrial Automation Systems, they are used to minimize the welding time by 50%.[6] These robots have become more affordable and are designed to be easily used and relatively trouble free. The growth of these robots in the United States is shown in Figure 2.

Finishing Operations

Spray painting is one of the largest finishing applications for industrial robots. Conformance with safety and health regulations and the high degree of consistency obtained makes spraying a natural application for industrial robots. Automo-

Figure 2 Arc-Welding Robots Sold in the US

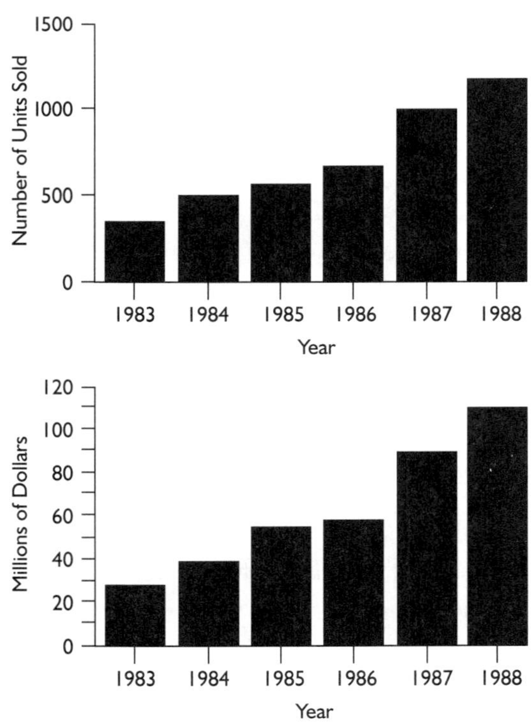

Source: T. Halligan, "Automated Welding Trickles Down to the Small Shop," *Production* (February 1986): 47–51.

bile and major appliance manufacturers are among the prime users. For example, major savings are realized at GM plants by using spray-painting robots. According to a GM vice-president, about 70% of the paint used in manual spraying operations is deposited elsewhere or sucked into giant exhaust systems that protect employees from fumes.[7] Major savings are also achieved in spray painting at General Electric's Hotpoint dishwasher factory in Milwaukee. At this factory, a single painting robot achieves improved coating uniformity and saves 20% of paint in one year.[8] Spray-painting robots are also becoming popular within durable goods industry for spraying stain, sealants, and rustproofing materials.

Assembly Operations

Speed, efficiency, and accuracy are the most important criteria for employing an assembly robot. Common assembly operations handled by robots include picking, placing, connecting, fastening, soldering, wiring, and part inserting. They are used in the automobile industry, and their use for electronic circuit board assembly has been growing rapidly. At IBM, robots are used to insert disk drives into microcomputers and snap keys onto electronic typewriter keyboards.[9] General Electric Co. is using robots to assemble the two major sheet-metal components of the Hotpoint refrigerator.

Materials Handling

Handling heavy loads can be dangerous, tiresome, and fatiguing work. Industrial robots are excellent for performing such arduous tasks of materials handling as machine loading and unloading and material transfer. The increasing popularity of flexible manufacturing systems is increasing the demand for materials-handling robots. Materials handling will account for almost one-third of the robot market within a few years.

These robots are used in every industry. For example, Reynolds Metal Co. uses a pallet-loading robot at its plastic plants for moving stacks of boxes to the automatic pallet wrapper to be wrapped in Reynolds plastic film. The use of robots has saved $1,360,000 per year in direct labor and further savings in packaging materials and has boosted employee morale in the company. The combination of robot pallet loading, pallet overwrapping, and carton forming resulted in increased plant productivity by 34%.[10]

Other Applications

Industrial robots have been introduced into the manufacturing environment in almost all operations, from machining to materials handling and storage. For example, robots are used in forming operations (e.g., founding and heat treatment). In foundry applications, robots are preparing molds, ladling materials into the molds, and removing the casting. In heat-treatment operations, robots place the material in the furnace, then remove the material and place it in a location for controlled cooling. Doehler Jarvis, a metal fabricator, uses robots to load and unload die casting machines, trim parts, and ladle molten metal.[11]

Robots are ideal for performing such machining operations as turning, drilling, milling, and shaping. They perform these operations by either moving the process tool around the work piece or by moving the work piece under the process tool. For example, at a General Dynamics plant in Fort Worth, it takes only three hours for an industrial robot to drill 550 holes in the vertical tail fins of an F-16 fighter. The company claims a fourfold increase in the output in drilling and routing of sheet metal used for wing panels.[12]

Inspection robots are used for measuring tolerances and detecting defects in assemblies. Packaging robots are widely used in the food industry. To achieve lower costs, improve quality, and gain greater sales and profits, Whitman's Chocolates Co. is investing $20 million to increase the use of robotics in its plants. It is considering simple pick-and-place industrial robots to automatically pack chocolates in paper cups in a multilayer, multiposition box. Robot installations help to maintain the consistency of the printed weight of the box and improve product quality and visual appearance.[13]

Bin picking is an increasingly popular field for robot use. This consists of selecting a single object from a bin of objects. Current industrial storage techniques place storable items in bins and the bins in an industrial rack assembly. Industrial robots move in and out of the rack assembly to remove and replace bins.

ECONOMIC BENEFITS AND JUSTIFICATION OF INDUSTRIAL ROBOTS

The use of industrial robots yields both direct and indirect benefits. Because one robot can usually perform the work of three to five humans, the cost of direct labor is reduced. Robots can work in a less comfortable environment than humans and provide savings in the costs associated with accommodating human workers (e.g., light, heat, and safety). The rising costs of human labor are making robots less expensive in comparison. More important, the unit price for robots continues to drop as a result of increases in their production rate and improvement in technology. Figure 3 compares the hourly costs of a robot with human worker hourly costs from 1960 to 1995. The continuing decline in robot unit costs in the face of ever-increasing labor wage rates for manual operations is the major reason market forecasters are predicting a significant increase in the purchase of robots not only by big manufacturers but by smaller manufacturers as well.

Other economic benefits of industrial robots include:

- Improved productivity. Robots can perform tasks under hazardous conditions more quickly than human workers. In some applications, such as arc welding, productivity can increase several hundred percent.
- Improved efficiency. The staff required for repetitive and monotonous tasks can be reduced. Robots perform a task precisely and repeatedly without fatigue or boredom. A robot is never sick or on strike, never demands higher wages or vacations, and never complains to the management about the work environment. For these reasons, industrial robots are becoming a more favorable choice for jobs involving exposure to radiation, dust, and other contamination.
- Improved product quality. The consistency of industrial robotics operation reduces defects and scrap materials.
- Improved quality of work. The human staff, no longer obligated to perform menial and hazardous tasks, can dedicate time to more important duties. Improving the efficiency and productivity of the staff creates room for better service for customers.
- Improved competitive position. Robots can help a company reinvent its factories and build a manufacturing system that concentrates on producing a high-quality product at a reasonable cost. After two decades of neglect, the factory is reemerging as the focal point of corporate strategy, and the automated factory is viewed as the key element in global competitiveness.

ROBOTICS IMPLEMENTATION

The problems of implementing industrial robots and other manufacturing automation projects are usually underestimated. Meticulous planning enables management to take the steps necessary to minimize the impact of possible pitfalls. To ensure successful implementation of robotics, the following process should be carefully considered.

Evaluate Current Processes

The current environment should be evaluated and the needs and priorities of the manufacturing firm should be carefully analyzed. Identifying these needs is essential in making a robotics application more successful.

Determine Long-Range Objectives

Management should perform an internal analysis of the organization in terms of its long-term

Figure 3 A Comparison of Robot and Human Cost

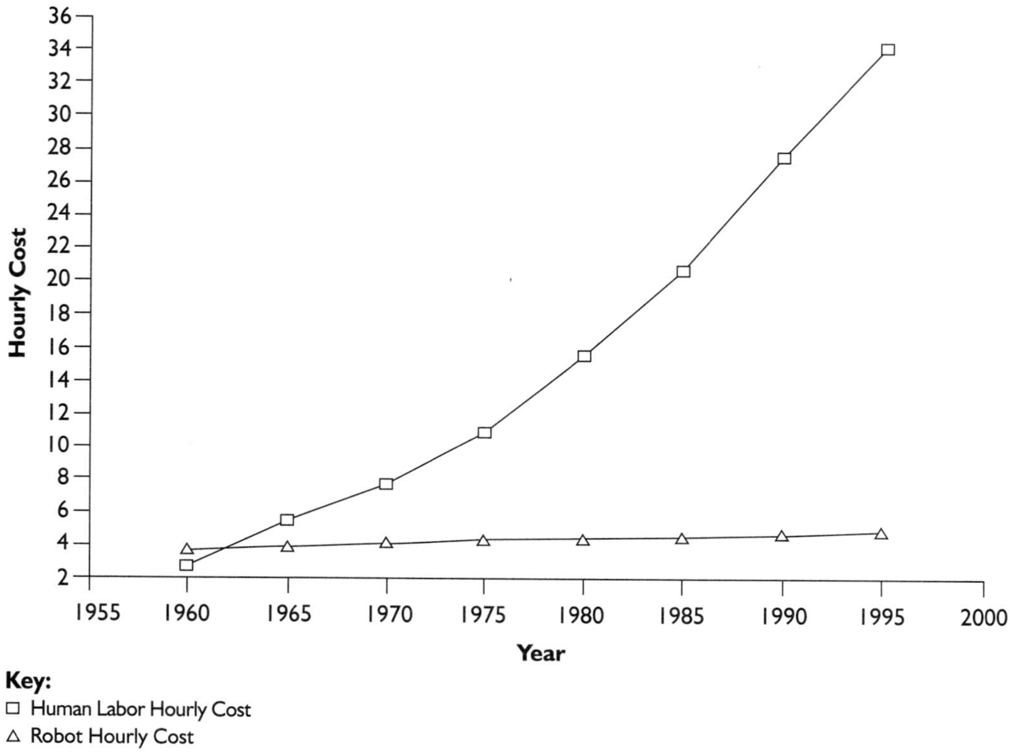

Source: J. Engleberger, *Robots and Automobiles: Applications, Economics, and the Future,* Technical Report 8000379 (Society of Automotive Engineers, February 1980).

goals and objectives. Robotics plans should take these goals into account. Strategic goals and business priorities must be identified before any effective robotics program can be designed.

Obtain Reliable Information

To ensure successful implementation, the organization should obtain accurate cost and operating information from suppliers, competitors, and professional organizations. Many manufacturing companies that are attempting to introduce robotics are having difficulties in quantifying costs and benefits, especially as they relate to the economics of integration and system support needs. The costs of hardware, software, programming, systems planning, maintenance, and labor are fairly easy to determine. Intangible costs and benefits (e.g., the cost of retraining personnel replaced by robots and the effects of robotics on productivity improvements and quality) are more difficult to identify.

Involve Appropriate Personnel

The right personnel should be involved in the planning as well as the installation process to reduce management resistance and work force skepticism. The use of multidisciplined project groups is needed to bring the necessary technical skills and experience to the project team to assess improvement opportunities, critical issues, and potential barriers. It is important that employees be a part of the planning process to improve their perception of the changes in the working environment caused by robotics implementations. Training must also be planned and provided before, during, and after the installation stage.

KEY FACTORS TO FUTURE SUCCESS OF INDUSTRIAL ROBOTS

Several factors are contributing to the success of robotics in manufacturing. The first is the robot itself, with future advances promising to deliver:

- Lower cost
- Smaller robots with increased ability
- Higher intelligence
- Improved sensory devices

The second factor is the robotics firms themselves. As the use of industrial robots becomes more feasible, producers must take advantage of improvements in robotics technology while anticipating future and meeting current demands.

The third factor pertains to the following environmental factors:

- Shortage of workers in certain areas.
- Government regulations regarding a safe working environment.
- Increased national and international competition, which requires a more efficient and more productive working environment.

Current trends in the progress of robotics indicate that the technology is moving ahead. First-generation robots were manipulator-based; currently, they are controller-based. Future robots will be AI- and sensor-based. Analog motors are being replaced by less costly and more reliable digital motors in robots. Control is also moving from analog to digital. The electronics is also getting more sophisticated, moving from hardwired to software. Moreover, robot control systems are becoming highly specialized computers that have real-time control capabilities and communicate with other elements in a manufacturing system.

Continued improvement in a robot's monitoring system is increasing productivity by tracking a robot's repeatability, trajectory, and accuracy. The monitoring system consists of several light-emitting diodes (LEDs) attached to selected points on the robot, cameras to detect light emitted from the LEDs, and a programmable vision system that accepts input from the cameras. The system inspects components and products for conformity to specifications and detects faulty materials early in the manufacturing cycle so productivity and quality are improved. According to some forecasts, more than 90% of all industries' inspections will soon be done by robot vision rather than by human vision.

The robots of the future will have smarter and more reliable end effectors—that is, the grippers or special tools at the working end. Smart grippers will be created by adding sensors. Common sensors include vision or proximity sensors to indicate location and orientation and quality transducers to measure excess force. Of all the sensors, vision is probably the most useful. A vision-based robot is more flexible, inspects much faster, and ensures a more reliable inspection system than humans, who can suffer from boredom and fatigue.

Robotics technology has started to overcome robots' limitations in sight. At its most sophisticated level, equipped with camera "eyes," a robot can capture an image, analyze its important features, interpret the results, and then decide what is the appropriate action to take. Robot vision must examine a range of visual characteristics, such as color, dimensions, and texture. So far, robot vision has been limited to performing best in highly structured environments, such as assembly lines, where a certain pattern can be recognized and the robot decides what to do based on the pattern. Robots' sight should improve further as more research is done.

Another area of future robotic development is sense of touch. Using a general-purpose "hand," a robot will be able to imitate a human hand. A recent development is an artificial skin consisting of piezoelectric firm, which transforms physical pressure into electrical impulses. Electrical signals activate a computer system programmed to interpret different types of touch.[14]

Future robots will be smarter and more mobile. They will be directed by a few simple words or even by a hand gesture.[15] In the future, robots will provide the needed labor for such potentially hazardous jobs as mining, firefighting, inspecting waste dump sites, cleaning up radioactive spills, and guarding warehouses. Other industries that potentially will use robots for hazardous duties include the textile, pharmaceutical, food processing, health care, and construction industries. It is anticipated that in space, robots will significantly speed up the assembly of such large structures as NASA's projected space station for the 1990s. Robots will also be used to staff permanent space stations.

First-generation robots were designed to do physical work. The robots of the future will have intelligence. This is made possible by incorporating AI programs (i.e., expert systems) with robots to develop robots that can do more advanced tasks. With the help of AI, robots will be able to plan sales programs, interpret documents, and diagnose mental and physical illnesses. Expert systems are already doing the work of maintenance planners, geologists, and mathematicians; it is only a matter of time before these functions will be integrated into robotics systems so that robots perform these tasks. Using AI technology, the new generation of robots will be able to cope with new situations in their environment, compare them to situations they have encountered in the past, generate a set of alternatives, and choose the best possible alternative.

With capabilities that approximate human sensing, feeling, seeing, and thinking, robots will be capable of doing many more things than they currently can do. With future robots becoming stronger, lighter, faster, more intelligent, and less expensive, there is a greatly increased opportunity to place these robots in a variety of manufacturing environments, from giant corporations to smaller manufacturers.

NOTES

1. G. Bock, "Limping Along in Robot Land: A Once Hopeful U.S. Industry Goes Awry" *Time* (July 13, 1987): 46–47.
2. D. Considine and G. Considine, *Standard Handbook of Industrial Automation* (New York: Chapman and Hall, 1986), 260–300.
3. R. Eckelmann, *A Study of the Competitive Prospects of the U.S. Robotics Industry* (Princeton N.J.: Petrocelli Books, 1984), 117.
4. "Material Handling to Spur Robot Growth," *Modern Materials Handling* (July 1986): 9.
5. G. Bock.
6. T. Zambetis-Jones, "Robots Reduced Welding Time 50%," *Appliance Manufacturer* (November 1987): 86.
7. "Agony Now May Mean a Brighter Tomorrow for U.S. Auto Makers," *Wall Street Journal* (February 3, 1981).
8. J. Engleberger, *Robots and Automobiles: Applications, Economics, and the Future*. Technical Report 800379 (Society of Automotive Engineers, February 1980).
9. G. Bock.
10. E. Brown, "Robotic Pallet Loading at Reynolds Save Money and Muscle," *Industrial Engineering* (February 1987): 24.
11. G. Bock.
12. G. Bock.
13. W. Kuzio, "Whitman's Samples Robotic Technology." *Candy Industry* (April 1985): 44–48.
14. J. Nash, "Ouch! Oooh Cut It Out!" *Time* (February 9, 1987): 78.
15. G. Bylinsky, "Invasion of the Service Robots," *Fortune* (September 14, 1987): 81.

QUESTIONS

1. What type of manufacturing applications can benefit most from robots? Explain why.
2. Suppose you are planning to introduce robotics technology in your manufacturing facility. How would you go about quantifying the expected benefits of your robots?
3. What are the major problems of implementing robots in a manufacturing organization? What steps can be taken to minimize these problems?
4. Suppose your firm decides to manufacture robots for the home. You have been appointed Director of Marketing. What salable applications would you recommend? Describe at least three applications and their potential market.

4.5 Effective Supply Chain Management

TOM DAVIS

Logistics is a term that comes from the military. It refers to obtaining, maintaining, and shipping goods so that they show up where they are needed at the right time. In business, one of the main aspects of logistics involves managing the supply chain so that goods reach the customer on time.

The supply chain extends from obtaining raw materials, through moving the resulting products through the factory and the warehouse to the retailer's shelf.

Davis describes the supply chain for a particular manufacturer, Hewlett-Packard (HP). The key problem faced by HP (and many others) is uncertainty. To combat uncertainty, inventory is increased, thereby increasing cost. Davis shows that by modeling the supply chain, including modeling uncertainty (statistics does have a use!), supply chain management can be made more efficient.

Supply chain management is a particularly good example of systems thinking. The supply chain is an integrated system and it cannot be made more efficient by optimizing it locally.

SUPPLY chain management is a frequently encountered phrase these days, as managers strive to improve factory performance. The trouble is that all too often the real meaning is lost. Instead, a casual observer might interpret the activities at the factory as evidence of an intensive effort to improve supplier management.

Good supplier management, while praiseworthy, does not constitute good supply chain management without a concurrent effort to manage the rest of the aspects of delivering products to customers. In this article, I will present a complete supply chain management methodology. This approach, developed at Hewlett-Packard, will enable a manufacturing operation to better manage its supply chain, ultimately improving customer satisfaction levels while reducing overall costs.

Hewlett-Packard has successfully used this methodology and is making efforts to implement the practice of good supply chain management to all its divisions. HP identified the need to improve its process for manufacturing and delivering products to customers as profit margins suffered pressure from increasing competition. Other factors have contributed to a renewed focus, namely:

- More instances of multisite manufacturing, where several independent entities are

Source: Sloan Management Review (Summer 1993): 35–46.

involved in the production and delivery process.
- Increasingly cut-throat marketing channels, such as independent computer dealers.
- The maturation of the world economy, with heightening demand for "local" products.
- Competitive pressures to provide exceptional customer service, including quick, reliable delivery.

HP's methodology has led to major changes in the way it does business. In the process, though, we have also discovered some appealing side effects. The proper execution of this methodology leads to improved teamwork and cooperation among employees, especially those normally separated by either business function or geography. We've also discovered that this methodology leads to greatly improved customer focus, in addition to better relationships with suppliers.

In the following pages, I will introduce HP's methodology. First, I will identify the problems inherent in supply chain management. Then, I will describe some key insights and briefly describe an analytical tool that we have developed to support a rational analysis of real supply chains. Finally, I will close with representative case studies describing HP successes with "real" supply chain management.

RESPONDING TO UNCERTAINTY

Large manufacturing companies are "hostage to complexity," according to Lew Platt, HP's president and CEO. The nature of the complexity is evident in a review of material flows for a complicated product. Multiple suppliers ship to manufacturing sites with varying regularity. There, subassemblies and final products are made by complicated and somewhat uncertain processes. Products are then shipped to direct customers, OEM customers, and "internal OEMs" (downstream manufacturing divisions of the same company treating the first factory's "final" product as but one of their many component materials). The scene is further confused by the wealth of transportation options available: planes, trains, trucks, and ships. And, of course, multiple carriers convey products to customers spread across the globe. The permutations (most of which are actually used) defy proper management.

The real problem with such a confusing network is the uncertainty that plagues it. This uncertainty—observed on a daily basis as late deliveries, machine breakdowns, order cancellations, and the like—leads to increased inventories. In fact, inventory exists more or less as simple insurance against uncertainty. In the case of a single factory, this uncertainty, while bothersome and costly, is relatively easy to overcome with properly sized inventories of raw materials, work in process (WIP), and finished goods (FGI). Using simple statistical techniques, it's straightforward to figure how much to hold so that, say 95% of the time a customer will find in stock the product he or she wants, despite the uncertainties in the factory process.

However, the problem is much more complicated when one considers the whole network. Uncertainty propagates through a manufacturing network. Consider the following case: A supplier of raw silicon to an integrated circuit (IC) manufacturer delivers a bad lot. Without properly sized safety stocks to buffer the production line against such an event, the IC manufacturer in turn delays its deliveries to one of its customers, a computer manufacturer. The shortage of parts shuts down the computer line, and shipments to computer dealers across the country are late. A customer walking into his neighborhood computer store discovers the machine he wants is out of stock. However, the opportunistic salesperson sells him a competing machine available then and there. For want of an on-time silicon delivery (perhaps many months earlier), a sale is lost.

Now that's an extreme example. Nearly every manufacturer carries some inventory to protect against such situations.[1] The real difficulty is knowing how much to hold and where to hold it, for the propagation of intrinsic uncertainties is mathematically complicated. Because previously there was no clear, analytical way to calculate a proper inventory level (and because there was no clear insight into other processes in the supply chain), firms have traditionally relied on a combination of experience and intuition. This has been their only recourse against the vagaries of their supply chains.

In the face of increasing pressure to reduce inventories and simultaneously improve customer service, the different organizations in a supply chain find themselves working at cross-purposes. This is true when the organizations exist side by side within the same operation as well as when they are physically remote, independent companies.

What often happens is this: One organization will reduce its inventory to reduce cost. Typically such reductions hurt service. To build up service again, heightened pressure is put on suppliers (which might be the line running just across the aisle) to improve *their* performance. If this can be achieved only by increasing the supplier's own inventory, the downstream operation's inventory reductions might be completely canceled out. In many instances, a "broken" supply chain will have substantial stock held at one point to enable another node in the supply chain to skate by with minimal stock. Overall, there is too much stock held in the supply chain system. Reapportioning the stock would reduce the overall inventory investment and improve end-customer service levels. Unfortunately, most organizations are designed to create winners and losers; working together for *system* optimization receives little more than lip service.[2]

Only analytical techniques—or darned good luck—can precisely tune a supply chain. In our experience at HP, even the best-run supply chains could reduce their overall inventory investment by 25% by more cleverly distributing stock within their systems. More typically, 50% reductions are possible.[3] With the proper tools, understanding and improvement will come from the following three steps.

- Benchmarking current performance. It is essential that the managers of a particular inventory network how just what performance is possible, given the existing set of operating characteristics (order review periods, forecast accuracy, etc.). Many of the potential savings come simply from the "tuning" process of adjusting an existing system for better performance. Possible metrics include inventory investment and order fill rate, among others.
- Controlling uncertainty. A company can make great strides by understanding the relative impact of different sources of uncertainty in the system and by then working to reduce (or avoid) the impact they have. As in the computer example, it is important to measure the indirect effect of uncertainty on downstream or upstream nodes in the supply chain. Most managers contending with supply chain issues do not adequately address this.
- Planning changes. In addition to adjusting the parameters that describe an existing network, it is possible, with the right tools, to study in advance the benefits (or costs) of sweeping changes to the inventory network. New policies and practices might lead to great reductions in cost and/or improved performance. Without an adequate analysis tool, opportunities for change might be lost for want of a credible argument.

MODELING SUPPLY CHAINS

When we first started investigating the problem of modeling Hewlett-Packard's worldwide inventory networks, we found ourselves grappling with three fundamental questions. They stemmed from our interest in studying a difficult academic problem and from our desire to provide value to the company by providing appropriate solutions to real problems. First, could we develop a simple, generic framework to describe the supply chains we faced? Second, could we accurately model the propagation of uncertainty up and down the supply chain? Finally, could we create a modeling approach to support strategic decision makers and not simply provide yet another tool for solving day-to-day operational problems?

As our work progressed, we found that the answer was yes to each of the questions. What started as questions evolved into the three key elements of our work. They represent our greatest insights to the problem of managing supply chains. We repeatedly use a single, generic model to capture the salient features of each node in a supply chain. We carefully account for the upstream and downstream effect of uncertainty introduced at any node. And we answer questions posed by strategists, not tacticians.

Model Simply

From an analytical point of view, a supply chain is simply a network of material processing cells with the following characteristics: supply, transformation, and demand (see Figure 1). This model applies at many levels. One can view a factory in this way just as easily as one can view an individual process step within that factory. In both cases, the central manufacturing process—whether it's called "automobile assembly" or "spot-welding station 109a"—requires raw materials from some supplier external to the process. Those materials are then transformed in some manner that adds value, creating a stock of finished goods (though a JIT process might have an empty stockpile). Finally, there is demand from external customers for the goods produced. Because of the strategic nature of our work, we avoid diving too deeply into the internal workings of an individual factory.

Figure 1 A Typical Supply Chain

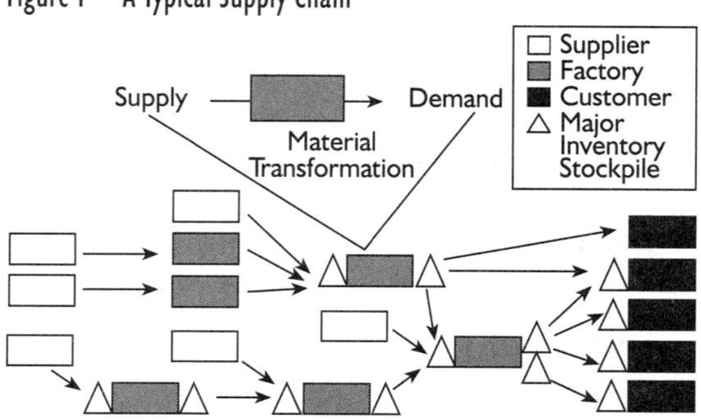

An important aspect of our approach is that we model material handling and distribution functions in the same way as more traditional "manufacturing" processes. They follow the simple rule we apply: materials arrive, value is added, and products exit. In a warehouse, goods arrive in a truck or on a lift. The distribution team adds value by sorting the deliveries, breaking down pallets as needed, and racking the materials. The distribution operation then fills orders for those finished goods. That spot-welding station might even be one of the customers.

Include Uncertainty

The reason we keep inventory is simple. Inventory is insurance—protection against life in an uncertain world. To meet our objectives for customer service, we keep a little extra material around (in what we call "safety stocks") so that service won't be adversely affected when something in the upstream process goes wrong. Understanding the impact of variability in the system is the goal of modern inventory control theory.

In fact, these are three distinct sources of uncertainty that plague supply chains: suppliers, manufacturing, and customers. To understand fully the impact on customer service and to be able to improve performance, it is essential that each of these be measured and addressed. The systemwide impact of each source of uncertainty must also be understood. Figure 2 illustrates how the three sources of uncertainty in the system combine to affect deliveries to customers.[4]

The first source of variability that leads to holding safety stocks is supplier performance. This is where naive supply chain considerations start and end. Your supplier quotes a lead time, but—alas!—he has a hard time getting you his materials exactly when he says he will. A storm delayed a delivery. A machine broke down. One of his suppliers was late. Any number of things could account for the late delivery, but in the end, it remains that sometimes those shipments are late (and, of course, sometimes they're early).

Figure 2 Uncertainty in the Supply Chain

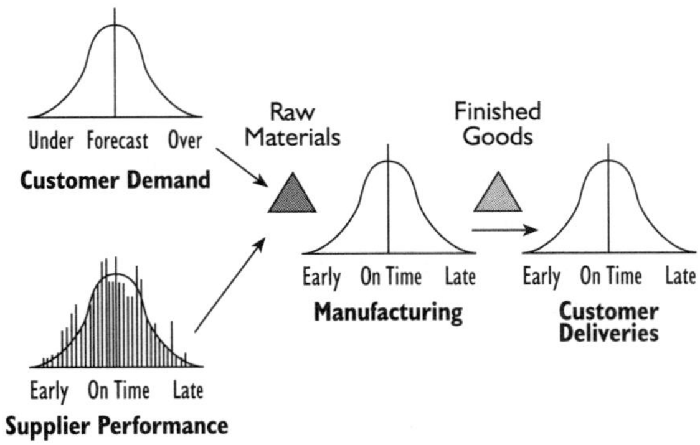

Over time, the vagaries of each supplier's delivery performance can be tracked. Each supplier's on-time performance, average lateness when late, and degree of inconsistency (as measured by the standard deviation of lateness) are the data to maintain. These data are immensely valuable, for the characteristics of each supplier tell you how much extra stock you must hold to keep your line up and running reliably. Supplier performance—and the factory's response, in the form of safety stocks—help determine downstream customer service.

The second source of variability comes in the manufacturing process itself. Here, a host of familiar problems can stop the flow of materials off the end of the line. Machines can break down (just as they can for your supplier), another project can tie up a key worker, or a computer foul-up can route materials to the wrong place.

In measuring process performance as it applies to overall supply chain performance, the key metrics are frequency of downtime (for the entire process, not just this or that machine), repair time, and the variation in the repair time. Note that again we're focused on a probability distribution of performance, for that is the nature of uncertainty. Sometimes the line runs for a week without shutting down. Other times, it seems we can't keep it running for more than an hour or two without a major shutdown. The reliability of the process—or its lack of reliability—is one of the three principal determinants of downstream customer service and the inventory investment to achieve that service.

Customer demand marks the third and final major source of uncertainty in the supply chain. Depending on the factory's location in the supply chain, this may reflect irregular purchases by a fickle public. Or it may stem from irregular orders from an industrial customer responding to its own up-and-down demand.[5] In any event, true build-to-order factories are relatively rare, and most factories keep some stock, filling orders from inventory as needed. The more variable the orders, the more stock required to reliably meet customer demand. Knowing average demand and the variability of that demand makes setting inventory goals more scientific and less seat-of-the-pants.

Unfortunately, week-to-week demand variability is often not the only problem. Because of long lead times, factories build stock in an attempt to meet *forecast* demand. Hewlett-Packard, for instance, often contracts for raw materials eight months or more before the final product will reach finished goods inventory. Thus, customer service in August depends on HP's ability to anticipate demand the preceding January. The effective variability in this case is greatly compounded from the actual variability we will experience come August.

We can measure historical order performance and attempt to use that information to predict future orders. Market research can help too. But such a strategy is often little more than a shot in the dark when introducing a new product. With ever-shortening product life cycles, our exposure to this final element of variability in the supply chain occurs more frequently.

Act Strategically

Customer service starts and ends with the customer. The customer feels the full effect of the complex, interacting uncertainties in the supply chain. In Figure 3, we show what we've coined as the "uncertainty cycle" to capture this effect. The forecast—based on historical customer behavior—starts things rolling. Materials are purchased

Figure 3 The Uncertainty Cycle

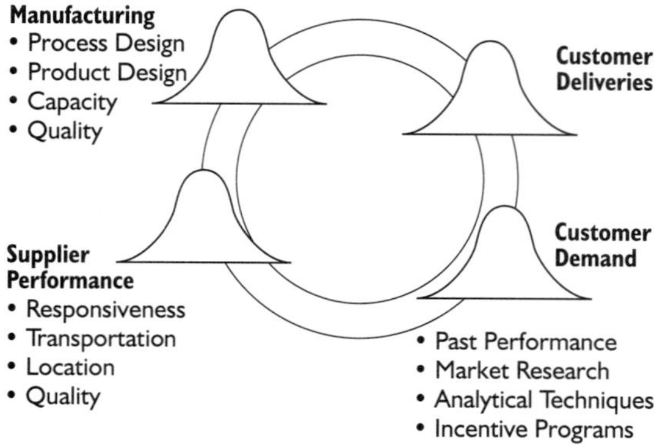

and delivered by suppliers. Then the factory takes its turn, producing the finished goods. Finally, the early forecasts are realized in the form of hard orders, and products are shipped to customers. This cycle repeats itself up and down the supply chain. It also repeats over time.

Strategic initiatives can lessen the impact of uncertainty. First, we can reduce the uncertainties using a standard total quality control approach. We can improve forecast accuracy through advanced analytical techniques.[6] We can investigate more reliable transportation modes. We can encourage suppliers to perform more reliably.[7] And we can change product designs to stabilize manufacturing processes. The costs associated with these changes can be balanced against the savings of reduced supply chain inventories or improved customer service.

Unfortunately, we cannot reduce all uncertainties. But other initiatives can redesign the supply chain to reduce the impact of the uncertainties. For example, opening a new factory can put production closer to key customers. I'll return to this point shortly with some specific examples from HP's experience.

A sound supply chain modeling methodology can address long-term problems. Focusing on the uncertainties in the system and the way they propagate up and down the chain can lead to new operating policies and objectives. Using tactical tools like transportation models, scheduling algorithms, or fast MRP programs can fine-tune the performance of the system.[8]

USING A STRATEGIC MODELING TOOL

At HP, we have developed a decision support system that allows us to model supply chains analytically. The tool captures the spirit of the model characteristics described above, though the mathematical details are not appropriate for this discussion.[9] As we explain to our internal HP customers, the tool is of little value unless one grasps the underlying methodology of supply chain modeling. Supply chain performance can be improved substantially without even switching on a computer. Nonetheless, there are a few points related to the use of the analytical tool that are important.

Valid Input

As noted earlier, there are many performance measures, like supplier lateness and the associated variability, required to adequately model a supply chain. Ideally, this data would be collected routinely. After all, these are fundamental measures. However, we have yet to work with an organization that has the data readily available.

While organizations typically imagine they can collect the necessary data within just a few weeks, it invariably takes six months or longer. Perhaps the single greatest reason for this is that businesses have not taken a careful look at uncertainty in their operations. This is a major failing on their part. Not only have they been ignoring one of the critical drivers of their performance, they force themselves to wait while collecting enough data points to ensure an accurate representation.

Another common problem with collecting the right data stems from the breadth of the supply chain itself. Because the typical supply chain comprises many organizations, data must come from many disparate locations. To ensure success of the modeling effort, it is necessary to obtain sufficient commitment of resources to ensure accurate, useful data about each of the links in the supply chain. Even within a single site, cross-functional cooperation is necessary.

Accuracy of the available data can present a problem too. Often the collected data paints a false image of an operation due to data entry errors, inconsistent collection procedures, and system incompatibilities. The breadth of the supply chain can compound the accuracy problem, for two sites often define their terms somewhat differently. A consistent, careful approach is needed in collecting data on the supply chain.

Finally, the collected data must be valid. Usually, some degree of abstraction is necessary. For example, it would be unwieldy to include every stock-keeping unit in a model that is used for strategic purposes only. Thus, some parts may be left out of the model completely (e.g., common nuts and bolts or passive electronic components). Others may be aggregated and represented in the model as a single, "generic" component, like "mechanisms." The summarized characteristics of the aggregated parts must be checked against expert opinion to see if they represent the situation fairly. We often use the "worst case" to get a conservative model.

Iterative Analysis

Usually, managers propose changes in operating policy and then carefully tally the costs of implementing the changes. However, even the best ideas often rot on the vine because of the difficulties inherent in quantifying the benefits associated with change. That is, typical decision criteria like net present value are not easily calculated; advocates are forced to rely on persuasive, qualitative arguments.

The tool we use at HP measures the benefits (or costs) of changes in supply chain policy. This fills a gaping hole in the strategic planning process. Specifically, our tool can tell the inventory requirements of a given supply chain, given the appropriate operating conditions and customer service objectives. Uncertainties throughout the supply chain are accounted for, and stockpiles are sized and valued accordingly.

We use our analysis tool in an iterative manner, as depicted in Figure 4. The process allows us to investigate numerous possible changes before the fact and explore thoroughly the strengths and weaknesses of each. The insight we gain from one analysis can lead to refinements in the next. In many ways, our analysis is akin to the use of CAD/CAM software in a product design lab. Simulation detects many flaws before the expense of a physical prototype.

Useful Output

It is important that any analysis tool provide information to facilitate decision making. That is, the product of the analysis must be *useful*. We look at a number of factors when reviewing the output of our supply chain models. However, we remain focused on balancing cost and customer service. Examples of output that we find most useful are shown in Figure 5.

We use two key measures of customer service. For off-the-shelf items, the line-item fill rate is usually appropriate. This measures the fraction of demands filled immediately. On average, the more inventory kept on hand, the higher the achieved fill rate. That explains the general shape of the curve in part A in Figure 5. Two different policies—two alternative means of shipping products from the factory to the distribution center, for example—can lead to different cost/service curves. Policy 2, in this example, provides equivalent service at a lower inventory cost than Policy 1. The question remains, of course, whether the cost of implementing Policy 2 exceeds the anticipated benefits.

The second measure we use is the so-called order aging curve in part B in Figure 5. This describes the more general situation by looking not only at off-the-shelf performance but also at the likely delay if delivery is not immediate. It might

Figure 4 Using Analytical Models Strategically

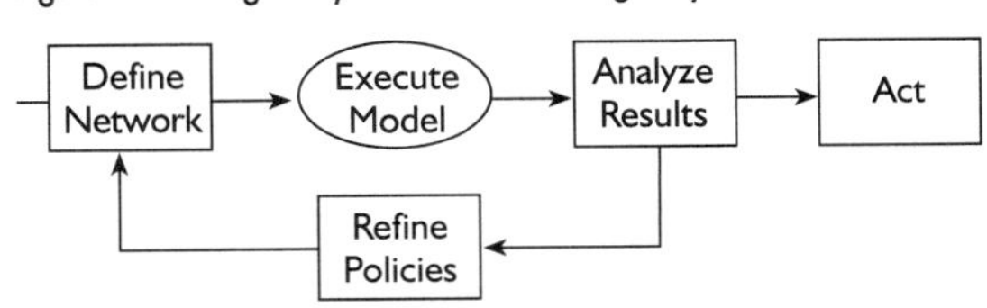

be helpful to think of a build-to-order environment. In the case illustrated here, Policy 1 has the higher proportion of on-time deliveries (0 days late), but late orders may linger on the books for weeks. Implementing Policy 2 would lead to slightly diminished off-the-shelf delivery, but the bulk of all orders should be filled promptly. Policy 2 inflicts less variability on the downstream customer. In real terms, Policy 1 might represent the case of a factory with quite a bit of FGI, but unreliable suppliers. Policy 2, in contrast, sees less reliance on FGI but more dependable suppliers or sufficient stocks of raw materials to compensate.

It is crucial to interpret properly analysis results like those shown in Figure 5. The mathematics of inventory management theory focus on long-term, average performance—yet an individual order can only be early, on time, or late. Just as factories experience variations in supplier performance or process uptime, there *will* be periods where customer service is high and other periods where it is low. The key is that the analytical curve represents an "efficient frontier"—the best possible performance given the operating characteristics. While it is possible to perform worse than the curves (i.e., results above the line-item fill-rate curve), better performance is impossible over an extended time. Thus, the curves presented as examples suggest average performance over the long haul, a goal to strive for. As they say, "your mileage may vary."

Despite a theoretical nature, output such as this is useful. Often performance is not adequately measured in these ways. Especially lacking is historical data on the variability of delivery performance. Simply discussing new ways of measuring performance—and the reasons for setting different performance targets—can lead to improvements in overall supply chain performance.

Benchmarking with data on historical performance is important to validate the results. If the model doesn't reflect reality, then it is dangerous to suggest its use to facilitate decision making. On the other hand, once validated, the model can be used with confidence on extensions from the base case. In the supply chain arena, we generally plot performance on a monthly basis. The goal is to match as closely as possible the predictions of the analytical model, as they do in Figure 6.

So how can there be points *below* the curve? Doesn't that curve define the minimum required inventory to meet customer performance goals? Yet Figure 6 shows that in October 1991, the service level was 98% with two and a half weeks of inventory. Theoretically it would take four weeks of inventory to achieve that level of performance; two and a half weeks buys you only 92%. The ex-

Figure 5 Useful Analysis Results

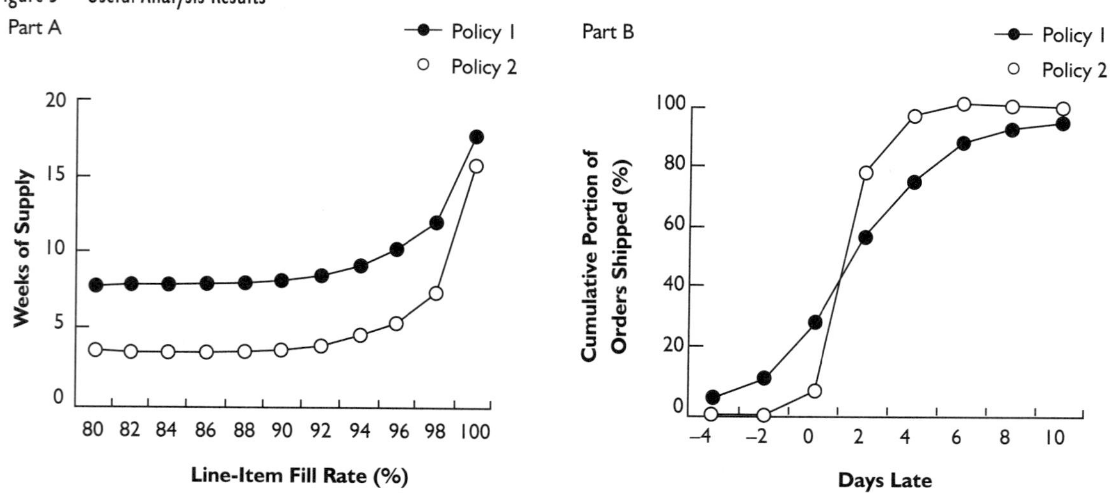

Figure 6 Benchmarking Historical Results

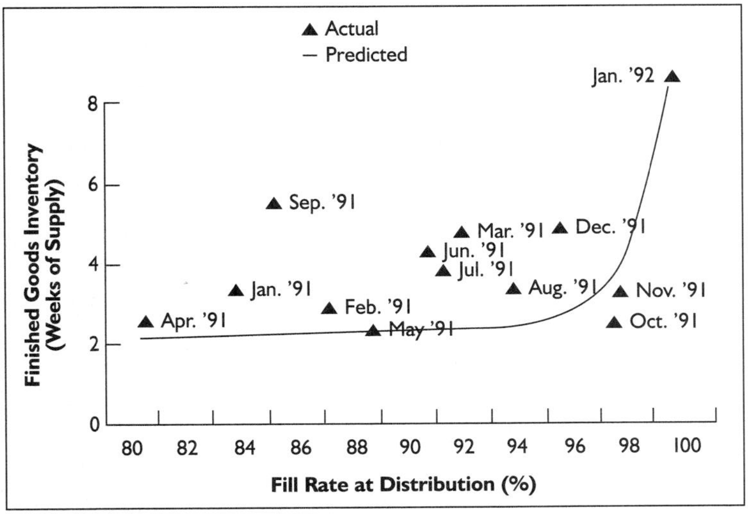

planation gets back to the matter of long-term averages. Generally, a site holds too much inventory, and just as we see in Figure 6, most of the actual results fall well above the theoretical curve. Operational difficulties of one sort or another explain why performance strays from the theoretical. Exceptionally high orders, for example, might lead to a stockout situation and poor customer service results. A slow order month, though, will lead to good results. The usual amount of inventory leads to higher than average fill rates simply because there were fewer customers than usual.

CASE STUDIES

I have described a framework for modeling supply chains in the preceding pages. But what good is it? Next I'll describe a few cases of the successful application of this modeling approach. The first is about the manufacture and distribution of one of Hewlett-Packard's most successful products, the DeskJet family of thermal ink-jet printers. The second case describes the situation HP faced when considering a new, cooperative manufacturing venture with an Asian partner. In the final case, I present some results that clearly indicate directions for future work in this area.

Localizing Generic Products

The supply chain for the DeskJet family of low-cost printers has five physically separate Hewlett-Packard facilities that contribute to the manufacture and distribution of the printers; the same shop assembles printed circuit boards and completes printer assemblies (see Figure 7). The pipeline itself is nearly six months long.

The product managers wanted to reduce their susceptibility to variability in the supply chain. Under then-current policies, they had to keep excessive amounts of inventory, and that tied up cash they might otherwise have applied to other projects. The division met its customer service goals, but the managers felt that there must be a way to lower costs without sacrificing customer goodwill.

Figure 7 The DeskJet Supply Chain

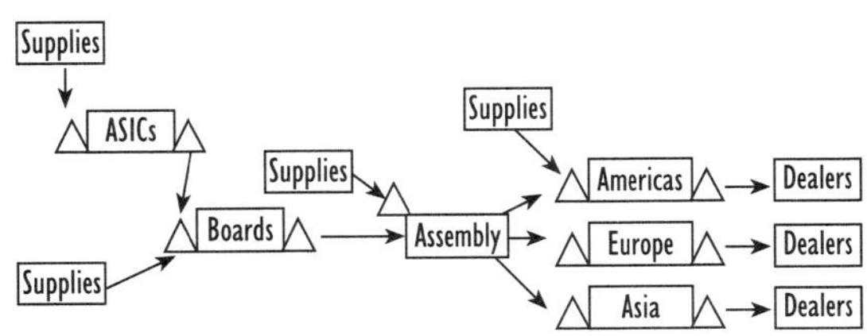

Under the existing policy, the factory "localized" products. That is, the central factory manufactured and packaged printers for each of the local markets served, adding a power cord and transformer to meet the plug and voltage needs of the country and a manual written in the proper language. The result was that the factory produced over a dozen variations of each seemingly identical printer. Each distribution center kept on hand a substantial stockpile of each printer/country option to fill the erratic pattern of orders received. This was especially troublesome in Europe, where there are more options produced for relatively small markets (e.g., Denmark).

The safety stocks required to meet high service targets were huge. Because the division chose to segregate the total market into its constituent components, it subjected itself to heightened variability—orders for Danish printers are much more unpredictable than the overall demand for printers in Europe. The question quickly became: What would be the value of risk pooling? Localization could be postponed to the distribution step; the factory could ship generic printers. The distributor would then combine the printers with the proper manuals and power cords only after receiving firm orders. This would greatly reduce the total safety stock (of generic printers).

We modeled the DeskJet supply chain to determine by how much switching to a generic printer would reduce the inventory investment. The results, shown in Figure 8, persuaded the distribution center to accept the plan. Under the old policy, and with a service goal of 98%, roughly seven weeks of FGI were required. Changing to a generic printer strategy would reduce the inventory requirement to about five weeks. The savings—worth over $30 million a year at current volumes—come from postponing the commitment of the generic printer to a specific model. Also contributing to the savings is the fact that stocks of generic printers are less valuable than a stock of an equal number of localized printers. That localization step adds value. Finally, the savings are not as great at lower service levels. High off-the-shelf service rates are relatively expensive.

Figure 8 Possible DeskJet Inventory Savings

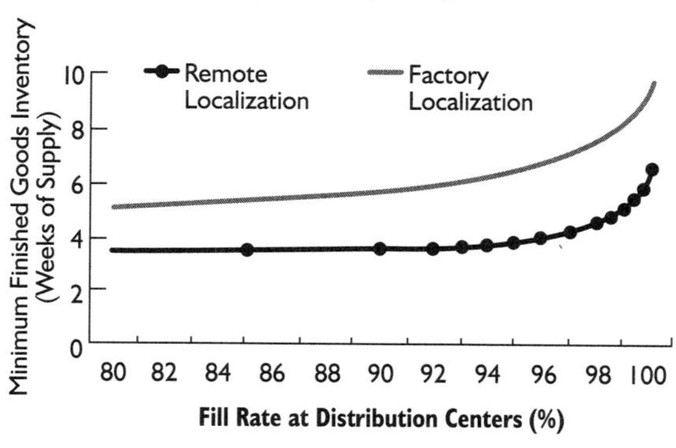

The savings from the reduced inventory (valued at a relatively high cost of capital because of the risk of write-offs due to short product life cycles) more than compensated for the cost of implementing the changes. Increased supplies of localization materials are necessary when postponing localization to distribution, but those costs are relatively small. The icing on the cake? Shipping costs dropped by several million dollars a year because shippers can pack generic printers more densely in bulk.[10]

Generic printers localized at distribution centers make sense in regions where there are many printer options. However, adding the manual and power cord is a slightly more expensive proposition at the distribution center than at the factory. This was a particular concern in the United States, where inventory savings for a generic printer are small, since virtually all U.S. volume is for the same model. An extension of the DeskJet study investigated the possibility of producing two printers at the factory—an ultra-low-cost unit destined exclusively for the U.S. market and a generic printer to serve the rest of the world.

Weighed against unit cost savings with this U.S.–only model was this supply chain consideration: with 100% generic printers, stockout conditions could be quickly remedied using a lateral resupply mechanism. If orders for a generic printer in Europe exceeded demand, the U.S. or Asia/Pacific distribution centers could send emergency shipments of the same printer in just a few days and restore high customer service quickly with little expense beyond air freight costs. With factory-localized printers, there would be expensive rework at the distribution centers too. The distributor would have to open a sealed box, change the cord and manual, and fill a new box. Labor and material costs would go up.

Under routine business conditions, this may not seem like a tremendous potential saving. However, consider the scenario of a new product introduction. Initial forecasts are for a lukewarm reception in Europe, while everyone expects the printer to be a big hit when introduced in the United States. We fill the pipeline accordingly. Then, against all odds, the forecasts prove to be exactly backward. Although having accurately projected worldwide demand, HP faces the unpleasant predicament of having orders it can't fill in Europe while the U.S. distribution center is stacked to the rafters with excess printers. With a generic printer strategy, those slow-moving printers could be immediately shipped to Europe. The alternative would be to wait and slowly let new production fill the pipeline to the European customers. With windows of opportunity for new products growing ever smaller, the ability to design the supply chain to protect against such unexpected circumstances is quite appealing.

As product life cycles shorten, Hewlett-Packard is facing more and more situations like this. The value of considering the problem in advance and devising a strategy using a supply chain methodology measures in the millions of dollars. In this case, generic printers reduce HP's exposure to forecast uncertainty. And when things do go wrong, they make recovery swifter and less costly, without sacrificing customer service.

Adding a Manufacturing Partner

In another Hewlett-Packard business, one of the factories had the opportunity to form a joint venture with a low-cost Asian manufacturing partner. This new partner would add an intermediate process step between two HP plants in lieu of a process handled at HP (see Figure 9). The perceived drawback was that shipping to and

Figure 9 Adding a Manufacturing Partner

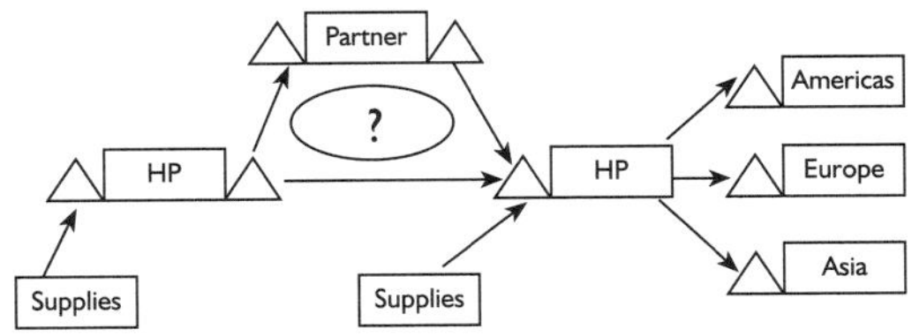

from the partner would take a month on a boat. Including the partner's processing time, the product lead time would increase by over three months!

The question was straightforward: How much additional stock would the distribution centers need if we added this intermediate step? At the time, the stock level was just over a month's supply of FGI in the U.S. distribution centers.

The result was staggering. With no other changes to the system, nearly five months worth of FGI would be necessary to maintain the same service level as before, far more than initially anticipated. The increase stemmed from the increased product lead time. Demand was anticipated to be quite variable. Introducing a three-month lag in the middle of the process greatly increased the product's exposure to changes in the marketplace.

These results opened our eyes to the value of rigorous supply chain analysis. While everyone involved knew that inventories would have to grow, we would have faced a terrible product shortage if we'd gone with our initial guesses. Moreover, the inventory situation forced us to take another look at our suitor's proposed financial arrangements.

We proposed several alternative supply chains. Since both firms had production facilities in Singapore, the most attractive called for colocating the two firms' factories. Doing all the work in the same place would virtually eliminate the onerous three-month delay, which was largely caused by shipping time. Our analysis gave us firm footing for negotiating a more attractive deal with our new partner. These results made the managers involved realize the importance of inventory in their decision-making process.

Interpreting Findings

Studying various Hewlett-Packard supply chains leads to some startling conclusions. As we gain experience treating the supply chains as systems, we build insight into their nature. However, the results shown in Figure 10 are downright unnerving.

These results come from a single HP division, but follow-on work shows that the figure represents the rest of the business as well. We have categorized the total supply chain inventory by its cause or source. Naturally, any supply chain has a certain amount of essential inventory. This stock satisfies the so-called conservation of mass laws. There is work in process in the factories and products in the pipeline from site to site. Also, periodic review systems lead naturally to a certain average stock on hand, even if there is no uncertainty in the system. At HP, these "natural" stocks make up about 40% of the total inventory investment.

The three sources of variability account for the balance of the stock. In HP factories and at the corporate level, we have exerted tremendous pressure on suppliers to improve their performance. Only a small percentage of the total stocks stems from late deliveries, etc. That is, if all deliveries arrived exactly when expected, HP's total inventory needs would drop just a few percentage points.

The same is true in manufacturing. HP processes are well-tuned. Moreover, the duration of process interruptions is negligibly short compared to the overall length of the supply chain. Thus, while a line may shut down fairly often for one reason or another, the global impact (from an inventory point of view) is not great.

The chief culprit is demand uncertainty. Over half of HP's inventories are protection against irregular orders. There are two partial explanations for this: First, demand is truly variable. No one can anticipate just when a big order might come in. Similarly, we have a difficult time anticipating just what price point will trigger a sales rush. This problem is partly natural, but improved forecasting ability would help alleviate the problem. If we knew better what total month-to-month demand would be (a forecasting problem), it would be much easier to meet the day-to-day changes in order volume (a true demand problem).

Second, we set high customer service goals. For example, for our computer peripheral products, we commonly shoot for an off-the-shelf line-item fill rate in the three sigma range. That is, we hope nearly all our customers will have their orders filled immediately. Much stock is necessary to protect against those rare instances when demand is very high. We don't always meet our goals, but operational problems or gross forecasting errors usually explain situations when we don't. New product introduction and the unanticipated popularity of some products has led to allocation of insufficient supplies.

Historically, we have spent large sums honing manufacturing processes and working with suppliers. However, in a supply chain perspective, the big reductions in inventory expense in the future will come from improving HP's ability to forecast demand and from designing products that allow better risk pooling. Continued efforts to improve product and process designs will reduce vulnerability to demand uncertainty, as with the DeskJet printer.

Figure 10 The Root Causes of Inventory

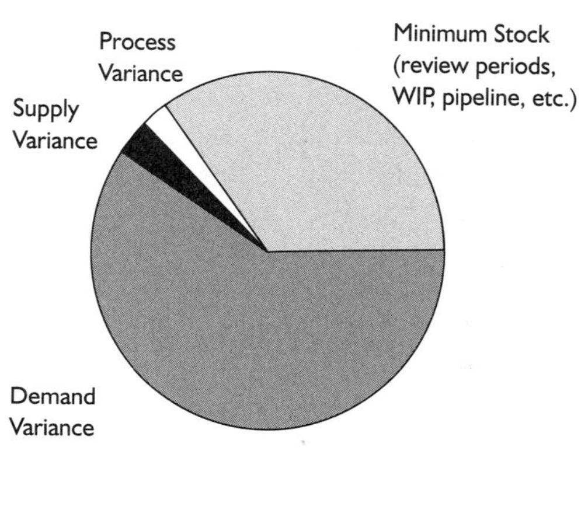

CONCLUSION

Sound data combined with the right quantitative techniques make thorough supply chain analysis possible. Without them, product managers must resort to qualitative approaches. Fortunately, there is a wealth of opportunity in most supply chains, even if using a simple, commonsense approach. Managers can implement a few obvious changes and best practices while collecting the first solid data on supply chain uncertainty. Then, once the data is available, they can fine-tune the system using a rigorous, analytical approach.

Table 1 outlines some of the ways to reduce or avoid uncertainty in the supply chain. Changes to product and process will stimulate reductions in the uncertainties plaguing each node in a supply chain, namely, supplier performance, manufacturing, and customer demand. These actions are not without costs.

The value of taking a systems view of the problem cannot be overstated. Organizational barriers introduce tremendous inefficiencies in supply chains. It is critical that all players in the business of getting products to customers consider their role in the objective of satisfying the customer. Strategic decisions on supply chain design can increase customer satisfaction and save money at the same time—the classic win-win situation.

Clearly, supply chain management should not be confused with supplier management. Supply chain management covers a far broader scope. Approaching problems with a systems view and a sound supply chain methodology can lead to great savings. At Hewlett-Packard, we have little doubt that hundreds of millions of dollars lie in the balance. Judging from our discussions with suppliers and customers, the same is true for other firms.

NOTES

1. Even JIT fanatics carry inventory of finished goods close to the customer and with suppliers. These stockpiles are small, though, be-

TABLE 1
Actions to Improve Supply Chain Performance

	Product	*Process*
Supplier Performance	Use common components and subassemblies in many products (to pool risk of stockouts). Follow industry standards (to increase part availability). Share information with strategic partners.	Reward good performance (based on *ship* date, not delivery date). Measure transportation performance separately. Subcontract inbound freight handling. Source locally (to shorten lead times). Review stocks more frequently.
Manufacturing	Lower tolerances. Pool engineering change orders. Use standard processes. Promote DFM, DFA, etc. Produce a generic product.	Remove bottlenecks. Size buffers appropriately. Reduce setups. Shorten cycle times. Introduce self-managed work teams. Install buffer capacity.
Customer Demand	Reduce product offerings and options. Design for localization. Customize products in software, not hardware. Manage delivery expectations (service requirements).	Adjust FGI safety stocks. Change transportation mode. Implement better data systems. Introduce improved forecasting techniques. Subcontract distribution operations. Build near customers.

cause the manufacturing process has been toned to reduce uncertainty in the first place and to recover quickly when something does happen. Paul Zipkin describes this as "pragmatic JIT." See: Zipkin, P. "Does Manufacturing Need a JIT Revolution?" *Harvard Business Review* (January–February 1991): 49–50.

2. Lee and Billington describe a number of ways in which a firm's supply chain can break down. This example corresponds to their Pitfall 11: Organizational Barriers. See: Lee, H. L. and Billington, C. "Managing Supply Chain Inventory: Pitfalls and Opportunities," *Sloan Management Review:* (Spring 1992): 65–73.
3. IBM reduced its U.S. spare parts inventory investment by half a billion dollars—a 20 to 25% reduction—by introducing an analytical tool to set stocking levels. See: Cohen, M. et al. "Optimizer: IBM's Multi-Echelon Inventory System for Managing Service Logistics." (*Interfaces*) 20 (1990): 65–82.
4. I depict uncertainty in the system with the familiar "bell" curve of the normal distribution. Raw data, shown in the underlying histogram for the suppliers in Figure 4.5–2, can be readily summarized in statistical form as the mean and standard deviation. These quantities determine the shape of the curve. While I've shown each distribution in the example to be the same shape, in practice some will be wider (more variation, or a higher standard deviation), and some will be narrower.
5. The vulnerability of upstream suppliers to variability in customer orders is wonderfully captured in the "beer game." This game casts players in the various roles of a beer distribution supply chain: retailer, distributor, wholesaler, and factory. The game is used at MIT to sensitize students to the importance of systems thinking. An ability to focus on system problems is critical to improving supply chain performance. See: Senge, P. *The Fifth Discipline.* New York: Doubleday/Currency, 1990.
6. Makridakis and Wheelwright provide a good introduction to the field of forecasting. See: Makridakis, S. and Wheelwright, S. "Forecasting: Framework and Overview." *TIMS Studies in the Management Sciences,* edited by S. Makridakis and S. Wheelwright, 1–15. Amsterdam: North Holland, 1979.
7. There should be more to negotiations with suppliers than price. While important, there are other measures of supplier performance that can have a greater impact on overall profitability than just a few pennies on the unit price. Hewlett-Packard's Corporate Procurement team posts, in public areas, the names of the company's best and worst suppliers as measured by delivery performance. Poor performers try to remove their names from that public record, while the top suppliers strive to maintain—or improve—their place on the list. For such a simple idea, this has had a remarkable impact.
8. Myriad examples of tactical planning tools appear in the literature. For example, see: Graves, S. "A Tactical Planning Model for a Job Shop," *Operations Research* (July–August 1986): 522–533.
9. Silver and Peterson describe the basics of modern inventory theory. See: Silver, E. and Peterson, R. *Decision Systems for Inventory Management and Production Planning.* New York: John Wiley, 1979. Nahmias provides another useful text. See: Nahmias, S. *Production and Operations Analysis.* Homewood, Illinois: Richard D. Irwin, 1989. Magee's timeless piece is a good introduction to the field. See: Magee, J. "Guides to Inventory Policy," *Harvard Business Review* (January–February 1956): 49–60.
10. The extension of the manufacturing process into the distribution center initially challenged managers accustomed to centralized control. They looked at the new approach with caution out of concern for issues like process control, quality, and operational efficiency. Because of different reporting lines for the factory and the distribution center,

high-level intervention was required to execute the change, which was widely acknowledged to benefit the company on the whole. Organizational obstacles such as this represent one of the major pitfalls of supply chain change management. Modeling, which facilitates data-driven decision making, is one of the tools that can be used to overcome these barriers.

QUESTIONS

1. How would you predict productivity gains from automated logistics?
2. What is the role of IT in the logistics operations?
3. Can manufacturing logistics techniques be adopted for retailing operations? Explain why or why not. Give examples.

4.6 The Role of IS in Manufacturing Automation

JAYANT V. SARAPH AND TOR GUIMARAES

Saraph and Guimaraes examine the computer needs of manufacturing from the information systems department's viewpoint. They argue that the changes in manufacturing change the way the information systems department allocates its resources and operates.

To make their point, the authors discuss the major components of computer integrated manufacturing (CIM). Their article ties together many of the techniques discussed earlier in this section (such as CAD and JIT) as well as others that have not been discussed explicitly. The value added by this article comes principally from the definition and description of the many systems that need to be integrated.

The large number of systems and their complexity imply that to implement CIM successfully, firms must apply the expertise and skills available in the information systems department.

COMPUTER technology is widely accepted as one of the major forces changing the way organizations operate. Among the many new developments and issues are the changing role of the information systems (IS) department and the chief information officer, the design and implementation of executive information systems, the deployment of technology to gain competitive advantage, and information resource management.

Although the many challenges that IS managers currently face have received much attention, the dramatic impact that changes in manufacturing automation are having on the IS function and its management have been largely ignored. This topic is important and should not be neglected by IS managers for several reasons.

First, the widespread use of computer technology throughout organizations has created islands of automation that must be integrated for these organizations to operate efficiently. Uncontrolled end-user computing can create significant long-term problems for many businesses. The manufacturing and other functional islands must communicate and share common resources.

Second, if computer technology is to increase the competitive advantage of manufacturing organizations, a smooth connection is necessary between the manufacturing and IS departments. Manufacturing automation systems and operations must be effectively integrated with their counterparts in other departments to obtain this strategic advantage.

Source: Information Systems Management (Winter 1991): 58–66.

Third, the IS staff in many organizations has supported the manufacturing function through development and maintenance of a variety of such manufacturing-related applications as order processing, inventory control, product demand forecasting, production scheduling, and manufacturing cost accounting. Quite often, this support has also included assistance in the selection, acquisition, and implementation of hardware and off-the-shelf manufacturing applications software.

In general, both manufacturing and IS have regarded this relatively limited cooperation to be critical to the success of the manufacturing organization. As organizations increase their use of computer technology, however, this relationship must be renewed and strengthened.

Fourth, manufacturing applications usually account for a large portion of a manufacturing organization's existing collection of applications. As a result, manufacturing often uses a large portion of the organization's IS resources. In the future, the changing manufacturing environment will require increased levels of automation and widespread use of computer technology to support multiproduct and multiplant environments and to produce information for various levels of management.

Therefore, the new manufacturing environment will require significant adjustments in IS's role and resource allocation and will increasingly demand a proactive IS view of manufacturing operations. This article describes the main technologies currently in use in manufacturing automation, discusses the likely impact of each on the IS department, and suggests ways IS managers can take effective actions in this area.

MANUFACTURING AUTOMATION ARCHITECTURES

Computer-integrated manufacturing (CIM) is considered by many to be a philosophy rather than just a technology. Several architectures, with various technological components, have been proposed to represent CIM.

Regardless of the specific architecture at hand, CIM is generally accepted as the highest level of integrated automation in manufacturing organizations. Currently, fewer than 100 firms throughout the world are at this level of automation, which may be because manufacturing organizations usually move slowly toward a CIM environment by adopting new manufacturing automation technologies selectively and in increments.

CIM generally involves implementing the automation of three major functions and their subcomponents in an integrated fashion (see Figure 1):

- Product design and engineering. This function is automated using technologies known as computer-assisted design (CAD) and computer-assisted engineering (CAE).
- Production planning and control. This function is automated using one or more systems (e.g., just-in-time [JIT] production, material requirements planning [MRP], and optimized production technique [OPT]) to plan purchase quantities, production quantities, and inventories for the future.
- Production. Automation of the production process (i.e., planning and operation) is known as computer-assisted manufacturing (CAM). Specific automation technologies related to CAM include a flexible manufacturing system (FMS), group technology (GT), computer-assisted process planning (CAPP), an automated guided vehicle system (AGVS), and an automatic storage and retrieval system (AS/RS).

Although there are many components and subcomponents to CIM, only some of these may actually be used by an organization. Depending on the organization's manufacturing strategy, executives usually select a manufacturing automation architecture to best support that strategy.

The manufacturing automation architecture specifies computer-based systems in each of these three CIM functions. As the level of auto-

Figure I Computer-Integrated Manufacturing

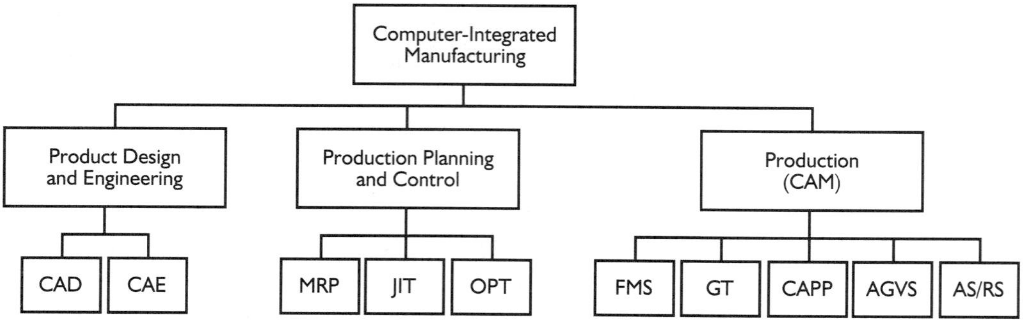

Notes: AGVS Automated guided vehicle system
AS/RS Automatic storage and retrieval system
CAD Computer-assisted design
CAE Computer-assisted engineering
CAM Computer-assisted manufacturing
CAPP Computer-assisted process planning
FMS Flexible manufacturing systems
GT Group technology
JIT Just-in-time production
MRP Material requirements planning
OPT Optimized production technique

mation and integration of these functions increases, a greater level of IS participation and support in implementing the organization's manufacturing automation architecture is required. In the following sections, the most important and well-known automation technologies under these three functions are described and analyzed in terms of their likely impact on IS management.

PRODUCT DESIGN AND ENGINEERING

Computer-assisted design and computer-assisted engineering are the primary technologies used to automate product design and engineering. These reduce costs by eliminating manual design and drafting, thus reducing the design and drafting time so that new products can be introduced quickly. In addition, CAD and CAE improve design quality by enabling engineers to test several design alternatives and improve design efficiency by using data bases of existing or standard designs of components and subassemblies.

Each organization has its own strategy for implementing CAD/CAE. Some have a stand-alone CAD/CAE operation; others integrate CAD/CAE technology with CAM. In the simpler CAD/CAE systems, there is usually one or more CAD workstations attached to a minicomputer or mainframe, CAD application software for design modeling and drafting parts, and related data bases. In more advanced CAD applications, product design and process planning are performed concurrently, which is accomplished by incorporating into the product design the features for which optimum shop processes exist in a predefined library.

Usually, the product design and R&D functions are responsible for the decisions involving the CAD/CAE system elements. The primary IS role is more likely to be as a consultant on the capabilities of various software and hardware products. When CAD systems are integrated with CAM, the migration of data from CAD to the CAM subsystem is accomplished by common data bases and standard or customized applications interfaces between the two subsystems, including the use of available CAPP software.

The direct impact of CAD/CAE on the IS department will be minimal in terms of the required changes to the existing portfolio of manufacturing applications. In complex applications of CAD/

CAE, however, IS support will be required in such technical areas as networking of CAD stations, data communications among geographically distributed locations, data communications from CAD to such systems as cost accounting and quotation systems, and in defining, developing, and administrating data bases required for CAD.

PRODUCTION PLANNING AND CONTROL

Three major information systems supporting production planning and control have emerged during the past decade—material requirements planning, just-in-time production, and optimized production technique. These systems require many data bases and interactive software modules used throughout the organization. They are the most information-intensive systems and will have a greater impact on IS than the systems supporting product design and engineering and the production process.

Material Requirements Planning

The MRP system evolved during the 1970s and is used for scheduling the production of parts, purchasing raw materials, and developing plans for appropriate inventories of raw materials and finished goods on the basis of the master production schedule. MRP evolved into a manufacturing resource planning system (also known as MRP II) by adding such functions as work center loading and capacity planning, manufacturing budgeting, cost accounting, and performance reporting.

Introduction of MRP or MRP II is likely to have a significant impact on IS. Some of the major modifications to the existing manufacturing applications include:

- New or modified cost accounting, general ledger, financial planning, and purchasing systems may be needed; these must be interfaced with the MRP system.
- Engineering change control systems must be integrated with the MRP system because MRP requires up-to-date bill-of-materials data produced by the change control system.
- Current production control and inventory control systems may need to be revised or replaced by modules within MRP software.

In addition to these changes to applications, the MRP system will require extensive data conversion, the development of new data collection and recording systems (e.g., bar-coding and laser scanning), increased computer capacity for processing MRP runs, substantial MRP-related user training, and a carefully designed plan to implement these required changes. The IS department will very likely play a significant role in these activities.

Just-In-Time Production

During the late 1970s, the JIT manufacturing concept and the associated kanban scheduling technique became popular in the United States after their success with Japanese manufacturers. The JIT approach to production is based on the idea that finished goods and components should be produced just in time to satisfy the need for them. Furthermore, the raw material should be received just before it is needed for production.

As result, JIT production eliminates most of the in-process, finished goods, and raw materials inventories. Because these inventories are kept small, any manufacturing process problems are quickly revealed and corrected. This allows a continuous improvement in the manufacturing process quality and productivity.

In a typical JIT application, efforts are made to streamline operations and reduce in-process inventories through automation, changes in layout, and changes in machines or tools. While these changes are being made, a compatible JIT production planning and inventory control system, such as kanban, is introduced.

In kanban, the production from each workstation is ordered by the succeeding workstation through tickets called *kanban*. The assembly schedule is short (i.e., three to seven days), and the components are manufactured according to the needs of the assembly requirements.

Because just-in-time production is a simplified approach, its implementation will have a substantial impact on existing manufacturing applications. Kanban tickets will replace the job orders and routing sheets. As a result, introducing the JIT approach is likely to eliminate the organization's existing shop floor control system.

In addition, because just-in-time production requires frequent deliveries of raw materials and immediate shipments of finished goods, both inbound and outbound freight systems may have to be redesigned and computerized. Current logistics and traffic applications may also need revising.

Purchasing applications are likely to change because under JIT production, vendors are evaluated on the basis of quality and delivery performance rather than price alone, long-term contracts are awarded to the vendors, and vendors deliver small lots more often. Current production planning and inventory control applications will be simplified because inventories are reduced to a near-zero level. The need for elaborate inventory control disappears.

JIT production relies on solving resource availability and process quality problems. To control process quality, a strong statistical quality control system or real-time process control system is necessary. In addition, just-in-time production requires a rigorous preventive maintenance scheduling system to ensure machine availability. The IS department will be involved in developing and implementing these systems.

Just-in-time production will cause changes in the existing cost accounting system because it calls for a flow-type production approach rather than a batch-type approach supported by traditional cost accounting systems. Just-in-time production is often used in conjunction with the production concepts of group technology and flexible manufacturing systems. The integrated nature of JIT, GT, and FMS makes their implementation a great challenge for the IS department.

In addition, some organizations may use JIT production in conjunction with MRP. In such cases, there is a strong need to integrate MRP and JIT systems through their shared data for manufacturing and production scheduling.

Commercially available software to support JIT production is limited; therefore, the IS department may be asked to develop the software. This may be a tall order for many IS departments, requiring the development of bar-coding systems, kanban scheduling systems, and the integration of MRP systems with JIT support systems.

Optimized Production Technique

OPT, a manufacturing planning methodology developed during the late 1970s, creates an optimum schedule and sequence of operations for manufacturing organizations' work centers, taking into account product demand, priorities based on delivery due dates, and work center capacities.

OPT is supported by algorithms that process data about production requirements and manufacturing facilities usually available from plant records. The system tests the existing work load to identify capacity bottlenecks, which it attempts to reduce through efficient job scheduling.

OPT implementation will require reworking the organization's cost accounting system to reflect accurate inventory levels, up-to-date product structure, assembly line routings, and setup and operational times. As with MRP, the data requirements for OPT can be overwhelming. Any organization not presently collecting such data is faced with a monumental new task of data collection, coding, and storage.

Currently, few organizations worldwide have adopted OPT. The promised benefits of OPT implementation on production process efficiency and output have yet to be demonstrated conclusively.

As with the other production planning and control systems, OPT will present its own set of challenges to the IS department and its manager. The organization may have to incur substantial up-front costs to collect and manage data required by OPT (e.g., data about bills of materials, operations routings, and setup times). Existing data bases must be made compatible with the OPT system, and solving compatibility problems can itself be a monumental task.

Unlike MRP, OPT is currently available from a few vendors and many algorithms used in OPT are proprietary. Therefore, the ability to customize OPT application programs is limited. Because OPT is an integrated production planning and control system, its implementation will affect many of the organization's existing applications, including cost accounting, order entry, general ledger, and inventory control.

PRODUCTION

The level of automation of the production process may vary considerably among organizations. At the highest level of automation, all of the CIM component technologies are completely integrated and can support the requirements of one or two production environments: repetitive (i.e., high-volume production) or small lot (i.e., high-variety production).

Table 1 lists seven automation levels and describes their corresponding significance for the IS department. At level 1 of the automation spectrum, there may be isolated automated workstations using robots or standalone numerically controlled (NC) machines at one or more times in the production process. Such automation usually requires considerable NC programming, but at this level, very little impact on the IS department is expected because the existing business systems remain largely unaffected.

At the next level of automation (level 2), clusters of computer numerically controlled (CNC) machines are driven by special-purpose computers. The introduction of this technology into the production process has little direct impact on IS operations and existing manufacturing applications because it is focused on the production process and does not usually affect commercial systems.

At the third level of automation, direct numerical control (DNC), a central computer is used to operate several manufacturing machines through programmable logic controllers. The manufacturing machine speed, feed, and job requirements are directly controlled by the central computer.

The architecture for this level of automation may be simple or complex depending on the configuration and the complexity of the manufacturing process. For reasons similar to those for levels 1 and 2, many business applications are likely to remain unaffected at this level of automation.

The fourth level of automation is computer-assisted manufacturing (CAM), in which an extensive network of computers using common data bases drives the manufacturing process. The fifth level of automation represents the integration of CAD and CAM. At levels 4 and 5, product design and the manufacturing process are linked through the sharing of parts, tools, and machine-capability data bases. Computer-assisted process planning (CAPP) automatically generates machine processing instructions to direct sequences of operations, thus producing parts designed by the CAD system.

The implementation of level 4 or 5 requires major investments in computer resources. As a result, various organizationwide technical and managerial issues arise. These issues include the selection and implementation of network communications protocols, applications development software, and data base management systems. Given the scope and large amount of corporate resources involved, it is important for IS, manufacturing, and other departments that depend on computer technology to cooperate in the design and implementation of the required infrastructure.

The last two levels of automation (6 and 7) are FMS and CIM respectively. In FMS, the manufacturing process itself may be highly automated but the level of integration is lower than in the CIM environment. For example, FMSs may or may not involve integration with CAD. With FMSs, collections of different types of numerically controlled machines are connected by a material-handling system and controlled by a central system.

Flexible manufacturing systems process a vari-

TABLE 1
Automation Levels and Their Impact on the IS Department

Level	*Type of Automation*	*Impact on IS Department*	*Overall Impact on IS*
1	Numerically controlled machines	Software programming of individual machines must be performed with the programming mechanism provided.	Low
2	Computer numerically controlled machines	Same as for level 1.	Low
3	Direct numerical control	Network programs may be necessary to upload and download data and programs from a central computer.	Low
4	Computer-assisted manufacturing and computer-assisted process planning	Machines and tools data bases must be created and integrated with parts data bases. Application programs are needed to supply process, loading, and unloading instructions.	High
5	Computer-assisted design and computer-assisted manufacturing (CAD/CAM)	Application programs are needed to transfer product design information to the CAM system.	High
6	Flexible manufacturing system	Integrate material handling and storage retrieval systems with computer-assisted process planning (CAPP). Group technology is used to create cell structure and processing sequences for families of parts. Group technology and CAPP should work together.	High
7	Computer-integrated manufacturing	In addition to the full integration of the systems in levels 1 through 6, factory transaction data is summarized at various stages of aggregation to produce management reports. AI and expert systems may be used, and most commercial applications must be changed.	High

ety of small-volume parts and differ from transfer lines, which are groups of production machines connected sequentially by an automated material-handling system used to produce high-volume parts. With an FMS, the manufacturing process, the production planning and control system, and the material-handling systems are highly integrated.

The group technology concept is usually associated with FMS and involves producing an optimum layouot of machine cells (i.e., groups of similar machines) to be used to process particular types of parts (i.e., part families). The various commercially available GT software packages have algorithms available to process data bases of parts, tools, and machines. These data bases are shared with such systems as engineering change control and the bill-of-materials updating system. An FMS is the automated production and material-handling system that implements the structure of machines cells designed through GT.

In conjunction with GT, FMS is supported by two other major types of systems: the automated guided vehicle system (AGVS) and the automatic storage and retrieval system (AS/RS). AGVSs are used to move parts, tools, and supplies along computer-controlled, programmed paths designed to support the processing of specific parts within and between FMS cells.

AS/RSs are used to store and retrieve the items carried by the AGVS to and from fixed storage locations. These two systems enable automatic parts retrieval, loading, unloading, transfer, and storage.

In CIM (level 7), the product design, manufacturing process, and production planning and control systems are highly integrated. Therefore, conversion to a CIM environment is likely to have a dramatic impact on existing manufacturing applications. Existing production planning and control systems will be replaced by components of the CIM software.

During the design of CIM, an organization has to plan for a hardware and software architecture that will integrate all of the automation levels. In the processing area, direct numerical control, remote diagnostics, and adaptive and laser control functions performed for cutting, measuring, and inspection operations are all supported by integrated software.

A successful CIM implementation requires much of the expertise available in the IS department in operating systems, applications development software packages, local area and wide area networks, and the resources to generate and distribute factory-level information for corporate management decision making. In addition, IS departments will be expected to lead the effort in modifying many of the existing business applications to suit the CIM environment.

THE IS MANAGER'S ROLE

On the basis of observations from several organizations advancing through the stages of manufacturing automation, several recommendations can be made to IS managers. IS departments must increase the level of awareness and participation in manufacturing automation projects on the planning board, getting involved from the start.

Employees in the IS department must develop technical and operational expertise in factory automation because they are usually asked to participate in projects involving communications networks, software development facilities, and data management facilities. The IS department should take on the role of collecting and sharing the organization's knowledge regarding the implementation of manufacturing automation technologies in various units of the organization.

IS management must be aware of and able to manage the impact of factory automation technologies on the organization's business systems, such as the inventory control, order entry, and cost accounting systems. Many of these systems will require significant changes as the level of factory automation increases.

Productivity improvements require increased organizationwide systems integration. The IS function must improve its technical and managerial ability to bridge the operational gap between departments, which will require greater knowledge of manufacturing operations as well as better data communications capabilities throughout the organization. Increasingly, applications development will have to focus on integrating islands of automation within the factory and between different functional areas.

Aligning IS resources and activities with the organization's manufacturing automation strategy will require a substantial shift in the organization's applications portfolio mix. Therefore, IS departments must develop the expertise and acquire the necessary tools to develop highly integrated manufacturing decision support systems.

In addition, IS operations must be able to support the design and implementation of knowledge-based (or expert) systems, which help manufacturing handle the complexity of working with numerous variables requiring judgment or inferential logic in a low-volume, multiproduct environment. In short, these expert systems capture the logic of the expert, support the design of products and processes, and control quality and production plans.

Project selection and development should no longer emphasize manufacturing economies of scale by automating highly repetitive tasks. IS support must include economies-of-scope applications that support the short-run production of multiple products.

CONCLUSION

This article focuses on the changing technologies in the manufacturing area and their impact on the

IS function. As the level of automation within the organization increases, significant changes in the IS role are required.

IS managers must accept the need to develop additional expertise related to factory automation. They must know how existing business applications will be affected by increasing factory automation, and they must understand the impact that various manufacturing automation systems will have on the organization's computing resources, including existing systems, IS department areas of expertise and staffing, and department funding.

Furthermore, IS managers must help the corporate learning curve by facilitating technology transfer among various divisions of the organization by centrally acquiring the disseminating information back and forth between business units, as needed. To the extent that IS managers are prepared to support the changing manufacturing environment, they will be able to contribute to their organization's survival and growth in the increasingly competitive global marketplace.

FURTHER READING

Ettlie, J. E., and Reifeis, S. A. "Integrating Design and Manufacturing to Deploy Advanced Manufacturing Technology." *Interfaces* (November–December 1987): 63–74.

Guimaraes, T., and Gupta, Y. "Personal Computing and Support Services." *Omega: International Journal of Management Science* 15: 467–475.

Hawley, S. "Computer Integrated Manufacturing." *DEC Professional* (May 1988): 42–46.

Hunsche, M. A. "The Three Challenges of Factory Systems Integration." *Infosystems* (April 1986): 72–74.

Turban, E., and Sepehri, M. "Applications of Decision Support and Expert Systems in Flexible Manufacturing Systems." *Journal of Operations Management—Special Combined Issue* (August 1986): 433–448.

Warner, T. N. "Information Technology as a Competitive Burden." *Sloan Management Review* (Fall 1987): 55–61.

QUESTIONS

1. In addition to acquiring some manufacturing knowledge, what other skills do IS managers need to support manufacturing automation?
2. How should the IS department interface with manufacturing in implementing and maintaining automated manufacturing systems?
3. Do you think a manufacturing automation project should be managed by someone in manufacturing or in IS? Describe the rationale for your answer.

4.7 Old System, New Life

PATRICK HOUSTON

This concluding paper of Section 4 is a case study. It describes how one small manufacturing plant turned itself around by using information to create a just-in-time environment. What is intriguing is that the firm did not require large amounts of new hardware. It used the information technology that it already had in a new way, and it empowered its employees by giving them that information.

DEBRA Miesner is engaged in a well-rehearsed and intricate dance involving pushes, swivels, twists, and turns. But she's not on a stage; she's on a factory floor, and the movement is performed with her hands. As if to a quick quarter beat, she's putting tiny springs, pins, wires, and screws into place as the first steps in creating a circuit breaker. In performing her number, Miesner takes cues from a nearby computer printout. By providing her with basic information such as customer name, product number, and due date, it serves as her musical score. She uses it to determine her workday routine—not at the behest of any boss, but on her own.

Miesner works in a 550-employee Eaton Corp. plant in Lincoln, Illinois. Besides circuit breakers, this small plant situated among the corn fields produces breaker boxes, safety switches, and meter centers—devices that usually reside in the basement or another unobtrusive corner of any building that uses electricity. With her dancing fingers and smudged printout, Miesner symbolizes the twin attributes of knowledge and power, forged to create a remarkable make-over in American manufacturing.

Four years ago, the tiny Lincoln plant represented the worst of U.S. manufacturing practices. Its processes were convoluted, its inventories bloated, and its jobs headed, in a manner of speaking, to the Mequiladora plants in Mexico. Now it represents the best. By implementing the concept known as just-in-time manufacturing, it has become a study in elegant and efficient simplicity. It reduced inventories by two-thirds, an amount equal to 14% of its $75 million in annual sales.

But intelligent decisions require knowledge and information. For Eaton's little plant in Lincoln, that meant sweeping changes within its information system—from an IBM System/38 bogged down by thousands of needless transactions, to a simplified system that could streamline the entire manufacturing process.

Lincoln's improvements didn't require big investments in new technology, though. Instead, the plant used a venerable, decade-old language to craft brilliantly simple applications, breathing new life into the ten-year-old System/38.

Eaton's Lincoln plant then did something extraordinary. It gave its streamlined, up-to-date information about customer orders and financial results directly to its workers. The plant proceeded

Source: *Corporate Computing* (November 1992): 160–167.

to recast its work force into dozens of cross-functional teams that use the information now at their disposal to make many of the decisions their supervisors once did.

What's even more remarkable about the change is that Lincoln is a 38-year-old facility, and aged plants are like old dogs. As a rule, it's impossible to teach them new tricks. But the Lincoln plant adapted so well to the new work mode—largely due to the insistence of a hard-headed plant controller named Al Houser—that it put itself on the vanguard of the emerging renaissance of American manufacturing. While overall productivity improvements for U.S. manufacturing have been less than 3% a year since 1990, Lincoln has racked up gains averaging nearly 6% a year—and expects to exceed 12% in 1992.

The three means toward that end are basic: simplify, simplify, and simplify. Especially when anyone starts thinking about replacing the plant's aging IBM System/38 with a new mainframe, or sticking terminals all over the plant floor. Houser, who never met an aphorism he didn't like, snaps off one of his favorites: "Obliterate, don't automate."

At Lincoln, workers now depend as much on daily reports as on any machine or tool. These reports contain the same information about finances, operations, and customers that American managers have traditionally hoarded as selfishly as Scrooge collected shillings. At Lincoln, Houser holds almost nothing back—not out of kindness but out of necessity.

PLANT BE NIMBLE

Whether you make circuit breakers or sell advice, being nimble in your market is the preeminent competitive advantage in business. "In every one of our markets, the pace of business decisions and product development is faster," says Alexander M. Cutler, who as executive vice president of operations oversees Eaton's 110 factories throughout the world. Moreover, he says, Eaton serves commercial customers—75% of Lincoln's products are sold through independent distributors to building contractors—whose expectations are similar to those of fast-food customers: They want it their way, now. The need to compress reaction time is forcing companies to whack away at management bureaucracies, pushing decision making down to the worker-bee level.

To make sound decisions, employees must know almost as much as managers do about their plant and its processes and customers. With automation, generating that type of information "is almost like turning on a fire hydrant," says Ed Lawler, a University of Southern California business school professor who studies employee involvement. There are two paramount questions, he says. They are: "What information do you need to get to employees?" and, "Will they know how to use the information that they get?"

In its approach to information and its new mode of operation, Lincoln joins an elite corps of plants, including Harley-Davidson's old Milwaukee works and Honda Motors' Marysville, Ohio plant. As persuasive as these success stories are, only about 10% of American manufacturing facilities have implemented just-in-time systems, with their characteristic work cells and reliance on up-to-date information, to the extent Lincoln has, according to Robert W. Hall, a director for the Association of Manufacturing Excellence. But according to Hall, this transition is inevitable, and the 90% that haven't implemented these processes most certainly will. "By the end of the decade, if you're not somewhat into these types of practices, you better have a sweet little monopoly, or you're dead in the water," says Hall.

In 1988, the Lincoln plant wasn't dead in the water, but it was a sitting duck. The concept that had dominated the manufacturing mind-set for the better part of the industrial age held sway over the plant's methods: To achieve economies of scale, you make as much as possible at the same time and in the same place.

At operations such as Lincoln, this concept was represented by a rigidly departmentalized batch-manufacturing process. Workers performed one task, all day, every day, as fast as they possibly could. They not only built many of the 12,000 parts that went into Lincoln's finished products, they built those parts in long production runs that created lot sizes in the thousands.

Worse, Lincoln didn't build these runs in reaction to customer orders; it set its workers loose based on forecasts divined by a Manufacturing Resource Planning (MRP) program. Or, as Houser puts it, the plant produced as much as it "hoped" it would use or sell. Typically, those thousands of subassemblies sat for months in a 50,000-square-foot stockroom, awaiting final assembly.

In addition, the process zigzagged like a rabbit tearing through a briar patch. Parts were ordered out of the storeroom. Labor was scheduled. To make sure the right parts reached the right departments—stamping, fabricating, plating, painting, molding, and welding, to name a few—everything had to be routed. And that was just the paperwork. This paper trail and the process it documented crossed paths as many as 15 times en route to final assembly.

To further complicate matters, employees worked under a piece-parts incentive plan. The more they made, the more they were paid. And that meant tracking who made how much of what—no small task in a factory that turned out more than 7 million circuit breakers each year. In the course of tracking labor, inventory, work in progress, and the myriad other elements of the process, Houser estimates that IS followed more than 6,000 reporting transactions a day. Even more mindboggling is that each transaction was entered by either a supervisor or an hourly employee in the plant. On average, each employee entered a transaction into the computer 11 times a day.

Like most other American businesses unchallenged by competition, Lincoln got away with these bald inefficiencies for a long time. But as the tide of general prosperity receded, the inherent shortcomings of these manufacturing systems were revealed. "What we grew to realize is that if you organized work in this old way, while it seemed to offer economies of scale, it had a hidden side effect," says H. Thomas Johnson, a Portland State University professor who specializes in quality management. "You added a lot of cost that was attendant on moving things from place to place, storing things, then designing and operating the information systems that were going to help you control all this stuff."

There were other problems, too. What if something was wrong with those big lots in storage? How would you know? Only when it came time to use them, perhaps after months of being made in the same errant way, would you recognize a lot with bad parts. There was an inherent flaw in the incentive pay system, too. When Lincoln assembly workers figured out a way to work faster, would they share it? No way. They feared Lincoln's industrial engineers would use that innovation to raise base production quotas.

Houser and other managers at Lincoln describe those days in one word: chaos. Delays, quality problems, and duplicated work made management an exercise in fire fighting. "You spent your whole day sorting through materials that had supposedly been delivered from the storeroom," recalls Jack Dowell, who oversees panel and meter center production. "And there was just a mountain of paperwork." Inefficiency was rampant. Production crawled to the point that at least 100 products a day would go on back order, even though the plant had 20% of its sales, which amounted to some $15 million, in inventory stacked up in its storeroom, or sitting in piles on the floor.

Of course, reality eventually set in. At Lincoln it took the form of some grim portents. A PPG Corp. plant, once Lincoln's biggest employer, closed down. For Houser, this amounted to a particularly painful reminder. His father had worked at the PPG plant. What's more, Lincoln's largest competitors, including much bigger Square D and General Electric, were forging ahead to recast themselves. Siemens, the big German firm, began expanding its presence in the United States. Among these behemoths Eaton's market share ranked a distant fifth.

It's no sin to be small. But in today's hypercom-

Figure 1 The Bottom-Line Benefits of Simplification

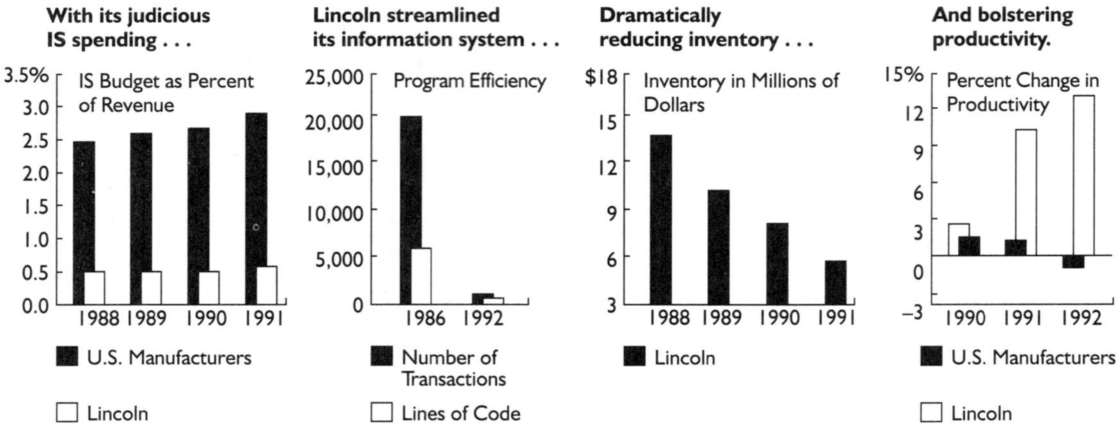

petitive environment, it is a sin to be fat and slow. You'll be squished like a fuzzy caterpillar trying to challenge the trucks rolling down Highway 55.

But the real call to change resulted from a visit by a division president. To deal with that very tangible, very quantifiable inventory problem, he created a task force and put Houser in charge. When Houser and a few other Young Turks at the plant began to look for solutions, they concluded that incremental fixes wouldn't suffice. The high inventories, they realized, were just a symptom of a bad system.

FROM PUSH TO PULL

With the assent of Lincoln's laid-back plant manager, Bill Kelly, the task force embraced the concept of just-in-time (JIT) manufacturing, which relies on a direct and continuous flow of materials—no more back and forth to a storeroom. Small lots of raw material roll off the shipping docks just in time, before a company runs out. That material is then processed to fill customers' orders, which are shipped onto waiting trucks by the next day. "The purpose of JIT is to cut throughput time," says Richard J. Schonberger, author of *World Class Manufacturing: The Lessons of Simplicity Applied*. What's more, a JIT plant ideally has no inventory or work in progress.

Its distinction from traditional processes also lies in the difference between *push* and *pull*. It's further distinguished from traditional processes by its reliance on actual rather than forecasted demand. Traditional manufacturing processes rely on MRP-based forecast planning to push work along. Parts or subassemblies are manufactured to meet expected demand just as Lincoln's were. A JIT plant, by contrast, relies on actual demand to pull the work through the manufacturing system.

Converting to the JIT concept meant reengineering everything at the plant—its physical layout, its pay incentives, its entire culture. The plant was physically reorganized, mainly to shrink the sometimes vast physical distances between consecutive production stages. The incentive pay system went out the window in favor of a universal pay scale based on a slightly higher hourly base wage for most of the workers. (Pay now averages $11.50 per hour throughout the plant.)

But the most significant change at Lincoln was cultural. Instead of working all day on the same mindless task, employees were organized into 50 teams, or cells, consisting of 2 to 20 people. Workers in each of these cells learned each other's jobs. Most of all, cell members began using their heads: They decided not only what to work on but who worked where, when, and for how long. They even held meetings to decide how to improve production. And they based all of their decisions on information presented on smudged green printouts, like those Miesner uses.

SIMPLICITY ITSELF

Once Lincoln decided to reengineer its work methods, Houser realized the plant had to streamline the information running on its System/38. Applications had to be simplified to match Lincoln's newly simplified work processes. Where to start?

The plant had an old and not terribly efficient manufacturing control system that included different modules for inventory control, materials resource planning (MRP), product costing, shop floor control, and an engineering database of parts. "We took the guts out of that package and transformed it," says Houser. "You wouldn't recognize the old system now."

Remarkably, only five IS staff, working entirely in RPG3, recast that ancient beast into a sleeker, trimmer information system. Where the inventory control module once comprised 6,000 lines of code, for example, it now runs on about 700. "And it is doing more than those 6,000 lines ever thought about doing," says Houser, justifiably proud of the IS staff's achievement.

In less than 18 months and for only $300,000, Eaton's small staff either rewrote old applications or developed new ones to support its JIT efforts. Once the incentive pay plan was sacked Houser just turned off the punch card system that tracked employee output.

Houser and his staff also devised an equally elementary method of tracking inventory called *backflushing*. Though it sounds more like a plumbing procedure, the term is descriptive. Rather than report on the flow of raw material and work in progress through the 15 or so manufacturing steps, the plant waits until a product is ready to go out the door. Lincoln's IS system has been programmed so that it knows the material content of every finished item. Gone is the tedious accounting of work in progress. And once an item is recorded as ready to leave the plant, the system automatically adjusts inventory downward. "I'm proud of the fact you don't see many terminals on the plant floor. It means our people aren't doing a lot of reporting."

Eaton's backflushing technology is finding its way into nearly all of the plant's homegrown systems, says Jan Taylor, the plant's IS coordinator (nobody is known as a supervisor now). Taylor and his staff wrote a new application for the shipping department that uses bar code technology to handle a host of functions. Now, Eaton's shipping dock prints bar-coded documents for each customer shipment. That single document backflushes inventory, creates the necessary accounting entries, generates customer invoices, and drives the computerized manifests for Eaton's freight carriers. Simultaneously, Eaton's divisional sales headquarters in Milwaukee receive the accounting information it needs in one transaction.

"Before, we had to go to a CRT and type the reports into our accounting system," says Houser. "Then we had to go to a different CRT connected to Milwaukee, and type it in again. And for each customer shipment loaded on a truck, which might be stacked with boxes for 30 different customers, we would have to type in a separate bill of lading."

Such streamlining has extended the useful life of the plant's ten-year-old System/38. "We're at the point where we're maxed out on the system," says Taylor. "We can't go any faster, and we can't get any bigger." The system has no room for additional memory. It could handle perhaps 10 more peripherals over its current 96. "I could go to an AS/400 tomorrow," he says. "But it's not going to happen." Partly because most of its applications are done in-house—Houser believes homegrown solutions are better suited to the plant's unique requirements—the IS department is running below its annual $336,000 budget.

So why not spring for an upgrade?

"I'd be the first to sign a requisition slip for a new computer, but you have to 'lead with the need,' " Houser says, flipping off another favorite aphorism. "There will be a day when we'll have a computer terminal at every workstation. But right now it's not justifiable because it's not going to improve productivity or customer service."

INFORMATION THEIR WAY

It's one thing to track information and something else entirely to present that information in a useful and relevant form. Workers have been quick to ask for information *their* way. In response, Houser, Taylor, and the plant team created the daily production report. From order entries downloaded nightly from Milwaukee's IBM AS/400, an applications program created by Lincoln's five-person IS group compiles a cell-by-cell report. Printed at 5 a.m. every working day, the report includes the customer's name, the product ordered, the quantity broken down on the basis of day-by-day due dates for the ensuing five days, the total due in the next five-day period, and the number of those same finished parts sitting in the warehouse. "We've given them what they need to become a fully functional business unit," says Houser. "All that we ask is that they meet the due dates. If the same product is due on Wednesday and Friday and they choose to make them all at once, that's OK."

The report is straightforward and accessible. Even a stranger to the plant can understand it. Miesner, for one, is far more comfortable with it as a printout—a familiar piece of paper—than she would be were it on a terminal. "I wouldn't know how to run a computer," she admits.

The report has been in constant evolution; it's taken a while to get the format finetuned. It was recently altered, for example, to include the tally of finished goods in the four Eaton distribution centers around the country that are fed by Lincoln. Factory-floor employees determine the report's look. IS coordinator Taylor says they actually look over his shoulder when he tinkers with it.

Tom Fabrizio, a senior consultant with Productivity Inc., which advises companies about ways to provide information to employees, says this is a laudable practice. "Information should not be 'done' to them," he says. "The user should always be the author."

Miesner, the leader of her work cell, keeps the report at her side throughout the day. Her eight-person work cell occupies a U-shaped table. Instead of having each member assemble a breaker circuit in its entirety, cell members complete the process in half a dozen steps and members switch tasks several times a day. Nor does the cell make the devices in long production runs. It produces them in numbers determined from a specific customer order contained in the report. The group makes as many as 35 different-size circuit breakers a day. On one particular day when two of its members were missing, cell members buzzed back and forth between stations as the breakers were assembled in six steps, tested, and then packed for shipping.

Even though their incentive pay system is gone, Miesner and her colleagues, by all appearances, work just as hard. Why?

For one, they're happier now that the system treats them as breathing, thinking human beings. "They like it better because they control their own destiny," says Bob Sanders, president of International Association of Machinists and Aerospace Workers Local 1165, which represents workers at the plant. Before, it was common to have eight or nine pending grievances. Now it's unusual to have even one or two, he says. Miesner has another reason, though. Thanks to the production report, she's not just turning out a product, she's doing it for a real customer whose name she sees. Says Miesner: "We know if we get them their orders on time, they'll keep buying from us, and we'll keep our jobs."

The system has other virtues as well. The breakers contain a subassembly consisting of two wires welded to small metal plates. Before, the welding was done in a department well away from the breaker room where Miesner works. Today only a walkway separates Miesner's cell from the welding operation. She orders only as many subassemblies from welding as her cell needs to complete an order. Miesner places the order with a welder either by writing it on a tiny blue Post-it note or by walking over and—how about this for simplicity?—just *telling* a welder how many she needs.

"IN TODAY'S BUSINESS NEWS . . ."

Houser says that as employees became more involved, they began asking for insights into the plant's overall performance, as you'd expect from the business owners of sorts they've become. In response, the IS group created a daily business report, which includes a tally of total plant inventories, a rundown of the plant's dollar sales for that day and for the month to date, and a list of items on back order. It also contains a breakdown of plant expenses along a dozen measures including labor, supplies, maintenance, scrap, shipping, and utilities.

Houser says the sales figures originate from Milwaukee with the nightly download. Technically, the trickiest part of the report is compiling the expenses. It required some in-house programming to fetch data from several different accounting applications such as accounts payable, payroll, working funds, and inventory.

Rick Ingram, 42, a brawny receiving-dock employee who's worked 26 years at the plant, watches these numbers closely. He can recite sales figures off the top of his head, following them with an unsolicited analysis of whether the plant is having a good month. Why is he so interested? Expressing a concern familiar among manufacturing workers who have watched their ranks shrink by the millions since 1980, he says, "If you've been here as long as I have, you like to know what the future holds."

The future looks pretty good. Since 1987, the plant's break-even point has come down by 18% to sales of $4.5 million a month. Despite a 14% recessionary decline in demand last year, Lincoln maintained respectable gross margins. "If we hadn't made these changes, we would not have remained profitable," says plant manager Kelly.

Productivity is up, too. Houser now tracks labor using only two broad measures. One is earned hours, or the total amount of labor that goes into every finished product, calculated on the basis of engineering-estimated labor standards. The other is total attendance. Comparing earned hours with attendance provides Houser with a perfectly adequate index of productivity. The more earned hours or real output to total attendance, the better. The trend has been very, very good: As a percentage of earned hours to attendance, productivity had risen from 46.2 in 1987 to 58.4 by July this year.

Along with being profitable even in hard times, the plant has become much quicker on its feet. It can fulfill an order in a day, down from the five days it took in 1987. The plant has become so efficient that if a customer calls in an order by 10:30 a.m., it will be shipped that day at no extra cost. Since the start of 1992 the plant has only once missed its goal of shipping 95% of its product on time.

"With their service and their ability to respond, they're probably one of the best manufacturers we work with," says Jack Mumford, a senior vice president with Platt Electric Supply, one of Lincoln's biggest customers. "Other suppliers could take a lesson from what they do."

What Lincoln has done in terms of process, technology, and information is simplicity itself.

QUESTIONS

1. What were the major problems that led Eaton to reorganize its manufacturing operations? What was the root cause of these problems?

2. What was the role of technology in the reorganization?
3. What general lessons can be drawn from Eaton's experience with factory automation?

5 Costs, Benefits, and Value

ALTHOUGH we know that appropriate use of information technology leads to large productivity gains, specific IT investment decisions are difficult to make and justify. Senior managers, faced with increased pressures for improved performance, are demanding solid proof of the expected costs and benefits before making significant capital commitments. Typical questions are:

- Are we getting our money's worth from this technology investment?
- What projects are the best uses of our resources?
- Why does it cost so much and take so long to get the benefits?

These concerns are quite legitimate and well founded.

The purpose of this chapter is to review the techniques used for making IT investment decisions. It is generally agreed that a single global measure is not adequate for assessing the full value of IT investments. As you will see in this chapter, IT value is measured in many ways. The best measures are the ones linked to the specific type or purpose of the system.

Another factor that complicates investment decisions is the time lag between IT investment and the benefits stream. Investment costs are immediate and highly visible. Benefits are in the future and currently appear only as uncertain returns. This time lag creates particular difficulties in determining the value of IT projects. This difficulty becomes even more pronounced for infrastructure investments (such as wiring a building with fiber optics) which benefit everyone but cannot be measured against specific future income streams.

This chapter begins with two sections that present the nuts and bolts of making cost/benefit calculations. They are followed by two sections that describe the process needed to take these calculations back into the organization to obtain approval for change. The last two articles deal with the decision process involving competing proposals, some of which do not lend themselves to standard forms of analysis.

5.1 Determining Economic Feasibility: Four Cost/Benefit Analysis Methods

RICHARD T. DUÉ

To be useful, cost/benefit analysis should take into account the time value of money, the risk inherent in the development of a new system, and the costs and benefits throughout the system's life cycle. The exact method selected by a firm to assess an information systems project depends primarily on the criteria used by the firm for all capital projects. Note that the building of an information system is a capital investment to a firm, no different than investment in a plant or a piece of machinery designed to improve productivity.

In this paper, Dué describes four commonly used methods of establishing economic feasibility: payback, return on investment, present value analysis, and present value analysis plus risk assessment.

To make these methods real, Dué uses a large example that allows computing returns by each of these four methods. You will observe that the sample project is feasible by some methods but not by the last method.

THE analysis of the relationship of a system's costs and benefits is usually performed by a team of user, systems, and financial personnel during the feasibility study and again during the postimplementation review. The resulting cost/benefit analyses are used to inform management, users, and systems developers of the economic consequences of the acquisition and implementation of a new hardware or software system.

A system need not be cost beneficial—that is, its tangible dollar benefits need not exceed its development and operation costs—for it to be acquired or operated. In some cases, a new system might be acquired to meet requirements of government regulations, to offer a service provided by a competitor, or to support other systems in the organization that are cost-beneficial. Some systems are acquired for their intangible benefits—for example, for their contribution to an organization's image. The cost/benefit analysis should still be performed in these cases to help management identify and minimize the tangible costs of these nondiscretionary system acquisitions.

The analysis and comparison of a system's costs and benefits should be a simple and straightforward exercise. In practice, however, the analysis is often flawed by three factors. First, in many cases the time value of money (present-value analysis) is not taken into account in the

Source: *Information Systems Management* (Fall 1989): 14–19.

cost/benefit analysis. This oversight stems from a widespread misunderstanding of the concept of the time value of money. Because there are always more investment alternatives than there are resources to be invested, organizations expect to receive a return on their investment as compensation for committing resources to a particular project. The organization's management and owners hope that the return on the investment will be at least equal to (and possibly better than) the return promised by any of the other investment options.

The time value of money is the recognition that, to an organization with the opportunity to earn a return on its investments, money that is available today to be spent or invested (i.e., cost dollars) is more valuable than money that will not be available to earn a return until sometime in the future (i.e., benefit dollars). For an investment to be cost-beneficial, it must return more dollars in the future (i.e., benefit) than the amount of dollars spent in the present (i.e., cost) to account for this difference in value.

The second flaw in many cost/benefit analyses is the failure to account for the risks inherent in the development of a new system. These risks may stem from changes in the environment that are beyond the control of the organization, such as technological, economic, social, regulatory, and competitive change. These risks might also be attributable to conditions that can be controlled, such as project management, training, and competence of systems development personnel and users. In either case, these risk factors can reduce or even prevent the organization from receiving the originally required return on the investment. In many cases, however, the development and implementation risks are not taken into account in the cost/benefit analysis.

The third major flaw in cost/benefit analysis is the failure to analyze and report on the costs and benefits of a system throughout its life cycle. Because the systems development and operation process is faced with many ongoing risk factors, the cost/benefit analysis should be reviewed or repeated at the end of each major phase of the systems development life cycle and periodically during the system's operational lifetime until the system is replaced or abandoned. The failure to monitor a system's cost and benefits can result in the continuation of the development and operation of a system that is no longer cost beneficial.

ECONOMIC FEASIBILITY ANALYSIS

A system's economic feasibility is analyzed by studying the relative costs and benefits of the technically feasible system alternatives that could satisfy the users' requirements. The costs of both developing and operating the proposed system must be determined. These costs are identified and quantified by referring to other existing systems or by obtaining cost estimates from the marketplace, using the request for information (RFI) or request for product quotation (RPQ) processes. MIS managers must ensure that all relevant costs are included. The system model shown in Figure 1 can be used as a costing checklist. It illustrates that the following costs for development and operation of each technical alternative must be collected or estimated:

- Environment (overhead)
- Input and output (hardware)
- Processing (software and personnel)
- Data base (administration)

Figure 1 System Model

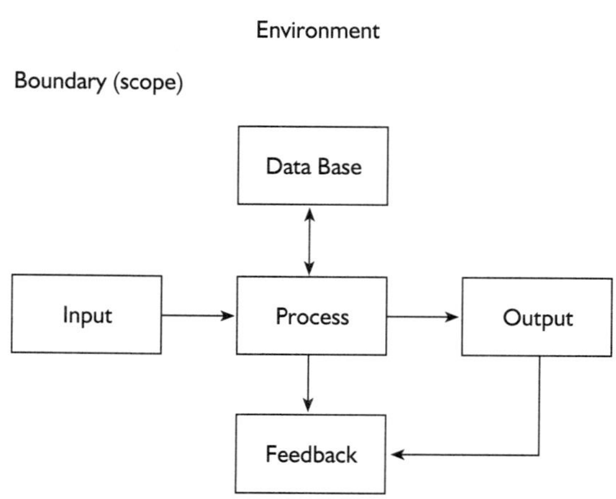

- Feedback (project management and maintenance)

The tangible and intangible benefits of the technical alternatives must also be determined. This is accomplished by comparing the benefit of having the system versus not having the system (the null-case approach) or by comparing the benefits of each of the alternatives to an existing or industry-standard system (the base-case approach).

Tangible benefits are those cost savings or revenue increases that can be measured and expressed in dollars. Intangible benefits cannot be quantified in monetary terms; they are the perceived gains to the organization if a particular technical alternative is implemented. Examples of intangible benefits include an organization's perception that its customers will view it as being modern and progressive, or that the organization will attract bright, capable employees, or that the organization will be using leading-edge technology. These perceptions may or may not be true and may or may not contribute to the organization's profitability. Nevertheless, they may persuade management to select a particular technical alternative over another.

One very effective approach to performing this phase of cost/benefit analysis is to involve experienced and technically competent systems personnel and users in the following way. The MIS manager should assign those systems personnel who will be responsible for the development and operation of the system to collect or estimate the costs of the alternative systems. In turn, the system's users, who will be responsible for justifying and obtaining the funding for the system, should be given responsibility for collecting or estimating the tangible and intangible benefits. This approach not only allows the technical and user groups to perform the tasks with which they are the most familiar, but commits each group to live with the results of its estimates.

After the costs and benefits are collected, the next phase of the cost/benefit analysis—the comparison of costs and benefits—is performed.[1] Four approaches can be used to compare costs and benefits (see Table 1); these approaches are described in the following sections.

Payback

The payback method of cost/benefit comparison involves the following steps:

1. Obtain the system's net annual tangible benefits by subtracting the expected annual operating costs from the expected annual tangible benefits.
2. Divide the total systems development costs by the average net tangible benefit.

The result is the number of years the organization can expect to wait before it recovers the amount of money spent in the development of the system. In the example in Table 1, the investment of 100 is divided by the average net tangible benefit of 38, to yield 2.6 years until the organization receives a return on its investment.

Payback is the most commonly used method of comparing the costs and benefits of systems because it is easily performed and understood. An organization can select a standard payback period (e.g., three years), which will be used as a cutoff point: any discretionary project with a payback period of more than three years will not be approved.

Unfortunately, the payback method of cost/benefit analysis is the least appropriate method of comparison of information system alternatives. Information systems can take months or years to develop, depending on project size, user and developer proficiency, resource availability, and stability of the development technology (see Figure 2). It is not until sometime after the system is developed that any benefits can be expected from the system. These benefits are usually limited during the early use of the system, increase over time as the users gain confidence and proficiency with the system, and then decrease as the system gradually becomes obsolete (see Figure 3).

The payback method does not take into account that current dollars are spent on the devel-

TABLE 1
Comparison of Economic Feasibility Methods

	Years of Operation							
	1	2	3	4	5	6	7	*Total*
Net annual tangible benefits	25	50	75	50	40	25	0	265
Present-value discount factor (35%) (formula in text)	0.74	0.55	0.40	0.30	0.22	0.16	0.12	
Discounted net tangible benefits (line 1 × line 2)	18.5	27.5	30	15	8.8	4	0	104
Probability of receiving benefits (past experience)	0.90	0.80	0.75	0.70	0.65	0.60	0.50	
Expected net tangible benefits (line 1 × line 4)	22.5	40	56.2	35	26	15	0	195
Discounted expected net tangible benefits (line 5 × line 2)	16.6	22	22.5	10.5	5.7	2.4	0	80

Total development cost (in dollars)			= 100
Average investment	Total development cost	$\frac{100}{2}$	= 50
Average net tangible benefit	Total net annual tangible benefits / Years of operation	$\frac{265}{7}$	= 38
Payback method	Investment / Average net tangible benefit	$\frac{100}{38}$	= 2.6 years
Return-on-investment methods	a. Average net tangible benefit / Investment	$\frac{38}{100}$	= 38%
	b. Average net tangible benefit / Average investment	$\frac{38}{50}$	= 76%
Present-value analysis method	Total present value of net tangible benefits (line 3) less investment	104 − 100	= +4
Present-value analysis plus risk assessment method	Total present value of expected net tangible benefits (line 6) less investment	80 − 100	= −20

opment of the system, but because of the time value of money, less valuable dollars are returned as benefits in the future. In addition, the payback method does not take into account the risk that the organization may not receive the required tangible benefits from the system.

Return on Investment

The return-on-investment (ROI) method involves the following steps:

1. Obtain the net annual tangible benefits of the system by subtracting the expected annual operating costs from the expected annual tangible benefits.
2. Divide the average net tangible benefit by the total systems development cost.

The result is the annual percentage return on the amount of money spent on the system. In the example in Table 1, the average net tangible benefit of 38 is divided by the investment (100) to yield an annual rate of return of 38% on the system investment.

An alternative calculation of ROI can be made by dividing the average net tangible benefit by

Figure 2 Investment Profile of the Process-Driven Systems Development Life Cycle

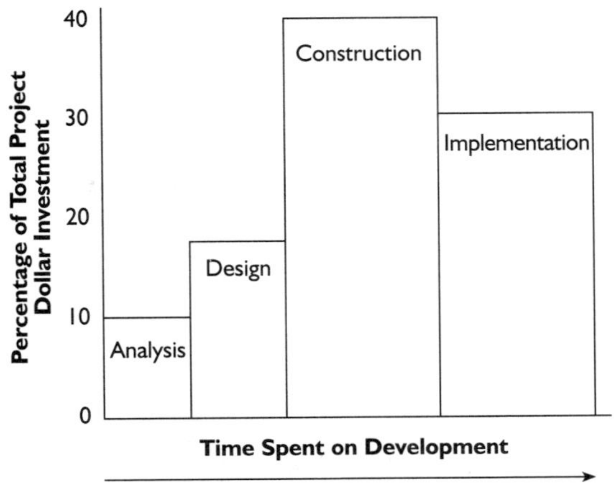

Figure 3 System Benefit Profile

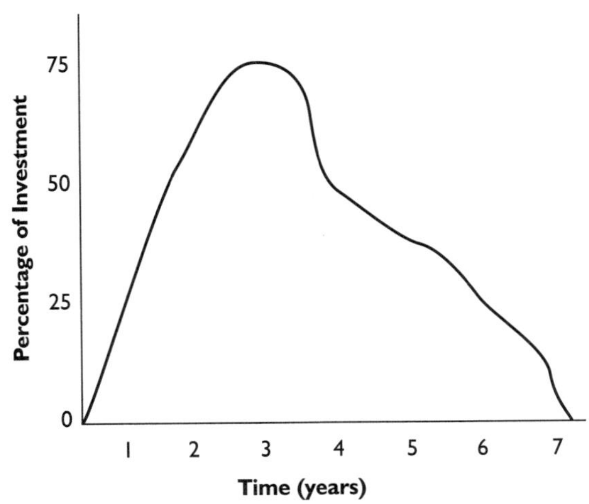

one-half the total systems development cost. The rationale behind this alternative calculation is that the development costs are being paid back throughout the life of the system, so that on the average, the organization's investment in the system is only one-half of the total development cost. In the example in Table 1, the average net tangible investment of 38 is divided by the average investment of 50 to yield an annual rate of return of 76% on the system investment.

The return-on-investment method is simple to use, but it does not take into account the time value of money or the risk of not receiving the benefits required from the system. It also understates the actual return on investment because it does not take into account the recovery of the development cost during the system's lifetime. On the other hand, dividing the average net tangible benefit by one-half the investment overstates the actual return on investment.

In the past few years, organizations have usually set the minimum annual return on discretionary investments between 30% and 50%. Some management accountants advocate setting the cutoff minimum return at the organization's after-tax cost of capital (which could be as low as 7% or 8%). The reason most organizations use a higher cutoff percentage is that capital is often not the limiting resource to new investment. The limiting resources for most organizations are time, management, and personnel. By setting a high minimum rate of return, an organization is forced to concentrate its relatively scarce resources on the most profitable projects. The high minimum rate of return also ensures that an organization will continue to experience a high rate of growth because it is not diluting its overall rate of return on investment with lower yielding projects.

Present-Value Analysis

The present-value analysis method involves the following steps:

1. Obtain the net annual tangible benefits of the system by subtracting the expected annual operating costs from the expected annual tangible benefits.
2. Discount the net tangible annual benefits during the life of the system using the following formula:

$$\text{Discount factor} = \frac{\text{Net tangible benefits}}{(1 + i)^n}$$

where i is the organization's minimum expected rate of return and n is the number of years in the future the tangible benefit will be received. If a stream of benefits is projected, the calculation should be repeated for each year in the stream.

3. Subtract the total development cost of the technical alternative from the total of the discounted net tangible benefits. The result must be greater than zero if the new system is to be cost-beneficial. In the example in Table 1, the total of the discounted net tangible benefits (line 3) is 104, whereas the total development cost is 100. Subtracting 100 from 104 yields +4: therefore, the system is cost-beneficial when the present value of the system's benefits is taken into account.

Present-value analysis is the preferred method of accurately determining the economic feasibility of an organization's capital projects. Unfortunately, the time value of money or present value is not a concept easily understood by most people. Another shortcoming is that this method does not take into account the risk of not fully receiving the required benefits of the new system.

Present-Value Analysis Plus Risk Assessment

This method involves the following steps:

1. Assign a probability of receiving the system's net tangible benefits for each year of the system's expected lifetime. These probabilities can be obtained by reference to similar systems or by reference to the organization's past history of success in systems implementation.[2] Multiply each year's net tangible benefits by the probability of actually receiving the benefit to determine the expected net tangible benefits.
2. Perform the present-value calculations (as shown in the previous section of this article) on the stream of expected net tangible benefits to obtain the discounted expected net tangible benefits.
3. Subtract the total development cost of the technical alternative from the discounted expected net tangible benefits. The result must be greater than zero if the new system is to be considered cost beneficial. In the example in Table 1, the total of the discounted expected net tangible benefits (line 6) is 80, versus development costs of 100. Subtracting 100 from 80 yields −20; therefore, the system is not cost-beneficial when the effects of the present value of the system's benefits and the risk of not receiving the benefits are both taken into consideration.

Present-value analysis plus risk assessment provides the truest picture of the economic feasibility of new system development alternatives. As the example in Table 1 shows, a technical alternative that appears to be economically justified with a 2.6-year payback and a rate of return somewhere between 38% and 76% is only barely justified with the present-value method and is not economically justified when the risk of not obtaining the benefits is taken into account. The example points out the importance of applying present-value analysis and of knowing and controlling the risk associated with the development and operation of the organization's information systems.

CONCLUSION

Unfortunately, the most appropriate method of analyzing the economic feasibility of systems development projects—present-value analysis plus risk assessment—is seldom used. The time value of money is a concept difficult for many managers, users, and systems personnel to understand. When the present value of the benefit dollars is understood and calculated, however, the MIS manager still needs to assess and apply the negative effects of the risks of developing and operating the system in order to gain a true picture of the desirability of a particular system investment. Because the payback and return-on-investment methods ignore present value and risk, the application of these methods can result in the misallocation of the organization's scarce staff, technology, and capital resources by biasing investment toward system projects that actually have low or even negative returns.

Data-driven systems development methodologies, end-user computing, and prototyping ignore feasibility assessments. They seem to follow the philosophy that if a system works, it must be technically feasible, and if someone is paying for it, it must be economically feasible. Failure to perform systems development and operation cost/benefit analyses in the data base, end-user computing, and prototyping environments results in the misallocation of the organization's scarce resources.

Finally, for a cost/benefit analysis to be effective, the assessment must be made throughout the lifetime of the project to halt poor development projects, to direct additional funding to successful investments, and to signal the end of the economic life of operational systems.

NOTES

1. R. N. Anthony and G. A. Welsch, *Fundamentals of Management Accounting* (Homewood, IL: Richard D. Irwin Inc., 1977).
2. R. T. Dué, "Stressing Project Accountability via the Project History Manual," *Computer Security, Auditing and Controls* 14 (1987).

QUESTIONS

1. What are the major strengths of each of the four cost-benefit analysis methods?
2. What are the major weaknesses of each of the four cost-benefit analysis methods?
3. How should intangible benefits be treated in making investment decisions?

5.2 Methods of Quantifying the Value of Office Automation

MATTHIAS SCHUMANN

The previous article by Dué focused on methods of computing the payback from information systems projects. These methods were generic. In the present paper, Schumann focuses on quantifying the benefit side of the cost/benefit relations. The work is founded in activity-based costing and is illustrated with representative values for office automation tasks.

Schumann argues that the benefits come in two ways: reduction of nonproductive work time and increases in work on higher-valued activities. The analysis examines both kinds of benefits. It determines the mix of tasks that are now performed and computes the current costs. It also forecasts how work would be done with office automation in place. These analyses are not trivial. They require measuring current processes and making estimates about future processes. If major expenditures are planned, it is well worthwhile to go into the detail called for by Schumann.

INVESTMENTS in information processing applications are subject to much closer scrutiny than they were at the beginning of this decade. The right investment made at the right time is often the critical factor in an organization's economic growth. Investigations into the costs and benefits of these investments serve two purposes: they provide the criteria to choose among alternatives, and they result in better control of project revenues.

Some published reports describe pilot projects and the results of cost/benefit analysis of their implementation. Analysis techniques and empirical results can be applied to other projects in other organizations. However, when applied to another organization, they provide only a first-cut justification of a project. Each organization's technical and managerial structure is different, affecting project justification. Therefore, the structure of a specific organization must be taken into consideration in the analysis.[1]

This article is concerned with the effects of office automation and the justification of this technology, which is no simple task. Office automation produces tangible and intangible results. A seemingly intangible result in one area of the organization can produce a tangible effect in another. However, an analysis of investments in a manufacturing application, for example, is not so elusive. A typical result of such an investment can

Source: *Information Systems Management* (Fall 1989): 20–28.

produce a return on investment (ROI) of 20% or higher. In most cases, the payback period for manufacturing applications is less than three years. Automated office systems, on the other hand, yield less than 20% and may even result in negative ROIs.[2]

This article presents an overview of cost/benefit analysis methods suitable for office automation. Special emphasis on methods that provide quantification of effects make the decision making process less complex for MIS managers.

CLASSIFICATION OF OFFICE AUTOMATION APPLICATIONS

For purposes of financial justification, three types of office automation technology are identified:[3]

- Substitutive applications. Replace manual processes (e.g., electronic mail systems and integrated text systems).
- Complementary applications. Not only automate manual processes but also offer additional support functions (e.g., spreadsheet and business graphics programs). Complementary applications improve the decision making process both qualitatively and quantitatively.
- Innovative applications. Yield competitive advantage (e.g., the automation of forms processing). Computer support that improves the communication and coordination between technical and administrative personnel accelerates the ordering process, which results in improved customer service, more orders, and ultimately, increased revenues.

Another dimension, the organizational level, must be considered in any analysis of an office automation investment. A cost/benefit analysis must distinguish between the individual level, group or department level, and enterprise level. The impact of an application varies at each level. Examples of the effects that an office system application has at these three levels are as follows:

- Effects on the individual level:
  Shorter working times (e.g., work preparation times, word processing, communication).
  Improved quality (e.g., fewer disturbances, reduction in necessary storage, improved communication).
- Effects on the group or department level:
  Shorter run-through times.
  Earlier availability of information.
  Better access to information.
  Improved quality (e.g., clarity, readability).
  Acceleration of coordination processes.
  More flexibility.
- Effects on the enterprise level:
  Flexibility
  Competitive forces
  Image
  Customer service

Quantification of benefits at the individual level is easier than it is at the enterprise level. Office automation systems are integrated with other products. Benefits can be accrued when information from one component can be used by another component.[4] Information is made available faster and is more reliable. These benefits are more often seen at the group level of the organization. Installation of an electronic mail network can be viewed as an investment in a backbone technology supporting future growth in communications applications. They are not easy to quantify but must be included in the analysis.

Finally, the introduction of office automation technology is often the cause of organizational changes. Only organizational changes that are a direct result of the introduction of the technology must be reflected in a cost/benefit analysis. Those that would have occurred for purposes of simplification or that are otherwise independent of the technology should not be included.

QUANTIFICATION OF BENEFITS

The cost of introducing new applications is determined by up-front hardware and software expenses and maintenance expenses incurred during the life of the application. Compared to benefits, the cost items are more readily identified but difficult to estimate. The problem becomes even more complex with regard to the benefits which must also be identified and quantified.

Before discussing a method of quantifying benefits, the three types of applications—substitutive, complementary, and innovative—are discussed in terms of how they link with the quantification methods. Substitutive applications have the most impact at the individual level and are most readily analyzed. Saving potentials for a single employee can be determined, and the ensuing productivity improvements can be identified.

Complementary applications also realize improvement in productivity, but the effect here is basically in the quality of the work produced. The effects reach different organizational levels of employees. The method to determine the benefits must consider the effect at these different levels and must be able to calculate a value with and without the benefit of office automation. Complementary applications bring the risk that new activities will occur or that the new tools will be used even when there is no benefit to the organization (e.g., a graphics package might be overused). In such cases, the hidden costs of the application must be determined.

Office automation applications can be integrated with other applications. The benefits of the integration is most obvious at the department or enterprise level. For example, the flow of information and the processes driving the flow have nodal points at various levels. Any approach to quantify benefits must include an analysis of the process chains within the organization. Innovative applications involve an assessment of alternatives and the risks associated with each. Most concern with these applications resides at the enterprise level.

Determination of Workplace-Related Savings

Potential savings through the use of office automation can be achieved for several reasons:

- The time to perform a task is reduced (e.g., with electronic mail).
- Performing a task is simplified (e.g., automatic text correction with a spell checker).
- The frequency of performing special tasks is reduced (e.g., control tasks, recording of information).

Only seldom is reduction in personnel a result of office automation. Although labor costs can be reduced by decreasing the number of personnel in the mail room or typing pool, for example, the major benefit of office automation comes from an increase in productivity.[5]

To measure savings in the workplace, an analysis of the different employee categories performing administrative tasks is presented. The method is supported with actual measurements. The following five-step framework, based on an activity decomposition, is useful to support the evaluation.[6]

1. Grouping of employees into different employee categories (and subcategories) based on their activities; for example:
   - Managers. Top and middle management.
   - Professionals. Technical, administrative, and scientific, at both senior and junior levels.
   - Administrative. Technical and general administrative areas.
   - Clerical. Typing and word processing.
2. Decomposition of activities at each level (e.g., managerial, professional, and clerical) and the percentage of time spent on each activity (see Table 1).
3. Evaluation of the task breakdown of the different activities for each employee level.
4. Estimation of the office automation savings potential related to these tasks.
5. Derivation of possible time savings, productivity improvements, and total savings quantified with labor costs.

TABLE 1
Sample Work Profile for Middle Managers

Managerial Work	
Representation	5%
Coordinating information	25%
Creating documents	4.5%
Reading and responding to mail	2%
Planning	3%
Communicating	2%
Giving instructions	10%
Exchanging information	10%
Total	61.5%
Professional Work	
Performing calculations	3.5%
Creating documents	9%
Reading	1%
Coordinating with others	10%
Handling mail	3%
Total	26.5%
Clerical Work	
Storing and retrieving information	5%
Retrieving external information	4%
Miscellaneous	3%
Total	12%
Tasks Reorganized for OA Support	
Create documents	17%
Handle documents	5%
Read/work on	9%
Attend meetings/travel	42%
Communicate orally	24%
Miscellaneous	3%

Basic values for the savings can be evaluated from published studies and personal experience. Uncertainties combined with the calculations and the individual aptitude and attitude of the employees to embrace the new technology can be taken into consideration to yield pessimistic, optimistic, and likely estimations (see Table 2). These values represent subjective probabilities. The use of probabilities allows the generation of a risk profile for the estimated benefits for each employee level. With decreases and increases in the saving percentages, it is also possible to reflect the transfer of tasks.

Table 2 shows a savings potential of between $4,248 and $17,649 per year. The following list shows examples of productivity increases for single tasks. The values are based on an empirical case study analysis:[7]

- Spreadsheet calculations—60%
- Business graphics—50%–60%
- Word processing—30%
- Transfer, delivery of documents—90%
- Handle documents, storage—50%

How can these savings be interpreted? In Table 2, a middle manager can see a likely savings of 45% in the time spent on handling documents, or 2.25% more work time can be spent on more important activities (i.e., transfer of tasks).

This method is only a first step of the total analysis. Only qualitative results on an individual level are evaluated. Integration effects and quantitative results are not included. Experimentation in this area has shown that office automation systems cannot be justified without additional procedures.

An Activity Price Model to Reflect Changes in the Work Profile

An alternative approach calculates charge-out prices or cost values for work activities to describe the changes of the work profile for different employee classes as a result of office automation. The methodology described, though based on previous work,[8] uses different calculation techniques. The model is based on the assumption that each employee, in the course of accomplishing goals, performs a spectrum of tasks. In this example, the tasks are described as managerial, professional, administrative, clerical, and nonproductive. Each employee category performs a different amount of each of these tasks.

Table 3 shows a work activity breakdown structure for different types of employees. Based on a 2,000-hour work year, Matrix 1 reflects an employee/activity profile without office automation and Matrix 2, with office automation. Percentages are determined from work decomposition methods described previously. The row items re-

TABLE 2
Work Savings with Office Automation for Middle Managers

	Pessimistic Estimations		*Likely Estimations*		*Optimistic Estimations*	
	Total Percentage for the Task	*Percentage of Total Work Hours*	*Total Percentage for the Task*	*Percentage of Total Work Hours*	*Total Percentage for the Task*	*Percentage of Total Work Hours*
Create Documents	4	0.68	6.5	1.11	14	2.38
Handle Documents	25	1.25	45	2.25	50	2.50
Read and Work on	2	0.18	4	0.36	6	0.54
Attend Meetings and Travel	5	2.10	15	6.30	25	10.50
Communicate	2	0.48	10	2.40	15	3.60
Miscellaneous	1	0.03	2	0.06	3	0.09
Total Saving Potential		4.72		12.48		19.61
Cost Valuation per Year		*$4,248*		*$11,232*		*$17,649*

Source: H. Koch (see note 6).

flect types of employees (i.e., employee classifications) and the column items reflect types of work activities.

Ideally, an office automation system must cause a shift toward the direction of higher valued activities as well as reduce the nonproductive work time. The lower portion of Table 3 shows that these benefits were realized for the middle manager in the example.

The percentages in Exhibit 3 are then used as coefficients of the different work activities (see Table 4). The sum of these values is equated to the hourly cost of that employee category. A similar equation is generated for each employee category, and the system of equations is then solved. What results is the value per hour of each activity. The starting point of the price determination is the quantitative lowest-paid employee category. This is because the market value plus the nonproductive value portion is paid for this type of work.

Table 4 shows the results of the system of equations solved first without office automation and then with office automation. The task values are lower after automation is introduced. For each employee category, the time differences resulting from the introduction of office automation (derived in Table 3) are multiplied by the new cost values per hour. The summation of these products for each category is the quantification of the benefits.

To illustrate the quantification for one level of employee, the calculations in Table 4 have been performed for middle managers. The same result is seen if the initial matrix is multiplied by the cost differences (i.e., task costs without OA and task costs with OA). Further investigations could produce separate effects based on work content for reduction in nonproductive work.

The model is based on three essential assumptions.[9] First, the salary structure of the employees is commensurate with the quality and the scarcity of the executed work. Because of the influence of the age structure of the workers, this assumption is not always valid at the individual or group level but becomes more valid as the focus moves to the whole organization.

Second, the actual number of employees corresponds to the actual work load. It reflects neither free time nor overtime. Office automation would increase their free time and reduce their overtime.

Finally, higher valued work is more worth-

TABLE 3
Activity Breakdown with and Without Office Automation

Matrix 1 Breakdown Without Office Automation

Assumption: *2,000 Work Hours per Year*

	Managerial Work		*Professional Work*		*Administrative Work*		*Clerical Work*		*Nonproductive Work*	
Middle Managers	49%	980	17%	340	9%	180	7%	140	18%	360
Professionals	5%	100	57%	1,140	11%	220	11%	220	16%	320
Administrative Staff	0%	0	5%	100	58%	1,160	27%	540	10%	200
Clerical Staff	0%	0	0%	0	8%	160	84%	1,680	8%	160

Matrix 2 Breakdown with Office Automation

Assumption: *2,000 Work Hours per Year*

	Managerial Work		*Professional Work*		*Administrative Work*		*Clerical Work*		*Nonproductive Work*	
Middle Managers	52%	1,040	20%	400	11%	220	5%	100	12%	240
Professionals	6%	120	66%	1,320	11%	220	8%	160	9%	180
Administrative Staff	0%	0	8%	160	60%	1,200	27%	540	5%	100
Clerical Staff	0%	0	0%	0	8%	160	86%	1,720	6%	120

Work Times for Middle Managers with and Without Office Automation (in hours per year)

	Without OA	*With OA*	*Difference*
Managerial Work	980 hours	1,040 hours	+ 60 hours
Professional Work	340 hours	400 hours	+ 60 hours
Administrative Work	180 hours	220 hours	+ 40 hours
Clerical Work	140 hours	100 hours	− 40 hours
Nonproductive Work	360 hours	240 hours	− 120 hours

while. There is more time to spend on making decisions, which results in an increase in the quality and number of these decisions.

Practical experience has shown that the increase in time made available by the use of OA is not necessarily spent on higher value work. The execution of clerical activities can be shifted to a higher paid employee. This results in higher costs to complete a task but can also result in time savings. Considering the three assumptions and this reverse shift of activities, the activity price method should be used in conjunction with other techniques.

Measuring Hidden Costs

Hidden costs are often observed and attributed to the use of office automation. Unnecessary mail is created and distributed through electronic mail. Graphics packages and spreadsheet programs are overused or misused.

Positive and negative factors can be used to more accurately evaluate productivity changes on the individual level. From experience, these factors range from 5% to 20%. For example, the $11,232 benefit per year in Table 2 can be re-

TABLE 4
Determination of Work Values with and Without Office Automation

System of Equations 1: Work Costs Without Office Automation

Middle Managers	Cost = \$45/hour
Professionals	Cost = \$35/hour
Administrative Staff	Cost = \$25/hour
Clerical Staff	Cost = \$18/hour

		Results:
Middle Managers:	$0.49(\times 1) + 0.17(\times 2) + 0.09(\times 3) + 0.07(\times 4) = 45$	×1 = \$67.62/hour
Professionals:	$0.05(\times 1) + 0.57(\times 2) + 0.11(\times 3) + 0.11(\times 4) = 35$	×2 = \$46.01/hour
Administrative Staff:	$0.05(\times 2) + 0.58(\times 3) + 0.27(\times 4) = 25$	×3 = \$30.51/hour
Clerical Staff:	$0.08(\times 3) + 0.84(\times 4) = 18$	×4 = \$18.52/hour

System of Equations 2: Work Costs with Office Automation

		Results:
Middle Managers:	$0.52(\times 1) + 0.20(\times 2) + 0.11(\times 3) + 0.05(\times 4) = 45$	×1 = \$63.31/hour
Professionals:	$0.06(\times 1) + 0.66(\times 2) + 0.11(\times 3) + 0.08(\times 4) = 35$	×2 = \$40.38/hour
Administrative Staff:	$0.08(\times 2) + 0.60(\times 3) + 0.27(\times 4) = 25$	×3 = \$28.04/hour
Clerical Staff:	$0.08(\times 3) + 0.86(\times 4) = 18$	×4 = \$18.32/hour

Cost Effective for Middle Managers per Year with Office Automation

	Time Difference	*Costs with Office Systems (per hour)*	*Productivity Advantage (per year)*
Managerial Work	+60 hours	\$63.31/hour	\$3,798.60/year
Professional Work	+60 hours	\$40.38/hour	\$2,422.80/year
Administrative Work	+40 hours	\$28.04/hour	\$1,121.60/year
Clerical Work	−40 hours	\$18.32/hour	−\$732.80/year
Nonproductive Work	−120 hours		
Value Advantage of Middle Managers with Office Automation per Year			Total Amount: \$6,610.20/year

Key:
×1 Managerial work
×2 Professional work
×3 Administrative work
×4 Clerical work

duced by 20%, but on the basis of existing test installations, a much more detailed analysis is possible. Also, the analysis shows that additional education in the use of new technologies is necessary to overcome these problems.

To identify hidden costs of an application, MIS managers should:[10]

- Select sample uses for the new system (e.g., specific reports, letters, graphs).
- Divide work load into groups on the basis of complexity and determine how much time is necessary to obtain the output with and without OA.
- Evaluate the savings in time and costs.

- Develop a scoring scheme to classify how the output supports the organization's goals (e.g., −2 = not useful and +2 = very useful).
- Score the selected reports and letters (preferably using an independent group).
- Quantify the value of both the positively and negatively rated products.

In addition to the cost values for the selected output, this method provides an increase/decrease factor that can be applied to the total evaluation of office automation.

Analysis of Process Chains

Although this analysis of hidden costs only considers individual activities, the combined and far-reaching effect of combined activities must also be considered. Specifically, changes affecting interfacing personnel and interacting processes must be analyzed. To do this, such processes and the interfaces must be identified. Potential savings can be realized by the elimination of transfer procedures, the reduction of control processes, and the availability of more accurate information on a more timely basis.

The integration of processes yields other benefits also. Information already recorded can have many and various uses. For example, a customer information system can be created without making a separate effort to record the pertinent data.[11]

With the advent of office systems, not only do procedural changes take place, but the execution of tasks can be shifted from one employee type to another. The transfer of editing and correcting tasks from clerical personnel to administrative or professional personnel is one example. As mentioned previously, work is done by higher paid personnel; however, elapsed time to complete the work is reduced. The benefit of this time savings must be compared with the increased cost of labor. These questions can be addressed by performing a break-even analysis.

Evaluation of Qualitative Improvements

Often, the innovative application of office automation systems produces intangible results. Benefits most often cited are faster decision making, more accurate information, improved customer service, and better corporate image.

An organization's image is mostly determined by the external environment. The major determining factor is the customer. Although it is difficult to measure an improvement in image, one factor forming this image is customer satisfaction and customer service. One effective way of improving customer satisfaction is to expedite the ordering and estimating process of special orders. In the following example, an organization uses office automation to accelerate customized order processing and thereby improves its customer service. How is that quantified?

Initial estimates of results of the acceleration must reflect the savings realized from the transmission, transfer, and verification of information. On the basis of these changes in communication, marketing research tries to estimate the effect of this acceleration on sales.[12] Different alternatives can be considered. It can be assumed that the behavior of the competition will remain unchanged. Subsequent analysis can include the reaction times of the competition and different values of projected sales. Other possibilities must also be examined: What will happen if the investment is made at a later date? Or, perhaps the competition already has the edge, and it is too late for the investment in office automation.

For example, an organization with a projected sales volume of $80 million a year has an average time of 10 working days to process an order. Often, the order is placed under rushed conditions. The organization's competition needs about the same amount of time to place the order. Office automation can reduce the coordination problems between the technical and administrative departments. Is there an advantage to filling that order sooner? If so, how can office automation help beat the competition? The marketing department has forecasted the impact that earlier processing with office automation will have on sales volume. Table 5 shows the expected increases in sales for two, three, and four days' earlier processing.

TABLE 5
Marketing Forecasts for Expected Sales Increases with Office Automation

Orders Processed 2 Days Earlier		*Orders Processed 3 Days Earlier*		*Orders Processed 4 Days Earlier*	
Sales Increase	*Probability*	*Sales Increase*	*Probability*	*Sales Increase*	*Probability*
0.5%	20%	2%	25%	3%	20%
1%	60%	3%	50%	4%	50%
2%	20%	3.5%	25%	5%	30%

Let's perform the calculations with the following probabilities:
20% probability of a 2-day acceleration
60% probability of a 3-day acceleration
20% probability of a 4-day acceleration

$$\begin{aligned} &0.2(0.20 \times 0.005 + 0.60 \times 0.01 \times 0.02) \\ &+ 0.6(0.25 \times 0.02 + 0.50 \times 0.03 + 0.25 \times 0.035) \\ &+ 0.2(0.20 \times 0.03 + 0.50 \times 0.04 + 0.30 \times 0.05) = 0.02765 \end{aligned}$$

The expected sales increase is therefore 2.765%. On the basis of the yearly sales volume of $80 million, this amounts to $2,212,000 per year. It can be shown that a sales increase of 2% ($1,600,000) or higher can be achieved with 84% certainty.

Justifying investments aimed at improving the quality of the product (in this case, information) has become an important cost/benefit issue. Most MIS managers believe senior managers have to determine how much money they would like to invest for the higher quality; that is, the additional value that contributes to the cost justification.[13] But senior managers need and demand to know what value they will get for their investment. How do MIS managers quantify the value so that senior managers can make a decision based on some expected payback?

There are methods to quantify the results of advertising campaigns in the marketing area. It should be possible to do the same for office automation improvements. Scales that allow such measurements must be derived. Methods similar to those described for innovative applications seem appropriate.

QUANTIFICATION OF THE EXPENDITURE

Office automation usually requires a large capital investment in new hardware and software, but most of the investment is in "backbone technology," which provides a basis for other existing or future applications. For example, the installed office system network can also be used to connect a mainframe with a workstation serving any other centralized application. Or, office system workstations are used as remote terminals. Other potential uses of the hardware and software must be considered in the cost/benefit analysis. Particularly if office automation is combined with substitutive investments, comparative calculations must be done.[14]

Justification can be done at various levels of expenditure. For example, ROI can be calculated with investment expenses that are used exclusively for the office automation application. ROI can also be calculated for a larger initial investment in hardware and software that will be shared by several applications. This stepwise process not only presents a more realistic scenario but also focuses on different aspects of the issue.

EXPANSION OF QUANTIFICATION ATTEMPTS

There are two approaches to expand quantification methods. The first approach, the portfolio approach, tries to avoid the quantification and determination of the single application, office automation.[15] Justification is based on the argument that tangible benefits only occur from bottom-line activities and their related technologies. In a comparative situation, the office technology would lose against bottom-line applications. Therefore, a portfolio approach that combines several well-quantified applications with those yielding good but less quantifiable benefits is recommended. The portfolio will allow a distribution of the risk, which can be assessed with a scoring checklist that assesses the risk and economic benefits of an application (see Figure 1). The approach can be described as a method to select or evaluate a technology package for the total information processing plan.

However, the problem is selecting the best possible portfolio. First, the benefits of the single application must be identified. The difference is that it is easier to justify risky applications when they are combined in a portfolio with secure technical projects.

Figure 1 Sample Risk/Benefit Matrix for Office Automation Application Portfolios

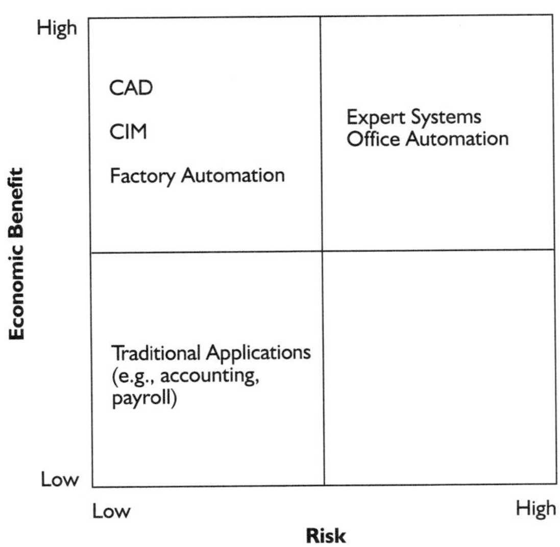

With this second approach to quantification methods, three primary questions must be answered when a cost/benefit analysis involves the assessment of improvement for which no immediate monetary value can be attached:

- Which values (i.e., measurements) should be selected?
- How should the values be described or measured?
- How can quantitative and qualitative values be combined?

On the basis of these questions, the results can only be subjective.

An example is a method that is based on such quantitative values as efficiency and effectiveness.[16] In addition, employee satisfaction plays a role in the justification. This method tries to evaluate employee satisfaction before and after the introduction of office automation. It is still unclear, however, how these subjective assessments (which could be made using a scoring system or survey) are combined with the quantitative values. Here, a scaling system that combines both measurements and determines the relationship between benefits and employee satisfaction is needed.

CONCLUSION

This article shows that a variety of methods are available to justify the investment in office automation. Difficulties exist in selecting a workable combination of methods and in determining the required information. Currently, MIS managers are faced with the challenge of developing and improving methods that allow the straightforward inclusion of qualitative benefits, because it appears to be qualitative benefits that will have far-reaching effects and will bear much weight in any decision making process.

NOTES

1. K. Crowston and M. E. Treacy, "Assessing the Impact of Information Technology on Enterprise Level Performance," *Proceedings of the Seventh International Conference on Information Systems*, San Diego (1986), 299–310.
2. A. B. Crawford, "Corporate Electronic Mail—A Communication-Intensive Application of Information Technology," *MIS Quarterly* (1982): 1–11.
3. M. M. Parker and R. J. Benson, "Enterprise-wide Information Economics (EwIE): Investment Evaluation Techniques for Information Technology," *IBM LASC Report G320–2782* (February 1986): 4.
4. P. Kapsales, "Before and After WP: An Office Automation Productivity Study," *Journal of Systems Management* (1986): 7–9.
5. D. L. Potak, "Function Analyzer of Office Automation (FAN), The Economics of Information Processing," *Management Perspectives* 1, ed. R. Goldberg and H. Lorin (New York: 1982), 221–31.
6. H. Koch, "Untersuchungen zum Nutzen-Kosten-Verhaeltnis der Bueroautomation unter Beruecksichtigung eines Unternehmens aus dem Anlagenbau" (Thesis), (Erlangen-Nuremberg, 1986).
7. P. Mertens, P. Zeitler, M. Schumann, and H. Koch, "Untersuchungen zum Nutzen-Kosten-Verhaeltnis der Bueroautomation," *Planung, Einsatz und Wirtschaftlichkeitsnachweis von Buerosystemen*, ed. H. Krallmann (Frankfurt-Main, 1986): 103–34.
8. G. Sassone and A. P. Schwarz, "Cost Justifying OA," *Datamation* (February 1986): 83–88.
9. P. G. Sassone, "Cost Benefit Analysis of Information Systems: A Survey of Methodologies," *Proceedings, Third International Conference in Enterprise-wide Information Management*, St. Louis (1986).
10. H. Brueckner, "Untersuchungen zur relativen Vorteilhaftigkeit des Einsatzes von Personal- und Universalcomputern in einem Unternehmen des Grossanlagenbaus" (Thesis), (Erlangen-Nuremberg, 1985).
11. C. E. Paddock, "An Assessment of Productivity and Operations Control as Motives of Office Automation," *Journal of Management Information Systems* 2 (1984): 76–86.
12. R. D. Smith, "Measuring the Intangible Benefits of Computer-Based Information Systems," *Journal of Systems Management* 9 (1983): 22–27.
13. W. J. Bocazany, "Justifying Office Automation," *Journal of Systems Management* 7 (1983): 15–19.
14. P. P. Chen, "Problems and Fundamental Issues in Cost Benefit Analysis of Office Automation Systems," *Emerging Office Systems*, ed. R. M. Landau, J. H. Blair, and P. Siegmann (Norwood, NJ: Ablex Publishing Corp., 1982): 223–30.
15. L. L. Gremillion and P. J. Pyburn, "Justifying Decision Support and Office Automation Systems," *Journal of Management Information Systems* 2 (1985): 3–17.
16. G. Schaefer and G. Wolfram, "FAOR-Methode zur Analyse und Bewertung von Kosten- und Nutzenfaktoren von Buerosystemen," *Handbuch der modernen Datenverarbeitung* 131 (1986): 54–65.

QUESTIONS

1. Why is it useful to classify different types of office automation applications?
2. What are the hidden costs of office automation? How can these costs be quantified?
3. What are the major challenges for justifying office automation? Briefly describe each challenge.

5.3

Making the Business Case for Change

PETER G. W. KEEN

Peter G. W. Keen is the co-author of Section 2.2 *Telecommunications: The Opportunity and the Challenge*, which described telecommunications technology. In the current section, he moves beyond the technology to the justification for undertaking projects. Be advised that this section is written in terms of telecommunications technology. However, the methods and ideas apply to technology-based productivity projects as well.

Keen categorizes various types of investments and shows how to make a business case for each type of investment. He divides business cases into four categories: operational moves, incremental moves, innovative moves, and radical moves.

These categories are of increasing boldness, complexity, and risk. He shows the different business cases that must be made to justify each type of investment, and the base case of "do nothing." It is important to recognize that the "do nothing" case must be considered. Often, the "do nothing" option is much worse than any of the proposals, even if the proposals do not clear the organization's rate of return criterion.

Failing to undertake a project can doom a firm to continuing to produce the proverbial buggy whip. Keen moves you beyond the nuts and bolts of the specific methodologies discussed in the previous two sections. He considers investments for which the traditional cost-benefit calculations are not appropriate or give answers that would ordinarily lead to rejecting a proposal.

Keen's advice is eminently practical. He tells you what information you need to gather and how to present the proposal so that you get "yes" as an answer.

Source: Competing in Time: Using Communications for Competitive Advantage (Cambridge, MA: Ballinger Publishing Company, 1988).

TRADITIONALLY, investments in data processing and telecommunications have been justified mainly in terms of cost displacement. This largely reflects a historical focus on automating clerical operations. It is also in part a reaction to the often uncontrolled growth in expenditures on data processing in the 1970s and the many unmet promises about the payoff. Senior managers are wary when invited by the information systems department to spend just a few million dollars more.

When telecommunications was almost entirely a telephone and telex utility, cost was the main issue. Service levels had to be maintained, but there were few "value-added" benefits to be gained from the available technology. As telecommunications became a force for innovations that affect business survival it became hard to justify the new investments in old ways. Obviously, though, those investments have to be systematically evaluated and justified. One cannot make the business case for major expenditures on telecommunications by vague talk about "productivity" or "competitive edge" nor underplay the very real risks involved in applying new technology to unproven business opportunities.

Vision and policy again are the starting points. There can be no one way to make the business case. Two extremes are to be avoided: forcing inventions and innovations to be put into a straitjacket of cost justifications and evading the very real problem of how to define business value when payoffs cannot be precisely forecast or benefits quantified.

An innovation like Merrill Lynch's CMA (Cash Management Account) is a radical move. The risks are high. The volumes and benefits depend on customer acceptance. There are few precedents to measure it against. No amount of cost-benefit analysis is likely to convince a skeptical or even a neutral listener. The case has to be made in terms of business trends and market opportunities.

The radical move has to be justified to and by the people at the top of the firm, and the time horizon they must address is a long one. The operational move requires a shorter, more budget-centered view, with decision makers at a lower level.

Rather than think in terms of "the" business case, we need to recognize that there is a range of business cases suited to the spectrum of telecommunications proposals, which goes from operational to incremental to innovative to radical moves.

- *Radical moves* explore an uncharted business, technical, or organizational area; involve major new commitments and risk; are an act of faith with no guaranteed payoff; and will have a strategic impact on the direction of the business if they are successful.
- *Innovative moves* involve a substantial increase in the level of commitment; build on current capabilities and experience; are based on a proven concept, though there are significant risks in implementing it; and provide substantial business value if successful.
- *Incremental moves* are the "next step" in an existing application within proven technology, involve an increase in scale or scope; have some degree of risk, and provide a measurable return on investment.
- *Operational moves* are a better way of providing an existing service or handling a standard operation; substitute new technology, equipment, or procedures for old; and involve limited, manageable risk; contribute directly to the "bottom line."

Making the business case involves getting the attention of the right people, marshaling convincing data, and persuading the people to make a choice. The people, the data, and the nature of the choice differ for each type of proposal. Figure 1 summarizes the sequence of steps.

In a radical move, what this article calls the "vision" is the business case. It sees communications as a means of refocusing key parts of the

Figure 1 Making the Business Case for Telecommunications

Classify the Proposal	**Identify "Panel of Judges"**	**Identify Data Needed to Make Case**	**Provide Initial Briefing/Education**
How radical is the proposal?	Who can make the decision, and whose support is also needed?	What competitive and/or financial information will be most convincing?	What does the panel of judges need to understand?
Radical Move: Communication as a means of refocusing major aspects of the business	Senior business management and corporate business planners	Competitive positioning and long-term business growth	Raise consciousness; focus on competitive context
Innovative: Support for key business activities in new ways	Business managers responsible and financial staff	Impact on revenues and profits	Highlight business options and implications for business planning
Incremental: Logical next step in extending or improving an area of business	"User": management and financial staff	Return on investment	Provide briefing on content of proposals; highlight tradeoffs
Operational: Improvement in cost, efficiency, or control	Financial and technical staff	Cost impacts	Provide detailed technical and financial vocabulary needed to evaluate proposal

→

Present the Options	**Analysis of Each Option**
• Do nothing	• Business benefits
• A few intermediate alternatives varying cost and risk	• Technology and technical risks
• "Ideal"	• Financial

→ **Decide**

business. It generally implies substantial potential benefits.

An innovative move might be a pharmaceutical distributor's building the infrastructure for managing inventories and putting terminals in the customer's store, in the same way as did McKesson and American Hospital Supply. The concrete business picture may well come from looking at what other firms are doing.

Incremental moves fit within the existing vision. For the grocery chain Publix, the decision to compete with Florida banks for control of EFT/POS was radical; the decision now to add a new bill-paying facility to its ATMs would probably be seen by its managers as incremental. Here communications adds a service or does something in a better way.

By definition, an operational move also fits within existing planning assumptions and organizational procedures. It does the same thing better. In the case of Publix, this might be an upgrade of the ATM equipment. Obviously, there is no point in even trying to justify a radical move if the vision is not clear and has not been successfully communicated by management.

CHANGING THE RULES OF THE GAME

Policy sets the organizational rules of the game. It defines authority, procedures, and planning assumptions—the constraints within which the business case must be made. It establishes territorial and political boundaries and, above all, delimits spheres of discretion.

The more innovative the move, the less likely it is to fit existing policy. An example is the proposal in a shipping firm to build an integrated communications capability that will handle all feasible aspects of document transfer, electronically link into banks' cash management systems to speed up the complex process of credit payment, and provide certification of ownership of goods. This intersection of electronic document management and cash management involves very complex legal issues and fairly complex technical ones. It costs about $500 to process the documents involved in a typical international trade transaction and a single week's extra delay in getting documents to the right place adds as much as $10,000 just in interest charges on a shipment of 200 cars or a tanker of oil, so that the potential payoff is huge. The business logic of such a proposal for electronic trade management is clear; however, such a radical idea—radical for all but the few firms in the industry that have already shaped such a vision—does not fit in with established lines of authority and responsibility or with capital investment and budgeting procedures.

The telecommunications manager is not expected to propose major changes in business and marketing. He or she is expected to submit investment proposals through the regular planning cycle, and expenditures will be allocated to operating units through a cost recovery formula.

In practice, in many firms the communications manager would not even try to present such a proposal. It makes no sense to stir up trouble by going around established procedures and by intruding on other people's territory. A firm's policy for telecommunications should make it acceptable for technical units to present proposals that have a strategic business impact.

Level and Mode of Justification, and the Panel of Judges

The business case is presented to a panel of judges—real people who have to be persuaded. Some of them are not part of the formal decision-making process but their opinions have informal weight. Before designing the presentation of the business case, one has to ask

- Who can *make* the decision; which group or individual must provide the authority, resources, and commitment needed to turn this idea into action?
- Who are the *opinion leaders* these decision makers rely on, to whom they turn for formal or informal review? These people can provide support and influence for the move.
- What *approvers* must one have? These are the people who can in effect veto the proposal, who must at least indicate their non-opposition if the proposal is to have a chance of being accepted.
- What kind of *data* is needed to make the case?

Generally, the decision makers will be a business manager or committee of managers, and the opinion leaders, respected planners or outside advisers, and the approvers financial and technical staff. The exact composition of the panel of judges depends on the type of move and the specific organization. As Figure 1 suggests, as one shifts down the spectrum from radical to operational moves, the relevant data needed to make the case become more specific and cost-focused. At the extreme of the scale of innovation, the key data presented to senior decision makers are trends and pictures from the industry and marketplace. At the other end of the spectrum, the relevant data are precise analyses of cost and volumes, to be reviewed by the people whose budgets and operations are most directly affected. This point is central. Appropriate data must be presented to appropriate people.

BRIEFING AND EDUCATION

Telecommunications and business strategy is a new field for everyone. Managers lack the experience and the training to handle it and most of them never expected to have to make decisions about applying communications. If they have not been briefed well before the proposal is presented and do not feel comfortable in taking part in the evaluation process, it is too late to change the situation. Education has to start early and be sustained.

The needs vary. For radical moves, the aim is to raise consciousness and establish a business focus; the workshop and reports should be filled with examples from industry and analysis of trends in the industry; the technology is related to these, not the reverse. Innovative business moves need a sharper focus and a framework for planning; the business relevance of major options and trends in the technology has to be highlighted; the goal here is to help business people feel they understand the key planning issues. Incremental moves are focused on the proposal, not the broader marketplace and technology; one needs to highlight the relevant choices and trade-offs. For operational moves the vocabulary and concepts needed to assess the proposal must be clearly defined; the focus is on briefing not education, on specifics, not concepts.

The steps described so far (see Figure 1) all precede the actual presentation and evaluation of the proposal. They are all too often neglected but are essential in establishing the context, criteria, awareness, knowledge, and vocabulary. Without all this being done, how can complex new applications of risky technology and business payoffs that go beyond cost displacement ever be adequately assessed?

By far the most common questions communications managers ask in discussions of strategic planning are: How do I get my senior managers to listen? and What do you do when they are only interested in the cost? The answer is to focus on vision and policy first, and build your education strategy well before you present your proposals. The questions can be restated to apply to senior managers: How should I be thinking about telecommunications? and What should I look at besides cost?

Presenting the Proposals

The uncertainties intrinsic in innovation make it essential not to get locked into presenting, and then having to defend, a single proposal. Equally, one must face up to and even highlight risk rather than downplay it. The more innovative the business application, the fewer the precedents against which to compare it. The greater the likely payoff, the greater the likely risk. The decision to make an innovative or radical move is a business decision about return and risk. The real issue is how far along the spectrum of risk and return management is willing to go, not whether to say yes to one specific combination of the two.

This is something too many data processing managers overlooked in the 1970s. They generally presented their senior management with a single proposal and rarely stressed the risks involved. This damaged their credibility and put them on the defensive. They had to respond to questions about risk with "yes, but" or discount the problem. To do otherwise meant the proposal being turned down.

A far better approach is to say "This is our preferred option. It has these risks and potential payoffs. Here is an alternative that has either less risk or less return. You choose." "You" means the panel of judges; the telecommunications manager has to give them a real business decision to make, not a go/no go response to a single proposal.

There is generally, for any one business opportunity, a wide range of technical options and levels of commitment. Obviously, it makes no sense to look at all the combinations. A much more effective approach is to narrow the analysis to three or four key alternatives:

- *Do nothing,* a real option, has to be evaluated systematically.
- *The "ideal"* alternative assumes the best of all worlds—things will work out as planned, market assumptions will be correct, the technology will work, etc.
- *A few options between these extremes.* One of them will almost certainly be the one recommended by the communications manager and should be flagged as such.

Based on these options the implications of changes in key factors can be analyzed from the start.

VALUE ANALYSIS

The formal proposal for any major investment must look at both cost and value. Generally, they are compared directly; however, until the value is established, *any* cost is disproportionate. Too often, as well, the costs are quantitative, predictable, and immediate and the benefits qualitative, uncertain, and deferred. Business inventions cannot be cost-justified, nor can direct and simple comparisons be made between cost and benefit.

Innovation is value-driven. There is a wealth of empirical data to support this point. The principles of value analysis are:

- Separate benefit from cost.
- Establish the benefit first (Ask: What is this worth?).
- View cost as a threshold (Ask: Would I pay $X to get this benefit?).
- Face up to qualitative benefits.
- Rank order "hard" and "soft" benefits.
- Define key indicators by which qualitative benefits can be evaluated.
- Use "bounding" to get a rough estimate of needed and likely benefits.

Often, in companies using telecommunications to gain a clear business advantage, cost displacement was not an expected or intended priority. Because automation in general, and telecommunications in particular, have traditionally been viewed in terms of cost, it can be hard to shift the perspective. Organizations are far more skilled in cost analysis than benefit analysis. The way to begin to change this is to learn to list, track (through pilots), and rank benefits.

THE DYNAMICS OF INNOVATION

Innovations are made by organizations on the basis of value. The early adopters differ from the late ones. They focus on a small number of benefits, which usually relate to a specific "felt need" with a definite time horizon. They are rarely interested in "office automation" or "productivity" but in getting the budgets in on time, seeing where they can improve short-term forecasting, or keeping the sales force better informed about inventories and prices. Innovators are not risk-takers. In fact, they tend to be skeptical about wild claims and do not respond to "technology push."

Many surveys show that most product innovations come from "demand pull"—the definition of a need in the marketplace to which the technology is then adapted. Technology does not create needs, though education about its capabilities and availability may. Quite often, knowledge about the technology is brought to the innovator by "gatekeepers," who bridge the two cultures of business and technical expertise; academics and planning staff often play this role.

Technical specialists too often think in terms of technology push. That is a solution looking for a problem. The approach to telecommunications and business strategy recommended throughout this book is to focus on the benefits in terms that are convincing to innovators—businesspeople who combine healthy skepticism about technology and open-ended promises with the willingness to put authority and resources behind a project once they are convinced. These are the people who can change the terms of reference for telecommunications. There are three ways to do so:

- *Focus on exemplars;* these are proven and concrete illustrations of benefits and opportunities.
- *Define the business vision* for telecommunications.
- *Focus on the benefit side* of the cost-benefit calculus.

This does not mean evading the issue of cost, but knowing where to put the emphasis. For radical moves, a strong focus on value is called for, with costs defined in terms of a floor: at least $X, probably $Y, and no more than $Z; once the benefits have been defined and management is convinced, then a follow-on study can look at the costs in more detail. For innovative moves, emphasize value but also be more detailed in indicating cost components and ranges. For incremental moves give equal weight to benefit and cost, but treat them separately and in that order. And for operational moves, be very precise about costs, since that is the issue here; the firm does not need to make the change and the real benefit is the cost impact.

TECHNOLOGY ANALYSIS

The risk inherently involved in information technology should be highlighted rather than smoothed over. As with investment in stocks and securities, high return generally implies high risk. With telecommunications, getting a strategic edge means doing something earlier than one's competitors, which may involve using a technology that is not fully proven. Value analysis establishes the return from the investment. Business managers can assess the business risk, the likelihood of getting the return, but cannot intuitively do so for the components of technical risk, which include the technology itself, the product, and the vendor. Detailed risk factors which apply to all three of these are:

- Performance. How likely is this to work as promised?
- Delivery. Will announcement dates and delivery schedules be met?
- Substitutability. If there are problems, can an alternative technology, vendor, or product be substituted without too much delay and cost?
- Regulation (especially key internationally). What is the likelihood of government (for example, PTT) regulation constraining acquisition, use, or even performance?
- Control. Are we dependent on the vendor or regulators if there are problems or changes?
- Costs. How reliable are our estimates of costs and cost trends? Are there indirect costs we cannot predict such as tariffs?

This list of risk factors is not complete, and the technical staff will have to go into far more detail in their planning and analysis. The goal here is not to produce an exhaustive technical analysis but to give the panel of judges a complete enough picture of the likely risks to be able to trade them off against the likely returns and to feel comfortable that they have been properly briefed.

FINANCIAL ANALYSIS

It is only after the issue of value has been established that cost can be brought in, and even for operational moves value and cost should be kept separate. This flies in the face of traditional approaches to cost-benefit analysis but is surely the only sensible way to handle innovative business proposals that rest on technology. Bundling costs and benefits together, trying to express uncertain qualitative benefits directly in terms of cost displacement or avoiding the problem entirely, simply muddles the issue. This in no way means one need not be rigorous about cost. The overall issue is handling uncertainty and establishing business value. This means be rigorous about value, be rigorous about risk, then be rigorous about cost.

Being rigorous about cost means focusing on cost dynamics rather than cost estimates. In most instances where a proposal for telecommunications introduces a new service, it is close to impossible to estimate volumes. In many instances, supply creates demand and companies consistently underestimate volumes sometimes because they previously overestimated them. An example is electronic mail.

Company A estimated demand would be fairly low; this was based on the assumption that 10–20 % of phone calls and memos would be shifted to the new medium. It turned out to be very high in one division because the sales force found it an entirely new and effective way of keeping in touch with the office when they were on the road. In Company B, use was far lower than predicted on the basis of a pilot because project users were charged for the service whereas in the pilot study it had been free. In Company C, use was initially low but took off rapidly when a new software package was installed which made it simpler and easier to use.

Cellular radio, voice mail, full-motion videoconferencing, home banking, network information services . . . There is a flood of new technologies and new services where supply may or may not generate demand, cost and ease of use may change customer response, or peoples' learning may open up new applications. In many instances the hope is that demand will be far higher than predicted—that the productivity gains from videoconferencing will lead to the rapid substitution of telecommunications for travel, or that distributors will enthusiastically adopt the dealer order entry system, or that the new all-in-one reservation system will be a winner.

For all these reasons, it will often be impossible to predict volumes and hence costs. The two key questions then are:

- How do costs change with volume?
- Would we rather overestimate demand and have unused capacity or underestimate it and not be able to meet demand?

The second question is a key business issue. It can be rephrased more specifically: We expect that about 200 customers will sign up; if we get over 1,000, we will be unable to accept them without significant degradation in service. If we get only 70, we will be operating at a loss. Which direction of risk do we choose—oversupply or undercapacity?

Costs do not necessarily (or often) increase directly with volume. There are many components which have a fixed capacity, and expansion involves additional capital investment. For example, a communications switch can handle a given number of terminal "ports"; when that limit is reached, new capacity must be added in multiples of ports. The dynamics of cost are thus shown in Figure 2.

The costs may be significantly affected by changes in tariff structures and rates. The use of private facilities and satellites may suddenly become less cost-effective when a foreign PTT or U.S. supplier switches to "volume-sensitive" tariffs or reduces costs for terrestrial circuits. This is

Figure 2 Cost Dynamics: How Fixed and Variable Costs Move with Volume

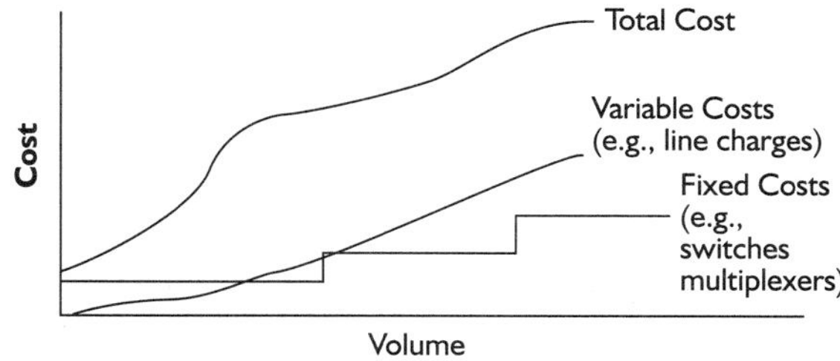

a growing problem; not only can we not predict volumes, but we may be wrong by a large margin in our assessment of cost trends.

The problem of projecting costs will not go away by ignoring them. We need to summarize them in terms of "If the capacity is X and the volume Y, the likely total cost is Z." The issue is not cost per se but net business value. For a firm that wants to use its communications capability to deliver products electronically, cost is only part of the calculation. But in each instance, the business questions are:

- What is the likelihood of particular traffic levels?
- What happens if the estimate is wrong? Would we rather err on the side of having too little or too much capacity?

Only management's panel of judges can choose the preferred risk direction. That is a central issue for telecommunications when the traffic has significant business value, or the relative costs are high, and in either instance volumes cannot be predicted. The telecommunications manager can highlight the technical risks and show the net business value for given traffic and capacity levels, but it is not his or her job to make the final decision, only to present the business case.

GIVE US A REAL BUSINESS DECISION TO MAKE

In practice, the business case is too often made in terms of cost and cost recovery, of single proposals, and of technical issues. That does not make sense, except for simple operational moves, and it locks out business innovation through telecommunications. Communications managers have to work through the stages outlined in Figure 1 and create a business-focused dialogue, as do business managers.

If there is no vision, how can existing policies be challenged and changed, so that opportunities for radical and innovative moves can be brought into top-level planning circles? If the policy for telecommunications cannot be changed, how can anyone step outside or around the existing justification process? If communications managers cannot influence the justification process, how can they find and convince the right panel of judges? If they cannot brief or educate the judges, how can they present their cases?

Senior managers can change policy and procedures far more easily than technicians. The strategy for making the business case presented in this chapter is the process that top managers can and should encourage, rather than the one that communications planners should maneuver to facilitate. The demands from senior executives to their communications staff then should be: "Highlight the uncertainties" and "Give us a real decision to make."

QUESTIONS

1. Why is it important to present alternatives rather than a single proposal?
2. What is meant by value analysis? How can it be used for justifying technology proposals?
3. Why is it often more important to focus on cost dynamics rather than on cost estimates?

5.4 Gut-Level Justification: As Valid as Cost Accounting?

LAWRENCE GOULD

Justifying projects based on conventional accounting and financial principles is the standard good business practice. Yet, companies can and do become non-competitive because they rejected office or factory productivity projects at some point in the past. Gould argues that in many cases analysis is not done. Firms recognize that they must introduce the new way of working because they will be overtaken by events or because they see it as an opportunity to leapfrog the competition.

Gould's article is primarily anecdotal, rather than being based on scientific method using a large data sample or a sophisticated model. Subjective articles such as this one are useful in bringing an opposing view to the forefront. It is up to you, the reader, to judge under what conditions, if any, the subjective approach should be used.

REDUCING costs is one reason why companies large and small implement automation and "reengineer" their organizations. The "softer," strategic reasons include improving product quality, reducing time to market, accessing data faster, and improving customer service.

Good business practices virtually demand a rigorous accounting of these reasons against cost. However, anecdotal evidence suggests that many financially based justifications are made at the gut level—especially by small manufacturers.

"Automation is expensive and it's often difficult to justify using conventional accounting rules," says Detlef Koska, a consultant at A. T. Kearney Inc. (Chicago, Illinois). "Some companies have rigorous guidelines. Others allow you to do all kinds of creative justifications. And with another group, their gut feeling says they want automation, that it will probably put them light-years ahead of their competition, and, therefore, they implement it."

Relaxing the rules on justifying automation is not an attempt to allow companies to implement more automated gadgets, fancier technology, and more wide-ranging Computer Integrated Manufacturing (CIM) systems. Instead, it is a way to make manufacturers more aggressive.

"Rigorous accounting rules can make manufacturers less aggressive than they could be," asserts Koska. "If they can't fully justify the automation project, then they wonder. 'Why even fill in the paperwork?' This attitude is counterproductive for the good of the company."

Small companies (generally with sales of less

Source: Managing Automation (April 1992): 52–54.

than $50 million a year) often look for spot solutions, says Robert Leach, partner in charge of the electronic consulting group for Andersen Consulting in Los Angeles, California. Their investments tend to be more tactical, the solutions more quantifiable. The piecemeal automation implementations and systems integration are more feasible, and more justifiable.

Firms in the $50 million to $400 million range typically realize that point solutions are not adequate. So, their automated solutions often cross organizational boundaries, thereby complicating the justification of them.

Those with sales above $400 million, says Leach, "typically think—and actually use—terms like 'strategy' and 'sustainable advantage.' The problem for these companies is in 'strategically aligning' expensive automation implementations with the overall strategic objectives of the company."

The more ambitious companies are usually more willing to take gut-level risks in justifying automation based on soft strategic criteria. "Companies that are interested in applying a particular new technology or that are interested in being among the pioneers realize that numbers might not exist," says John Detwiler, vice president of technology and development for Daxus Corp. (Pittsburgh, Pennsylvania), a supplier of data collection and acquisition systems. "They go out on a limb and do what feels right."

The extent of the cost-benefit analysis is also a function of the risks involved. "Small automation projects are often done with little justification," explains Robert Eiler, partner and national director of cost management for the accounting firm of Price Waterhouse in Cleveland, Ohio. "Major automation projects—completely automating a facility or developing new automated plants—require a fairly stringent justification.

"The number one criterion used to determine whether a company uses a lot of analysis seems to be what has happened in the past. If a company automated and didn't get a return, it will be a lot less willing to spend more dollars on automation without a fairly stringent justification. Companies that are successful in using automation to meet a strategic need are more confident they can pull off these types of projects without the finest of justifications."

For these companies, cost is a factor, but not the sole factor, when evaluating alternative investment proposals, says Martin Piszczalski, director of the Manufacturing Automation Planning Service at the Yankee Group in Ann Arbor, Michigan. "Most of these companies justify projects more by gut feel than they'd really like to admit. In some cases, the justification is a matter of pacifying the accountants."

One automation-related expense that defies a rigorous cost justification is retraining, notes Leach. "To achieve a directional change, such as a time reduction, you have to make an enormous investment in retooling people—namely, in retraining. No way can you get a direct tie between the time for training and the time you take people away from their traditional jobs in order to justify a return to the business based on that incremental investment you made in training. Yet we know that retraining is critical for getting the results you're trying to achieve."

Consequently, justifying automation often boils down to a management decision: Do you want it or not? Does the automation improve a process? If it does, that's reason enough.

"I don't think financial justification is dead,"

Figure 1 Most Common Metrics Measured by U.S. Manufacturers

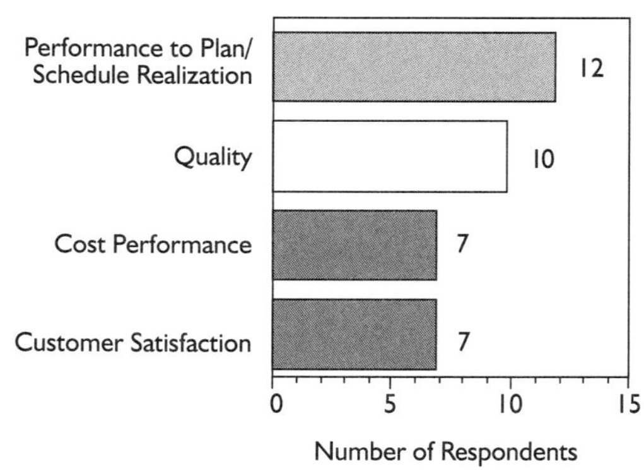

Source: The Yankee Group

says Eiler. "Companies implementing automation are probably requiring more justification now than in the latter part of the 1980s. It's probably taking a different direction."

In particular, the justification metrics have moved from strictly cost to a strong interest in, for example, time and quality, explains William Lehman, managing partner of manufacturing management consulting for Price Waterhouse in Oak Brook, Illinois. "While the real benefit of an automation project is in terms of dollars, one hopes for an improvement in the revenue stream. Therefore, you are betting that your strategic imperatives are right, and if so, you'll leapfrog a competitor or two, and the revenue stream will pick up to the point where the return on investment is shorter. The problem is that you can't prove this with a pencil like you can when you are doing engineering cost studies."

But some companies, through experience, have been able to "prove" the softer metrics. For example, IBM Corp. has computed that every 1% improvement in customer satisfaction leads to $257 million in additional revenues.

NO perfect, mathematical, scientific, and time-proven formula exists to crank out a justification in a clear and convincing manner. Even many of the people who have implemented automation—and justified it to management—will admit that. What many companies do is to identify where the pools of controllable dollars are. They identify expensive processes which might need to be reengineered. They then try putting a price tag on potential benefits. As a result, says Eiler, "they might not be able to get to the second or third decimal point of a cost justification, but they can surely get to the area of business that seems to give them the highest payback."

Unfortunately, cost-benefit analyses—whether based on cost or other factors—is fraught with politics. "I can poke holes in any justification if I'm against the project," admits Koska. "Conversely, if I'm in favor of it, I can show that my manufacturing engineers forgot half of the savings and here's a way to get the justification past the CFO. It depends on what kind of political campaign the automation sponsor is running."

Some companies have no choice but to implement some form of automation. "When something is a condition of doing business, you do it," says Roger Nagel, a professor of manufacturing engineering and operations director of the Iacocca Institute at Lehigh University (Bethlehem, Pennsylvania). "You have to spend the dollars to get the quality that will get you the business. If you don't, you won't have the business. For instance, many more companies are insisting that their suppliers be linked in some form of electronic data interchange."

That is exactly what drove GKN Automotive Inc. (Auburn Hills, Michigan), a supplier of constant-velocity joints to the automotive industry, to implement electronic data interchange (EDI) in the mid-1980s (*Managing Automation,* January 1990, page 42). "Our customers pretty much forced us into EDI," says Jay McLean, vice president of business development for GKN Computer Services Inc., the company's management information systems subsidiary. "It's a condition of doing business with our customers. It's not an option, so implementing it didn't need a lot of analysis."

Likewise, the strategic driver might be a result of a desire to sustain expertise. A company might be worried about losing its key people through attrition, so it might consider implementing automation not only to capture that expertise but to also generally streamline operations for a strategic advantage.

The justification for automation might be based on technology. For example, GKN justified implementing manufacturing resource planning (MRP II) because MRP II promised to manage the enterprise through one set of integrated modules. GKN's justification for the specific MRP II system it bought was based on the breadth of its functionality and features compared to competitive MRP II systems. And it chose the particular EDI package it did because it integrated well with the MRP II system.

Often, the justification comes after the fact.

Figure 2 Metrics for Determining Manufacturers' Highest Priorities

Source: The Yankee Group.

"Projects often get proven once you're in them because then the real numbers come out," says Norman Schmotzer, president of B&A Manufacturing Co. (Riviera Beach, Florida), a manufacturer of masonry drill bits. B&A installed an MRP II system in 1990 to give itself and its customers better product and cost information. "The next project is easier to justify because now you have a basis from which to justify added automation."

With automation already in place, softer metrics typically apply when justifying the addition of more automation. Those metrics, says Charles Sirro, director of business solution development at CAP Gemini America (New York, New York), an information-systems software vendor and consultant, include the need to move away from an old or unsupported technology or to increase the capacity of the information systems in line with business growth. The availability of third-party products on one computing environment may drive a change to that environment.

Compliance with a corporate information system standard is another reason why some firms automate or change systems. Likewise, corporate management may demand that remote sites be integrated.

Taking a softer—that is, gut level—approach to justifying an automation implementation is not for the faint hearted. It is fraught with risks.

Reducing the risks falls under the domain of classical project management. "The major risks are the same ones we've always had to deal with," says Lehman, "and they have relatively little to do with how well the hardware works. They are the organizational and management risks—those that don't get modeled too well."

Start by getting top management's buy-in. "The justification process has to involve developing a consensus among the top executives as to what they think the business can be and what an investment in technology is likely to accomplish," he adds.

ANOTHER way to justify an implementation is to view it in the context of an overall strategy and not as an incremental investment that will generate a particular return on investment revenue. For example, Emerson Electric (St. Louis, Missouri) took a year to develop its strategy, which resulted in product design and factory changes that took four years to implement. Leach says, "That's hardly a quarter-by-quarter correction."

Companies taking the softer approach to justifying automation, explains Detwiler, usually rely on the track record of the people they are dealing with, including the people in their own organization, the equipment supplier, and the systems integrator.

Benchmarking and evaluating the successes of other companies is another way of adding meat to a cost-benefit analysis. "Fundamentally, benchmarking is an attempt to not reinvent the wheel," says Lehman. "You'll find a company that has invented the kind of wheel you're looking for and see how they did it. Then you can use that wheel, use only half of it, or improve on it."

Most important, says Sirro, a company can reduce the risk of implementing automation by spending more time up front in planning, designing, and prototyping. These initial steps typically cost less than the last step: implementation. "If you spend the time and money up front in a comprehensive fashion, then the execution stage, instead of becoming a technical challenge, becomes a management challenge because you have addressed virtually all the technical complexities."

Money does talk. A company's health—in fact, its very existence—is based on balancing costs with cash flow. However, a company can spend more time talking about justifying a project than implementing it and being competitive. "The hardest part in justifying automation is making the decision to go with a project," says Schmotzer. "It's that first step that's the make or break."

In the end, the company's managers have to believe in the project and believe that it will really do the company some good.

"Decisions are very different from what they were 10 or 15 years ago," adds Nagel of Lehigh. "Now companies are making investments on which they don't have an option. The conditions of doing business have changed. You're being judged more on your systems, not just your products."

QUESTIONS

1. What are the limitations of cost-benefit analysis?
2. In what situations should managers use "creative" justification methods?
3. Gut-level investment decisions are considered to be risky. How can the risk be reduced without going to formal cost-benefit analysis methods?

5.5 Strategic Information Technology Investments: Guidelines for Decision Making

ERIC K. CLEMONS AND
BRUCE W. WEBER

Under appropriate conditions, information technology can confer advantage to a firm. Even when it fails to confer advantage, it may still prove crucial. Both concepts, competitive advantage and strategic necessity, confound traditional financial analysis.

As the strategic importance of information technology increased, the decision of where and when to allocate resources to information technology projects became more difficult and more risky. Executives are tempted by the opportunities for strategic impact, but struggle with the massive expenditures and uncertainties involved. Evaluating an opportunity and judging the strategic impact it will have is difficult. Since these analyses are done infrequently, they are usually ad hoc.

In this section, Clemons and Weber offer seven principles on which to base an evaluation of a strategic information technology venture. These principles are eminently sensible guidelines based on experience. They range from modeling the investment decision—through managing risk—to preparing for unanticipated upside and downside implications.

INTRODUCTION

The IT Gamble

New technologies and technology-driven opportunities are being developed and presented to managers at an ever-accelerating rate. Chances to join networks, forge electronic linkages with customers and suppliers, and develop and implement new systems architectures require careful consideration and analysis. The stakes and the possible returns are enormous.

In September 1988, AMR Corporation (parent company of American Airlines) announced that it would consider bids for selling off its highly regarded Sabre travel reservation system, in use in over 14,000 travel agencies. Wall Street put a

Source: Journal of Management Information Systems (Fall 1990): 9–28.

value on the system of $1.5 billion. At the time, AMR's market value was $2.9 billion, indicating Sabre's value in the 12 years since its introduction had grown to exceed AMR's core airline business.[1] And yet when American was first considering the $40 million start-up cost of the system, the decision to develop was not easy to justify (Quickel, 1989). AMR's chief competitor, United Airlines, had decided against wider distribution of its own travel agent reservation system, which was in use on a small scale. A full roll-out of United's Apollo system came several months after Sabre's introduction. The market share advantage gained by Sabre has continued, and its share is now 40%, compared to Apollo's 29% (Oram, 1989; Standard & Poor's, 1989).

The downside can be substantial as well. Chemical Bank and a Chemical and AT&T joint venture abandoned their disappointing electronic home-banking product, Pronto, in December 1988, and were estimated to have written off $70 million in assets associated with the system (Guenther, 1988). Despite aggressive promotion and technological excellence, Pronto attracted only 10,000 subscribers in six years, which was far short of expectations and well below the level necessary to break even.

Strategic systems initiatives such as these carry immense risk, yet may offer commensurate reward. They are difficult to undertake, and the decision to develop is rarely obvious in advance. Sabre's ultimate value was not readily apparent in the mid-1970s and the investment decision was neither obvious nor easy. Difficulty in justifying an investment for which the value now seems indisputable should not be surprising. User adoption, future benefits, and competitive impact are difficult to forecast. And the system itself may so radically alter the industry that prior assumptions are invalidated.[2] We have found that strategic IT programs often lack formal economic justification. We disagree, however, that an *act of faith* is necessary to justify strategic technology investments.[3] Principles exist, and many of the novel techniques that have been used to justify some of today's highly visible and successful systems are rational and can be applied to other IT investment decisions.

Plan for the Reading

This reading describes the IT investment decision problem, and proposes principles for evaluating and selecting IT opportunities in which to invest. Our guidelines are based on empirical studies, and on theoretical work on the valuation and timing of risky investments. We also consider the role of a company's unique resources in generating competitive advantage, and the role of strategic necessity in guiding IT investment decisions (Clemons and Kimbrough, 1987).

The second section of this paper will describe the general problem of making technology investment decisions. Companies are no longer able to justify investments in advanced technology solely on the basis of cost savings and tangible returns on investment. At the same time, these projects are considered essential for firms to maintain their competitive position. The third section examines the difficulties specific to strategic IT decisions that executives face. In the fourth section, we detail intangible benefits of IT that should be considered when making the investment decision. Often these benefits appear serendipitously midway into the project. In the fifth section, lessons from strategic IT successes are developed. In our studies, nontraditional justification methods have often proven more appropriate for making strategic IT commitments. The sixth section proposes seven principles for guiding the evaluation of an IT investment. These range from applying analysis tools, through identifying upside and downside implications, to recognizing multiple forms of risk that are inherent in a strategic IT venture. The final section presents ongoing research into active management of the risks of large-scale IT developments.

THE TECHNOLOGY INVESTMENT DECISION PROBLEM

Companies concede that they can no longer justify leading-edge automation projects in their factories, the development of large-scale systems architectures, or communications networks for their organizations, solely on the basis of return on investment. The direct-cost savings and higher production rates that once paid back IT investments in a short time are no longer sufficient to offset the high price of many of today's technology projects. Apprehensively, managers are using broad strategic considerations, such as enhanced quality, better customer service, and keeping up with the competition, to justify technology investments. Executives are considering alternative justification methods, and questioning the use of discounted cash flow (DCF) analysis. Unfortunately, generally accepted principles for making IT investment decisions are not yet available.

In this section we review evidence that managers struggle to evaluate technology investments. We review the strategic systems literature, and approaches to justifying strategic IT initiatives. Work in the early years of data processing management by Emery (1971) identified the critical role of intangible benefits in justifying systems investments at a time when the applications were largely intended to reduce costs and to automate paperwork and routine tasks. Emery noted that "the difference between a *tangible* and an *intangible* lies in the difficulty of estimating monetary value." He proposed methods for calculating the value of intangibles, such as the reduction in stockouts from a new inventory system. Subsequent research on measuring the intangible benefits of strategic systems has not kept up, and shortcomings still exist. The impassioned support of a "champion" (Beath and Ives, 1988), or convincing top management of less visible benefits of strategic systems (Lederer and Mendelow, 1988), may not be suitable for some of today's board-level decisions on $100 million IT infrastructure investments.[4]

Evidence

Recent studies have shown that executives are troubled by their firms' technology investment decisions. A 1988 survey of technology managers in UK banks found that 80% of the respondents feel technology has fallen short of their expectations, and more than half confirm that systems investments are not generating the anticipated improvements in productivity (Weil, 1988). IT applications have grown beyond supporting efficiency improvements. In seeking improved effectiveness and strategic goals, performance and return on an IT investment can no longer be quantified so easily. Instead, executives struggle to weigh intangible benefits and strategic impact against the often imposing hardware and development costs.

In a 1986 survey of how technology investments are justified in 67 Fortune 500 companies, 77.0% of respondent firms felt they will soon be making additional IT investments. However, 89.5% have *great difficulty* quantifying the intangible benefits of proposed systems, and another 61.2% find that the *uncertainty of benefits* makes justification methods hard to apply (Bennett, Reed, and Robertson, 1988). Clearly, executives are confronting challenging IT investment decisions, yet few appear to be well served by existing investment justification models.

Discounted Cash Flow Analysis

A broad debate has unfolded on the appropriate use of Discounted Cash Flow (DCF) methods for directing investment spending.[5] Managers recognize that many of today's large-scale technology investments—flexible manufacturing systems and integrated account processing systems in banks—are not easily justified on a cost-reduction basis (Holusha, 1986; Port, 1988). As we discuss in subsequent sections, this is because the costs of *not* making the investment are frequently ignored. Critics of DCF methods feel that an overreliance on discounting techniques has inhibited capital investment in American businesses, reduced research budgets, and harmed the competitiveness of U.S. industry (Hayes and Garvin, 1982).

Discounting's defenders reply that DCF analysis is based on sound theoretical principles (Logue and West, 1983). Rather than biasing against long-term investments, they contend that DCF is an objective framework for comparing the relative merits of investments with different cash flows (Brealey and Myers, 1981).

We feel that the evaluation of strategic IT programs is fraught with complexity that is not well handled by DCF analysis. The debate over the use of DCF analysis for technology investments obscures the current misunderstanding of how IT can provide strategic value. Rather than continuing the debate, our guidelines are techniques for augmenting—not replacing—formal financial analyses.

The Vanishing Status Quo

The trap of the *vanishing status quo* is real and prevalent. As management often neglects in its analysis of IT opportunities, the alternative to investment and development is rarely a continuation of existing industry conditions. Technological innovations have continually reshaped industries since the Industrial Revolution began in Britain in the 18th century (Landes, 1969). IT has increased the pace of change. Centuries-old traditions of floor trading in London's stock market were abandoned a few days after the Big Bang market reforms took effect in 1986 (Clemons and Weber, Spring 1990). Activity moved to a screen-dealing system that was better equipped to handle the demands of modern securities trading.

Executives making IT investment decisions must acknowledge the near certainty that technology-induced change will alter the structure of their industry. Continuation of the status quo, as the alternative to a proposed IT program, is a glaringly naive assumption. Barwise, Marsh, and Wensley (1989) identify *misstating the base case*—what will happen if the project is not carried out—as a primary flaw in most companies' investment decision making. They recommend spending more time considering the likely outcome of *not making* an investment; the company's position is likely to deteriorate, rivals will act, and market share and margins of firms that fail to respond will fall.

Strategic Systems

The strategic systems literature has focused on innovation champions and winning the support of the CEO as methods for gaining acceptance of hard-to-justify IT projects. Identifying opportunities and threats, and analyzing the financial consequences of competitive advantage and strategic necessity, have received less attention than they deserve. IT affects more aspects of a company's operations and has become a critical enabler of many business strategies. For instance, IT has:

- Formed distribution channels, reaching out to the organization's customers; McKesson Economost, AMR's Sabre, and BZW TRADE are examples (Clemons and Row, Summer 1988; Clemons and Weber, Spring 1990).
- Enabled tight linkages with suppliers; GM and Hewlett-Packard have extensive EDI programs with vendors.
- Spawned powerful cooperative partnerships; examples include MAC, a shared network of ATMs managed by Philadelphia National Bank (Clemons, 1990), and the "virtual" merger of SAS and Texas Air by linking their reservation systems and combining their flight codes.

In spite of the attention that several of these successful IT initiatives have received, managers have trouble evaluating the initial opportunities to use technology strategically, and have difficulty making investment decisions when these opportunities arise. The CFO of a Fortune 500 company said, "in today's environment we don't think the numbers really tell the story—many investments are required for survival or to stay ahead of the competition" (Port, 1988). Clearly, the decisions concerning where and when to allocate resources to technology programs have become far more risky and difficult.

Evaluating Strategic IT

Our approach to developing IT investment principles is to review the determinants of a strategic system's value, and use them as a basis for making investment decisions. Few companies that have spent heavily to create a competitive advantage through IT have been rewarded. However, those firms that fail to respond to competitive requirements, and assume the continuation of a comfortable status quo, consistently find themselves in deteriorating positions.

STRATEGIC IT INVESTMENT ANALYSIS: DOWNSIDE RISKS

Unfamiliar risks and difficulties confront executives deciding on strategic IT programs. Many firms pursue competitive advantage, but few systems actually provide it. Still, the benefits that result from strategic implementations are real, even if often unquantifiable and even if omitted in the initial financial analysis. A number of hard-to-quantify parameters have a strong bearing on a system's strategic impact, but these factors are often overlooked in the investment analysis. Our studies indicate executives must consider the following factors in the evaluation of strategic IT programs.

Replicability and Competitor Response

In general, IT capabilities are easily copied, and the benefits of an innovation (e.g., lower production costs, enhanced customer service) are not held exclusively by the innovating firm.[6] The technology itself is rarely a source of advantage. Instead, IT can enable other resources controlled by the firm to be exploited for advantage (Clemons and Row, 1987). Lacking patent protection, genuine first-mover benefits, or a truly sustainable edge, managers usually are not justified in forecasting competitive benefits from a system. The ability to duplicate IT capabilities readily means that any advantage conferred will most likely erode. In addition, since opportunities to apply technology are usually open to all competitors, rejecting a technology program leaves the firm vulnerable to other players.

Misapplication of Financial Models

In practice, many firms misapply discounted cash flow models to IT investments. Surveys of current capital investment practice indicate that (Hayes and Garvin, 1982; Kaplan, 1986):

- Benefits that are not easily measurable tend to be ignored and hence set to zero. The advantages of better quality, greater flexibility, lower throughput times, and experience with new technology are additional, but rarely quantified IT benefits.
- Financial analyses focus too narrowly on labor and materials savings; that is, cost displacement is emphasized over new products and enhancements to existing offerings.
- To compensate for high perceived risk, rate-of-return hurdles are set at arbitrarily high levels. True cost of capital is far lower than that generally applied in deciding on new investment projects.
- High hurdle rates set in order to compensate for risk have no theoretical justification and have a discouraging effect on technology investment.

Poor application of DCF techniques to risky, long-term, or strategic programs is capable of misleading managers, and misguiding firms in their use of IT innovations.

Industry Restructuring and Environmental Hazard

Because of IT's ability to alter the structure of an industry, environmental conditions (market size, margins, etc.) can change radically with the introduction of new systems, thus invalidating even well-constructed financial models. By creating new communication links and enabling coordination of businesses, IT brings about fundamental economic change. For instance:

- Big Bang, the sweeping deregulation and restructuring of the London stock market in 1986, was accomplished in large part by the introduction of a screen-based dealing system. The improved market efficiency, however, has had disastrous effects on member firms' profitability, causing losses that several years earlier would have wiped out the capital of most firms (Clemons and Weber, Spring 1990).

- The Chicago Mercantile Exchange, the Chicago Board of Trade, and Reuters are progressing in a joint venture, Globex, to facilitate computer-assisted, off-hours trading in futures contracts (Campbell, 1988). Available worldwide on 173,000 Reuters terminals, Globex could radically alter the markets' liquidity, and the role of floor brokers and other intermediaries that have had a long-standing presence in Chicago's derivative markets.

Anticipating the environment at the time of the investment decision is difficult and perilous. While predicting the future is always subject to error, technological advances make even the most expert forecast unreliable.

Long Lead Times

Information technology investments have long lead times. Before a full roll-out, an IT project typically passes through many phases of development, which include requirements analysis, coding, testing, modification, training, and limited implementation with close monitoring. Fourth-generation languages and CASE tools can shorten the timetable, but in most cases a strategic IT program will have a substantial lag between acceptance of the investment proposal, and the eventual time when benefits can be gauged. This delay introduces added uncertainty into assumptions made about market conditions and the actions of competitors.

In general, the future environment is highly uncertain at the time of the investment decision, and simply bounding the range of future cash flows may be extremely difficult, even for experts. To have projected Sabre's value correctly would have required accurate predictions of the airline industry's growth under deregulation, the adoption rate of travel agents, and American's competitors' response to paying a booking fee to a competitor. Even extensive scenario evaluation would not have yielded the full range of future environments to be considered in making the strategic commitment to Sabre.

Organizational Barriers

Several organizational barriers handicap management in properly evaluating strategic IT ventures. The size of the commitments is intimidating, and often requires board-level justification. NatWest Bank in the UK recently committed itself to the first part of a £500 million program to develop a new IT infrastructure for the bank and its 3,500 branches (Cane, 1989). Presenting the decision to the bank's executive committee forced senior IT managers to justify a massive technology program to a group unfamiliar with many of the risks of large-scale strategic IT developments.

As the costs and the number of risks and environmental uncertainties grow, IT decisions will not get safer or easier. As a result, paralysis can occur. Managements may have the misplaced belief that a technological panacea lies around the corner, obviating their need to act now. Promising new technologies that may threaten a company's existing products are often neglected. Also, it is simply difficult to deviate from existing practice. There is an observed incapacity of incumbents and leading firms to respond to challenges of start-up firms that apply new technologies and innovations that represent *generational* change rather than incremental change (Henderson, 1987).

STRATEGIC IT INVESTMENT ANALYSIS: UPSIDE OPPORTUNITIES

Several benefits of IT are often overlooked or not fully valued in traditional investment decision models. All too often, to maintain precision, financial analyses assign zero for the value of intangible benefits. Our studies indicate that less visible factors can significantly enhance the realized value of IT projects, and may prove to be the principal value realized from the investment.

Divisibility and Expandability

Systems are usually highly leverageable. Once developed, systems can be expanded and implemented in other parts of the organization with little incremental cost. There are few technical restrictions (and low marginal costs) for sending more bits to travel agents or hooking up another customer into an order entry system.

Systems can be designed to have flexible and readily increased capacity. A customer database developed by the sales department can be used in new product development, and summaries could be available for executive-level decision support systems. The output of a production planning system can be fed to an inventory and supplier ordering system to ensure that components are on hand at the beginning of a manufacturing run. The result is lower ordering costs and fewer stockouts.

IT has the capability to enhance a company's offerings and their convenience to its customers by expanding geographical reach or extending the hours of business. These enhancements can reach new segments of demand, and enable the business to grow in ways unforeseeable at the time of the investment decision.

Marketing In-House Systems

IT cost centers can become profitable ventures for firms that commercialize the systems they create. One writer points out that, "as IT takes up an increasing proportion of budgets, the attraction of turning parts of that investment into a revenue stream—of turning programs into products—is growing too" (Madden, 1990). A number of systems that were developed as in-house projects and received a favorable internal response have been offered as a product, or on a licensing basis, often resulting in generous returns on the incremental selling expense.

- First Boston has spun off a separate company, and is now licensing its proprietary systems development technologies—CASE tools—and its advanced global trading and clearance systems to competitors ("First Boston . . . ," 1988).
- British Telecom sold its Customer Service System (CSS) to Singapore Telecom for £13 million, fractionally offsetting its £500 million development cost (Madden, 1990).
- Mellon Bank developed sophisticated, large-capacity transaction and account processing systems in the early 1970s. Mellon became a leading provider of processing and other correspondent banking services to small banks (Salomon Brothers Research, 1985).
- SNCF, France's national rail body, recently purchased a license to use Sabre software from American Airlines for $160 million (Thomas, 1989).

The process of turning an in-house system into a marketable product can be difficult. However, the possibility of packaging and selling internally developed IT services and applications creates an opportunity for additional value from an IT investment. Many IT managers now recognize that any advantage from a system is generally short-lived, so commercializing a system is unlikely to result in the loss of competitive opportunities.

Timing Value

Timing value is often overlooked when evaluating IT investments. Technology offers management a choice of when to adopt an innovation. The ability to choose when to implement a system is valuable, and there are several approaches to the optimal timing of a technology implementation. Lee (1988), for instance, modeled the returns that an investment project generates as a stochastic process, and analyzed the value of the timing option—the ability to defer the project—contained in many technology investments. Using options valuation, he derived the optimal deferment policy and calculated a trigger-point value, the value of the project, which, when reached, should trigger its implementation.

Flexibility and Option Value

Although difficult to incorporate in traditional financial frameworks, quicker response to customer orders, greater customer satisfaction, and enhanced service quality are aspects of IT programs that affect market share and growth, and thus require explicit consideration in making an investment decision. Fine and Freund (1990), note that there is scarce "knowledge and experience in evaluating investments that enhance flexibility." They derive the added economic value provided by flexible production capacity when demand is stochastic, and note that this added value is rarely considered in investment decisions. Harrigan points out that flexibility—by enabling firms to overcome exit barriers in declining industries and unprofitable market segments—has strategic value (Harrigan, 1986).

LESSONS FROM IT SUCCESSES

IT Investment Decision Cases

In spite of the obstacles to evaluating IT programs accurately, and the resulting difficulty in making sound IT investment decisions, it can be done. We examined five strategic systems investments made with sound, rational methods (Clemons, 1989). The price of these investments ranged from $2 million to well over $100 million. The companies justified strategic technology investments by viewing IT as a source of change, and as a tool for strategic positioning; that is, several of the systems investments formed a platform for reaching a number of strategic goals. Measuring IT ventures solely on the basis of fallible estimates of future cash flows, omitting other benefits, was recognized to be misleading. The analyses these firms conducted prior to the launch of their IT ventures made use of traditional capital investment methodologies, but return on investment projections were not the primary criteria. They used evaluation methods that are rational, but not quantitative.

Strategy and Systems Decisions

As technology increases its strategic role, successful firms are making the evaluation of IT opportunities look more like the company's strategic planning process. Managers base strategic planning on the application of a number of different models and paradigms. Lederer and Sethi (1988) report that "organizations generally apply one of a number of methodologies in order to perform strategic IS planning," and may "select features of the alternative methodologies and in effect combine them to arrive at decisions." Brealey and Myers (1981) refer to strategic planning as "capital budgeting on a grand scale." The "grandness" is accomplished by taking explicit consideration of factors and of strategic risks and benefits not included in a standard NPV analysis.

Recognizing the Components of Risk

Risks in IT investments may come in forms with which management is unfamiliar. Successful companies recognize these alternative forms of risk, and take steps to hedge these risks or mitigate the impact of an IT project's failure. The first companies to switch over to automated payroll processing faced *technical risk*—there was uncertainty whether the technology was in fact capable of supporting the payroll processing needs. On the other hand, the development of electronic securities markets today presents entirely different types of risk. *Environmental hazards* arise; the electronically mediated market can so radically change the environment and business practices that forecasts are grossly unreliable estimates of what the response and the future environment will be. Given the unfamiliar risks, and the likelihood of error in *ex ante* financial valuations of IT projects, alternative principles for making IT investments are necessary.

GUIDELINES

We offer several guidelines below, which we believe to be useful in structuring the evaluation of strategic investments in information technology. We have not sought to validate them in a statistically significant sense. We have, however, observed them to be useful in numerous case studies, and in real implementations where managers had to make decisions and millions of dollars had to be committed. Inquiry through future case research can be used to support or undermine these hypotheses.

Guideline 1: Investment decisions can be made on a rational analytical basis, even when the numbers required for discounted cash flow analyses cannot be obtained.

Using discounted cash flow analyses to rank alternatives generally requires that reasonably good estimates be available for all relevant cost and income streams. However, even when true investment values are not obtainable, a ranking may still be possible. As noted, estimating future cash flows resulting from a strategic IT investment is complex and subject to significant error. An alternative is to perform decision-tree analysis. Identifying branches that characterize scenarios—for example, rapid or slow customer adoption, and limited or widespread competitor implementations—is often easier than valuing future cash flows. A decision tree may look as shown in Figure 1.

As can be seen in Figure 1, the threat of scenario C, in which customers adopt quickly, competitors act, and your firm does not invest, may be sufficiently compelling to justify investing. When properly structured, decision-tree analysis can frequently establish a ranking of outcome scenarios even when precise valuations are not available. It also surfaces assumptions and lays out the arguments for the viability of an IT program. In some cases the eventual realizations of individual parameters may be irrelevant to the decision. The ranking of outcome scenarios may not be affected within wide ranges of parameter values, and some choices may turn out to be dominated.

Guideline 2: Thresholds established by sensitivity analysis can be used as trigger points for fine-tuning a project once it is initiated.

Figure 1 A Decision-Tree Analysis for IT Investments

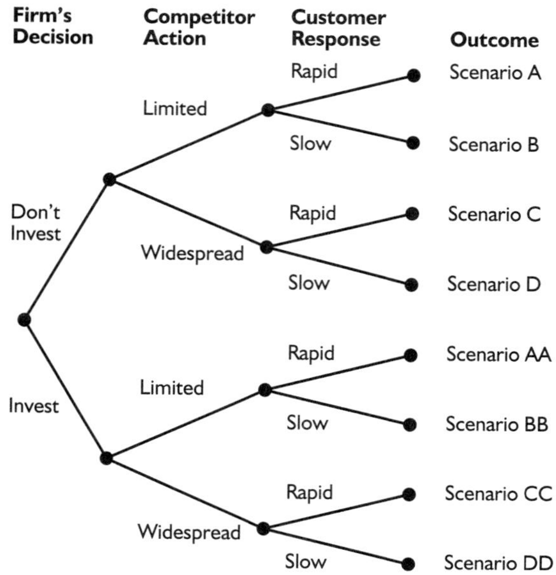

Sensitivity analysis is simply a procedure for calculating the consequences of mis-estimating parameters in a decision model. It entails using alternative estimates and examining their implications. As Leamer (1985) points out, a model's "conclusions are judged to be sturdy only if the neighborhood of assumptions is wide enough to be credible, and the corresponding interval of inferences is narrow enough to be useful." In making investment decisions, sensitivity analysis is used to:

- Determine input assumptions required for the viability of an investment
- Establish control ranges for changes in input parameters and constants over time for which the present investment decision remains optimal

An IT program, like any significant investment, should be subjected to sensitivity analysis. An investment that relies on a particular adoption rate by users or a certain completion date in order to be viable should be tested for its robustness to mis-estimation.

Furthermore, IT developments that follow a sequential plan allow systems efforts to be shut down early at substantial savings if prospects no longer appear favorable, or alternatively scaled up if conditions improve. While an IT project may be worth undertaking on the basis of present information, if a parameter value falls below a threshold indicated by sensitivity analysis, it signals the project no longer has a positive value and abandoning it ought to be considered.

In addition to testing for the attractiveness of a project when one input parameter is varied at a time, it is also valuable to consider scenarios, different but consistent combinations of variables. One reason for creating plausible scenarios is that inputs themselves may not be independent. For instance, the cost of providing third-party processing may be related to the size of the market. If demand rises it may be possible to charge less and thus attract more business. Also critical to assess in scenario analysis are possible unspecified variables, or *unknown unknowns.*

Guideline 3: Advantage results from unique assets and resources of the implementing firm.

The underpinnings of this hypothesis are fundamental economic postulates. If technology increases margins, we expect entrants and copying by existing players. As Teece (1987) has demonstrated, when there are limited barriers to copying—a "weak appropriability regime"—advantages will be competed away. Unless patents or trade secrets offer protection, the advantages from most IT capabilities are likely to have weak appropriability.

IT will likely have an asymmetric impact on different players. IT innovations will make some assets singularly more valuable. Since firms in the same industry often operate with different portfolios of assets, an IT capability even when universally implemented, can take two firms with equivalent profitability and radically affect their future margins and position in the industry (Clemons and Row, 1987). As a result, those players that develop or acquire the requisite cospecialized assets before IT hikes their value will be advantaged.[7] With access to *cospecialized assets,* a firm may be able to derive advantage that is not readily available to its competitors who lack these assets. Two examples are:

- In the drug distribution business, market share was the cospecialized asset whose value was greatly enhanced by the automation of the ordering and distribution functions. Systems such as McKesson's Economost required economies of scale to make their development and operation economically viable (Clemons and Row, Summer 1988).
- In third-party processing, Provident National Bank had developed resources inhouse that enabled them to win a large contract to handle Shearson's processing for its response to Merrill Lynch's Cash Management Account. In subsequent contracts, Provident enjoyed economies of scale and experience that drove their costs even lower (Clemons and Row, January 1988).

Thus, the value of an IT innovation will in part be determined by a firm's ability to exploit it.

If a fundamental impact of an IT innovation is to raise the value of a set of cospecialized assets, then one of the key decision criteria in considering a strategic IT venture will be which of these assets to buy and when to buy them.

Guideline 4: Several types of risk exist and must be recognized early in the evaluation of an IT development.

A number of forms of risks arise in undertaking strategic IT programs. Companies' IT investments have several dimensions of risk. As IT projects have become bigger and more complex, there are more ways they can fail. These risks need to be recognized, and in some cases, the risks can be hedged, through out-sourcing and joint development, or by investing sequentially and scaling the project up or down as uncertainties become resolved. Components of IT risk include:

Technical risk. The systems and the envisioned capabilities are not feasible with the current technology.

Project risk. The IT capability that a firm seeks to develop is too large or too complex, or overwhelms the staff's technical skills. It may be possible to bring in outside help, such as consultants or independent software developers, to reduce this risk.

Functionality risk. The firm may get the system design or implementation right—that is, according to specification—but still fail to realize the anticipated benefits.

- The design is wrong and systems analysis failed to assess the user needs accurately.
- The target moved in an unexpected way. A system has been built for what was originally wanted but the world and the business environment have changed; the system as designed is no longer appropriate.

Internal political risk. An IT investment can be undermined by vested interests in the innovating organization. Reacting to a perceived threat to their positions in the current environment, staff or divisions of the company may not cooperate in the development or may be slow to adopt an implementation.

External environmental risk. The source of the risk is unanticipated response from competitors or regulatory bodies. Responses from competitors include imitation and substitute products that are intended to prevent an advantage from developing. In addition, customers may bypass, or threaten to develop their own system.

Systemic risk. An extreme case of environmental risk, the IT program may so dramatically shift the environment that the advantages expected from it vanish. An IT project may precipitate a large-scale discontinuity in an industry, and so radically alter the industry and change prevailing economics that prior assumptions about market share, growth, and profitability are rendered obsolete.

Guideline 5: Technology investments may have option and timing value, and unexpected upside benefits.

Options theory can improve the evaluation of strategic IT investments. IT developments give firms valuable options and provide a platform for enhancing the bundle of assets and resources that make up the firm. For instance:

Similar to a call option, IT ventures frequently have significant upside potential. Several authors have pointed out that a technology investment can result in a residual value that is similar to a financial option that ends up "in the money" (Bowman and Hurry, 1987; Mitchell and Hamilton, 1988). For instance, an R&D investment may position the firm to move quickly into a new technology at a later time, or to respond to needs and opportunities created by a changing regulatory or business environment. Thus, a bank with an outdated and inflexible systems architecture might consider investment in technology to increase flexibility, and thus to facilitate responding to opportunities created by regulatory change, and to increased threats posed by new competitors. It might consider this investment to represent *buying an option on the future profitability of the firm;* in the extreme, this investment might even represent buying an option on the firm's survival. Thus, these investments might well be undertaken even when no specific requirements or threats are perceived.

Firms can cobble together option-like prototypes. The initial versions of Merrill's Cash Management Account and McKesson's Economost were initially rolled out as prototype versions (Clemons and Row, Summer 1988 and January 1988). When market response permitted more precise and accurate estimation of their impact, these prototypes were replaced by full implementations. Leasing or subcontracting to gain an IT capability is a second alternative analogous to a call option. If demand and benefits of the leased

IT fail to justify it (similar to an option expiring "out of the money"), the project is cut at minimal loss. In the early 1980s, Manufacturers Hanover built GEONET, a $300 million international X.25-based telecommunications network (Manufacturers Hanover Corporation, 1984). Actual volumes reached only 50% of the estimates, well below capacity, and well below the level needed for recovery of the system's costs. And yet, since virtually all of the technology was acquired from a third-party provider of packet-switching services, it is likely that initial capacity could have been leased. Necessary expertise would have been acquired, and the bank could have begun conversion of existing applications as well as the development of new applications to exploit packet-switching technology. This, too, is equivalent to acquiring a call option; if demand had materialized, the bank would have been in a position to respond rapidly with its own network.

Other sources of call option value from an IT program derive from the ability to scale up or down an IT capability in response to demand, and the ability to achieve scope advantages through a system that is capable of extension beyond its originally intended application. For instance, a bank that develops a corporate account management system may later use it to support trust department operations. In addition, timing options are often available that can improve the value of an IT program. Brealey and Myers point out "the fact that a project has a positive NPV does not mean it is best undertaken now" (Brealey and Myers, 1981, p. 95). Waiting may have benefits and may result in a greater project value.

Guideline 6: Downside risk exists in rejected IT programs, which then may become strategic necessities through another firm's initiative.

Since effective technology is generally widely adopted, firms rejecting IT programs because they are satisfied with the status quo should assess the position of ending up a follower in that technology. If a firm decides not to initiate the development of a system, some competitors most likely will. Also, the unit of analysis for analyzing an IT program needs to be chosen carefully. If development of an IT capability appears uneconomic, not developing while the rest of the industry does may be even less attractive.

Managers deciding on a strategic IT program need to decide on a clear *metric* for measurement of its value. The unit of analysis may not be clearly defined when an investment is intended to have a strategic impact. The desired strategic objectives could include increased market share, lower costs, new geographic reach, and/or business growth. The alternative to investment is unclear and hard to know in advance. It might be lower margins, or a necessary "catch-up" investment later. As a result, the decision to invest in a strategic system needs to be compared not to the status quo, but to the future environment, assuming competitors develop it and users adopt it. Falling margins and vanishing market share are usually the true alternative to a strategic IT investment. The negative NPV of an IT program may be less negative than that of not doing it.

Guideline 7: Cooperation may be the dominant investment alternative under conditions of strategic necessity.

Cooperation enables smaller and more specialized firms to counter scale weaknesses that they may face in developing and implementing strategic systems. If a strategic system relies on high volumes to be economic, the smaller players could be better off by initiating joint development efforts and providing a shared system. After Citibank (the most powerful player in the New York retail banking market) rolled out its ATM network in the late 1970s, smaller institutions banded together and formed the New York Cash Exchange to offer equivalent ATM service on a cooperative basis. Other motivations for cooperating can be to:

Enhance scale advantages. Had Delta been allowed to merge its travel agent reservation system with Sabre, the combined system would have controlled close to half of the market for travel agent reservation systems. Maintenance costs, marketing efforts, and the expense of any further enhancements would be spread over a wider user base.

Leverage existing assets. TRADE, BZW's system for stockbrokers to enter customers' buy and sell orders, has been enthusiastically adopted by brokers (Clemons and Weber, January 1990). In large part, the favorable reception came because the securities house had strong relations in place with the brokers, and the system did not try to bypass them, but was intended to enhance a broker's service to his or her customers.

Make up for resource deficiencies. SAS and Texas Air, two airlines with strong regional franchises in Europe and the United States, respectively, decided to share flight codes and link their reservation systems. In the hub-and-spoke operations made popular by U.S. carriers since deregulation, there is a certain minimum size or minimum number of spokes needed to operate effectively. The virtual merger increases the number of spokes each airline has available to it. In essence, SAS and Texas Air will now serve as a feeder system for each other, enabling them to compete more effectively with larger carriers that cover both regions.

ONGOING RESEARCH: MANAGING THE COMPONENTS OF IT RISKS

Potential for Active Management of IT Risks

There are different components of risk associated with strategic information systems projects, and different projects have different risk profiles. Additionally, these different components of risk have different attributes, and our investigations show they can be actively managed. Steps can be taken to mitigate the impact of failure, or to reduce the loss associated with failure. For example, when dealing with uncertain technology, one can explore alternatives, diversify the risk, and reduce the likelihood of significant loss if one technological route proves infeasible. Alternatively, in some instances managers can secure necessary services from a reputable third-party provider on a fixed-cost basis; if development proves to be more expensive than anticipated, the loss is borne outside of the organization. In addition, research by Sherer and Clemons (1987) developed an estimation procedure for the economic impact of software failures, one aspect of technical and project risk. Based on expected failure costs, resources can be appropriately allocated to the testing and review of critical program modules.

The components of risk we find most pervasive are *project risk* and *functionality risk,* and often they can be traded off against each other; actions taken to reduce one will often increase the other. This is most evident in strategic architecture projects, investments made to replace strategic IT infrastructure to cope with environmental change, especially after deregulation. Implementation through a centralized systems development group will frequently allow use of experienced and expert technical personnel, and can reduce project risk. Unfortunately, this can greatly increase functionality risk. In the unknown future environment associated with regulatory changes, future systems requirements are often uncertain, and the end users are in a far better position to evaluate and predict future needs than are systems personnel. More compelling, however, is the difference in incentives. The users will be unable to perform their jobs in the future if the systems requirements are not obtained correctly, or if the necessary flexibility is not incorporated to deal with uncertainty. However, meeting their needs can add substantially to the cost and the development time, and the systems group is judged by cost and adherence to schedule; thus their incentives make project risk more important to them than functionality risk. On one occasion, we heard the argument that, since the systems group almost invariably was able to deliver to the user exactly what had been originally promised, and at the scheduled time, there was no need to allow any other group to participate in systems development. While meeting deadlines is valuable, such a record needs to be measured in relation to users' needs that may go unmet by adhering to an inflexible deadline.

Actions are available to reduce project risk and functionality risk. User-sponsored development efforts can use outside experts, from consulting firms or from the firm's central systems group, to reduce the risk associated with complex aspects of the project. For example, expertise needed to manage the phased conversion from files to a shared database management system environment can be obtained on a contracting basis. Or, if development is through a central data processing organization, management can take steps to assure that the incentives of the systems group lead them to work closely with users to determine requirements and thus assure that the flexibility necessary to meet future needs is provided.

An Investment Analysis Mosaic

Investing in strategic IT programs is a high-stakes decision that carries enormous risk and large potential rewards. We have addressed the difficulties that management faces when making these investment decisions. How can the strategic impact of an IT opportunity be judged in advance? What factors distinguish an IT program likely to generate competitive advantage? We feel that the success of a large-scale IT introduction depends on the company's existing strategic plan, and the assets and resources of the firm, which any IT venture should seek to leverage (Clemons and Row, 1987).

Given the degree of uncertainty in development and implementation, and in the eventual environmental response, precise valuation methods such as DCF are frequently inadequate. In light of the vast sums of money and high levels of corporate vulnerability associated with strategic IT programs, we sense that many of the existing methods are biased dangerously toward conservatism and risk aversion. We have presented several procedures for evaluating IT investment programs, and have offered guidelines for making decisions to pursue strategic programs in IT. The credibility of these principles will be established through further experience, and ultimately through more formal empirical study. Due to the size of the commitments and the period of study required to determine the success of a large IT project, sample sizes will remain small. We feel that these early guidelines represent a significant improvement over methods generally in use.

NOTES

An earlier version of this paper was published in the *Proceedings of the Twenty-Third Hawaii International Conference on System Sciences.* IEEE Computer Society Press, 1990.

1. The reason for Sabre's value is profitability and prospects for growth. In 1987, Sabre revenues were $400 million, or 5.5% of AMR's operating expenses, but its $248 million profit accounted for 54% of AMR's operating income (Bradley and Smith, 1988). In 1988, costs of new technology lowered Sabre profits to $124 million, or 17% of AMR's total (Weil, 1988).
2. For instance, the proportion of plane tickets issued by travel agents has more than doubled to over 80% since the introduction of reservation systems.
3. In particular, we challenge the notion that "justifying organization-spanning networks whose benefits are uncertain . . . is in part an act of faith" (Rockart and Short, 1989), or that investment in technology is "an act of faith, a belief that the future will be as promising as the present" (Hayes and Garvin, 1982).
4. National Westminster Bank has committed a staggering sum, estimated by one journalist at £3 billion, to a single ongoing infrastructure investment (Cane, 1989).
5. DCF analysis is the most widely taught model for investment justification and applies a simple criterion to investment decision making: those projects that have a *positive* net present value (NPV)—that is, the present value of after-tax cash inflows exceeds the present value of after-tax cash outflows—should be undertaken, and all others should be rejected. Future cash flows are discounted at the appropriate cost of capital to make them comparable to today's costs.
6. In fact, under certain conditions, there may be

advantages to pursuing an *active follower* strategy (Clemons and Knez, 1988).

7. Cospecialized assets are those assets whose values are enhanced by a system or the implementation of any other technological innovation.

REFERENCES

Barwise, P., Marsh, P. and Wensley, R. "Must Finance and Strategy Clash?" *Harvard Business Review* (September/October 1989):85–90.

Beath, C., and Ives, B. "The Information Technology Champion: Aiding and Abetting, Care and Feeding." *Proceedings of the Twenty-First Hawaii International Conference on Systems Sciences* (January 1988):115–123.

Bennett, E., Reed, S., and Robertson, D. "A Survey: How Do You Justify an Investment in Technology." *Financial Executive* (September/October 1988):44–48.

Bowman, E., and Hurry, D. "Strategic Options." Reginald H. Jones Center for Management Policy, Strategy and Organization, Wharton School, 1987.

Bradley, C., and Smith, D. "Networking Reservations." *Network World* (August 22, 1988):22–45.

Brealey, R., and Myers, S. *Principles of Corporate Finance.* New York: McGraw-Hill, 1981.

Campbell, K. "Technology Wars Ravage Chicago." *Financial Times* (November 15, 1988):30.

Cane, A. "Nat West Plans £3 bn Spending to Update Computer Network." *Financial Times* (January 26, 1989):8.

Clemons, E. "MAC—Philadelphia National Bank's Strategic Venture in Shared ATM Networks." *Journal of Management Information Systems,* 7, 1 (Summer 1990).

Clemons, E. "Making the Investment Decision: Evaluating Strategic Opportunities in Information Technology." Decision Sciences Department, Working Paper 89–03–01, Wharton School.

Clemons, E., and Kimbrough, S. "Information Systems and Business Strategy: A Review of Strategic Necessity." Decision Sciences Department Working Paper 87–01–04, Wharton School.

Clemons, E., and Knez, M. "Competition and Cooperation in Information Systems Innovation." *Information & Management* (1988).

Clemons, E., and Row., M. "McKesson Drug Company: A Case Study of Economost—A Strategic Information System." *Journal of Management Information Systems* (Summer 1988):36–50.

Clemons, E. and Row, M. "Merrill Lynch's Cash Management Accounts: An Analysis of a Strategic Use of Information Systems." *Proceedings of the Twenty-First Hawaii International Conference on Systems Sciences* (January 1988):131–140.

Clemons, E., and Row, M. "Structural Differences Among Firms: A Potential Source of Competitive Advantage in the Application of Information Technology." *Proceedings of the Eighth International Conference on Information Systems* (December 1987):1–9.

Clemons, E., and Weber, B. "Making the Technology Investment Decision: Barclays de Zoete Wedd's TRADE System." *Proceedings of the Twenty-Third Hawaii International Conference on Systems Sciences* (January 1990).

Clemons, E., and Weber, B. "London's Big Bang: A Case Study of Information Technology, Competitive Impact, and Organizational Change." *Journal of Management Information Systems* (Spring 1990).

Emery, J. "Cost/Benefit Analysis of Information Systems." *The Society for Management Information Systems* (1971):16–46.

Fine, C., and Freund, R. "Optimal Investment in Product-flexible Manufacturing Capacity." *Management Science* (April 1990):449–466.

"First Boston, Kidder to Share Data Technologies." *Wall Street Journal* (July 12, 1988):42.

Guenther, R. "Chemical Banking, AT&T To Scrap Home Bank Service." *Wall Street Journal* (December 5, 1988):21.

Harrigan, K. "Strategic Flexibility." In *The Eco-*

nomics of Strategic Planning, edited by L. G. Thomas, III, 81–111, New York, Lexington Books, 1986.

Hayes, R., and Garvin, D. "Managing As If Tomorrow Mattered." *Harvard Business Review* (May/June 1982):70–79.

Henderson, R. "The Failure of Established Firms in the Face of Technological Uncertainty." Doctoral dissertation, Harvard University, December 1987.

Holusha, J. "Cost Accounting's Blind Spot." *New York Times* (October 14, 1986):D1.

Kaplan, R. "Must CIM Be Justified by Faith Alone?" *Harvard Business Review* (March/April 1986):87–95.

Landes, D. *The Unbound Prometheus: Technological Change and Industrial Development in Western Europe from 1750 to the Present.* Cambridge, MA: Cambridge University Press, 1969.

Lavich, K. "Winners in the Air Wars." *Fortune* (May 11, 1987):68–79.

Leamer, E. "Sensitivity Analyses Would Help." *American Economic Review* (June 1985):308–313.

Lederer, A., and Mendelow, A. "Convincing Top Management of the Strategic Potential of Information Systems." *MIS Quarterly* (September 1988):525–534.

Lederer, A., and Sethi, V. "The Implementation of Strategic Information Systems Planning Methodologies." *MIS Quarterly* (September 1988):445–461.

Lee, C. "Capital Budgeting Under Uncertainty: The Issue of Optimal Timing." *Journal of Business Finance* (Summer 1988):155–168.

Logue, D., and West, R. "Discounted Cash Flow Analysis: A Response to the Critics." *Productivity Review* (Summer 1983):233–241.

Madden, D. "Turning Programs into Products." *Financial Times* (March 1, 1990):12.

Manufacturers Hanover Corporation—Worldwide Network. Harvard Business School Case Services, Boston, 1984.

Mitchell, G., and Hamilton, W. "Managing R&D as a Strategic Option." *Research Technology Management Journal* (May/June 1988):15–22.

Oram, R. "U.S. Airlines Discuss Reservations Link." *Financial Times* (September 19, 1989), 22.

Parker, M., and Benson, R. "Information Economics: An Introduction." *Datamation* (December 1, 1987):86–96.

Port, O. "How the New Math of Productivity Adds Up." *Business Week* (June 6, 1988):103–114.

Quickel, S. "Management Joins the Computer Age." *Business Month* (May 23, 1989):42–46.

Rockart, J., and Short, J. "IT in the 1990s: Managing Organizational Interdependence." *Sloan Management Review* (Winter 1989):7–17.

Salomon Brothers Research, "Information Technology and Banking." New York, 1985.

"Aerospace and Air Transport." *Standard & Poor's Industry Surveys* (May 5, 1989).

Sherer, S., and Clemons, E. "Software Risk Assessment." *AFIPS Conference Proceedings,* vol. 56, Chicago, IL (June 1987), 701–707.

Teece, D. "Profiting from Technological Innovation: Implications for Integration, Collaboration, Licensing, and Public Policy." In *The Competitive Challenge,* edited by D. Teece (Cambridge, MA: Ballinger Publishing, 1987):pp. 185–219.

Thomas, H. "Express Service Before the Train Pulls Out." *Financial Times* (October 17, 1989):15.

"Wanted: Co-pilots for Reservation Systems." *Business Week* (April 9, 1990):78.

Weil, U. "Making Computers Pay Their Way." *Institutional Investor* (June 1988):18–19.

QUESTIONS

1. What type of investments are considered strategic?
2. What is meant by the "vanishing status quo"? How can managers avoid it?
3. Select a specific technology investment opportunity that can be considered strategic. How would you propose to justify the investment?

5.6 Managing Investment in Information Technology: Mini Case Examples and Implications

PETER WEILL AND
MARGRETHE H. OLSON

The subject of information technology investment is so important and the moneys involved are so large that this book devotes two sections to this topic. Not only is information technology the fastest growing segment of capital investment for many companies, it is also the source of possible opportunity costs for failing to develop needed information capabilities.

Measuring both the costs and benefits of information technology is difficult. It is not even clear what costs to include. For example, should telephone costs be considered? With the increased reliance on decentralized computing, it becomes all the more difficult to gauge expenditure for personal and departmental computing. Measuring benefits is more difficult than measuring costs because benefits are in the future and because the gains are often not related to simple work hours displaced.

To obtain a better understanding of the cost/benefit picture, Weill and Olson conducted interviews with six substantial firms. The objective was to understand how the firms define and manage their information technology investments. Based on these interviews and their understanding of the problem, they summarize the trends observed and recommend how information technology investments can be defined, measured, and tracked.

Source: *MIS Quarterly* (March 1989): 3–17.

INTRODUCTION

PRESIDENT:	I have been struggling with how heavily this company should invest in IT.
CONSULTANT:	Clearly it all depends on your strategy and associated niche, competitors, environmental turbulence and industry structure, technological profile of your firm, level of differentiation and vertical integration, and, of course, how you feel about it.
PRESIDENT:	Really?

Businesses are investing enormous resources in information technology (IT). A recent study (PIMS Program, 1984; Strassmann, 1985) finds that the average information technology expenditure in 1983 was 2% of revenues. Another survey (Diebold Group, 1982; 1984) finds that centralized management information systems (MIS) expenditures on average accounted for 1.44% of revenues in 1984. For a Fortune 500 firm with average revenues of $5 billion, this expenditure amounts to $100 million per year. Even more dramatic, two studies demonstrate that high-performing companies allocate a significantly greater proportion of revenues to IT expenditures than companies with lower performance (PIMS Program, 1984; Harris and Katz, 1988).

While it is tempting to conclude that so much IT investment increases organizational performance, there is little evidence that this is so. Frequently, the only spending guidelines managers have for IT investment are the spending levels of competing firms in their industry (based on percentage of revenues), or a flat increment on last year's IT budget. Unfortunately, therefore, much of this investment is based on blind faith that real returns will occur. Conversations such as the one reported above illustrate the quandary many managers face. Information technology is essential to their businesses, but inadequate guidelines exist for determining the appropriate level of investment.

The purpose of this article is to increase understanding of the basis for IT investment. We have performed six mini case studies of large companies in five different industries, addressing the questions of what firms define as IT and how they manage their investment in IT. Our goal was to gain insight into the management of IT investment and the relationship between IT investment and organizational performance. The cases led to a series of implications and recommendations for practitioners responsible for investment decisions.

THE PROBLEM

The general problem is: What is the impact of the level of IT investment on firm performance? This is a complex issue, as IT investment is only one of the many factors that affect firm performance (Lucas, 1975; Turner, 1985). The crux of the problem is whether investment in IT really makes a difference. This problem raises the following related questions:

1. Is there a significant, unique effect of investment in IT on firm performance?
2. What is the nature of the relationship between the level of IT investment and firm performance? Does increased investment lead to increased performance? Is there a level of investment that produces an optimum level of firm performance, after which further investment has a negative effect (Bender, 1986)? Alternatively, is there a diminishing return, where increased IT investment produces decreasing but still positive effects on firm performance?
3. Is the relationship between levels of IT investment and firm performance dependent on the industry in which the firm operates?
4. Is there a threshold investment that firms must make to survive?

5. Does the strategic reason for investing in IT make a difference? What effect does the objective for investment have on firm performance?
6. Do other factors that affect firm performance overwhelm the effects of IT investment levels?
7. Is it possible to predict the effect of a particular IT investment level on performance for a given firm?

While we do not claim to have answers to all these difficult questions, they are the basis for another major research study we currently are conducting. The purpose of this article, however, is to examine how practitioners responsible for IT strategic planning approach the investment decision. We report the results of six mini case studies, during which we attempted to answer the following specific questions:

1. How do firms define and conceptualize IT?
2. How do firms measure and track or otherwise manage their investments in IT?
3. How do firms infer or calculate return on IT investment?
4. Do firms in different industries approach IT investment differently?
5. What is the impetus for IT investment? At what level in the organization does the impetus originate? Is it top-down (driven by corporate strategy at the highest level), or is it bottom-up (driven by operational requirements)?

PREVIOUS STUDIES

A number of studies have investigated IT investment and its effects on performance. In this section we discuss the different approaches that have been adopted. The studies are summarized in Table 1. Some studies focused only on definition and measurement of IT investment, while others measured both IT investment and organizational performance and attempted to draw conclusions about the relationship between them. The results of the studies are summarized below.

What Is IT?

Each study uses a different definition of IT, making comparison of results difficult. Some studies only include management information systems and related personnel, consulting, external services, and personal computing expenditures within the centralized MIS budget (Datamation, 1985; 1986; Cahners Publishing, 1987; Diebold Group, 1982; 1984; Strassmann, 1985; Turner, 1985). One study broadens the definition to include expenditures for communications such as electronic mail, telephones, facsimile, and reproduction machines (Panko, 1982).

With the recent noticeable trend toward decentralization of expenditure in IT to functional departments, considering IT investment as equivalent to the centralized MIS budget is probably inadequate. However, since these decentralized expenditures are generally embedded within the functional department's operating budget, they are more difficult for surveys to identify. As a consequence, most surveys underestimate firm-wide IT investment.

Measures of IT Investment

Studies have used a variety of measures of the levels of investment in IT.

In its biennial surveys of major corporations over the last 10 years, The Diebold Group (1982, 1984) consistently uses three measures:

- MIS budget as a percentage of revenues.
- MIS staff as a percentage of total staff.
- Ratio of hardware expenditures to personnel costs.

Datamation regularly conducts a survey of the magazine's circulation base and the Fortune 1000 and 400 firms to assess centralized MIS investment (Datamation, 1985; 1986; Cahners Publishing, 1987). The measure used is MIS budget as a percentage of revenues. In a similar approach, Bender (1986) uses the ratio of IT investment to premium

TABLE 1
Previous Research on IT Investment

References	*Industry*	*IT Investment*	*Effect on Performance*
Bender, 1986	Insurance	IT expenses as % of premium income: average not reported	"Range of IT expenses as % of total expense of 15–20% seems to produce optimum results." More or less IT expense is associated with weaker performance.
Cron and Sobol, 1983	Wholesalers	Substitutes for investment (e.g., number of standard applications computerized)	Extensive IT users are either high or low performers.
Datamation, 1985; 1986; Cahners Publishing, 1987	All	MIS budget as % of revenue; average 0.57%	Not attempted
Diebold Group, 1982; 1984	All	MIS budget as % of revenue; average 0.95% (1971), 1.44% (1983)	Not attempted
Harris and Katz, 1988	Insurance	IT expense as % of non-interest expenses; average—14%	High performers spend more on IT.
Panko, 1982	All	$3,000 per year per office worker	Not attempted
PIMS Program, 1984	All	MIS budget as % of revenue; average 2%	High performers spend more on IT; average − 8%
Turner, 1985	Banks	DP expense as % of operating expenses: average—5.2%	"Unexpectedly no relationship"
U.S. Office of Management and Budget, 1985	Government	MIS expenses as % of total expenses: average—1.6%	Not attempted

income in the insurance industry. Strassmann (1985) refines the measure to be the fraction of management costs devoted to IT. He believes that the base of management costs is a more appropriate measure than revenues, since it is more directly influenced by management decisions.

Several studies rely on substitute measures rather than absolute investment dollars for comparison of firms within an industry. Cron and Sobol (1983) use three substitute measures:

- Computer ownership: owned/leased versus timeshare.
- Number of standard application areas computerized.
- Types of application areas computerized (accounting, management control, or environmental links).

In a study of the banking industry, Turner (1985) uses a number of measures, including "relative DP expenditures" (ratio of DP expenditures to total assets) and the number of standard functional areas having computerized applications.

All of these approaches have advantages and disadvantages. It is doubtful that any of them captures all of a given firm's investment in IT due to the increasing decentralization of IT investment and increasing ad hoc expenditures on IT (e.g., personal computers) from non-IT operational budgets.

In the 1986 *Datamation* survey, the MIS budget as a percentage of revenues for all industry groups combined was 0.57% (Datamation, 1985; 1986; Cahners Publishing, 1987). This figure is statistically biased (due to the large number of small sites) and underestimates MIS expenditures in large organizations. The actual figure for large MIS centers is more likely around .9% to 1.0%.[1] The Diebold (1982; 1984) survey showed average IT expenditures to be 1.44% of revenues and to have grown steadily over the six years represented by its surveys. Strassmann (1984) found that the average IT expenditure in 1983 was 2% of revenues. In the government sector, IT is a significant expense, which in 1982 was 1.2% of the federal budget and increased steadily to 1.6% in 1986 (U.S. Office of Management and Budget, 1985). In 1986, half this expenditure was on outside commercial IT expenses, such as computer service bureaus. Panko (1982) finds that, on average, firm IT investment was $3,000 per year per information worker.

The disparity in the results reflects the differences in defining and measuring IT investment. Comparison of these broadly defined investment levels among companies (particularly across different industries) is useful only as a rough guide. The macro investment percentages are taken out of context of firm strategy. In addition, the ability of a particular management team to effectively convert the investments into useful outputs (i.e., improved organizational performance) is not considered.

Measures of Performance

The determination of performance is highly problematic in this type of work. A variety of social and economic measures can be adopted (Zammuto, 1982). Many combinations have been used, often with more regard to convenience than appropriateness.

One view is to establish a single global measure of performance. In a study of the insurance industry, Bender (1986) uses the single measure of the ratio of expenses over premium income. Another view is that one measure of performance will not capture all factors that contribute to high performance, and, consequently, the use of any single measure is relatively arbitrary (Turner, 1985).

In a study of the impact of IT investment on financial performance for a group of wholesalers, Cron and Sobol (1986) use four measures:

- Pretax profits
- Return on assets
- Return on net worth
- Five-year growth rates

Other performance measures include market share (PIMS Program, 1984) and the ratio of net operating income to total asset base (Turner, 1985). Given the complexity of organizational performance, it is clear that no single criterion is sufficient and that a number of measures is needed (Zammuto, 1982).

Relationships Between IT Investment and Performance

A comparison of the results from this research demonstrates some of the difficulties of assessing IT investment and performance.

In his study of 58 savings banks, Turner (1985) concludes that "unexpectedly no relationship is found between organizational performance and the relative proportion of resources allocated to data processing" (p. 1).

Bender (1986) divides IT investment into components and finds only the expenditures on people, hardware, and environment to be significantly related to performance. In his study, investment in software is not significant. He concludes that in the insurance industry there is an optimum level of investment in information processing. Firms that invested below this optimum had lower performance, and firms that invested above this optimum added to costs unnecessarily. However, Bender assumed that all firms convert the investment in IT into measurable performance with the same degree of effectiveness. In reality, some firms are more likely to be effective at this conversion than others; thus the assumption of an industry-wide optimum level of investment is questionable.

Cron and Sobol (1983) find that firms in the warehousing industry that made extensive use of computers were either very strong or very weak financial performers. This finding appears to support the strategic planning literature that stresses the importance of strategic position and concurs with findings from PIMS (1984). The PIMS study finds that of the firms that increased IT investments annually, those with superior strategic position increased management productivity three times more than firms with inferior strategic position (PIMS, 1984; p. 35). The study concludes that if the firm has a strong strategic position, investment in IT will enhance performance. If the firm has a weak strategic position, significant investment in IT will be less effective.

A four-year study of 40 insurance companies (Harris and Katz, 1988) reveals that the most profitable firms are more likely to spend a significantly higher proportion of their non-interest operating expenses on IT. Any statement of causality in the study was not possible, leaving the question open of whether higher investment causes higher performance.

The findings reported above are sparse and somewhat contradictory. Where a relationship between IT investment and performance is found, the results cannot be generalized beyond the particular industry studied. The methods of measuring investment and performance and the units of analysis differ, limiting possible comparison across studies. The studies measuring both IT investment and performance ignore the time lag between the two.

A further weakness of these studies is that, in each, no distinction is made among different types of IT investment. Investments in operational IT (i.e., transaction processing and operational control systems) are intended to impact different aspects of firm performance than investments applied to systems for strategic competitive advantage (i.e., new products or services, executive information systems, etc.).

These conflicting findings are of little joy to practitioners faced with daily IT investment decisions. In the next section, we describe a series of six case studies to derive a series of recommendations for managers concerned with IT investment and organizational performance.

CASE STUDY RESULTS

We conducted six mini case studies to help us understand how firms define and manage their IT investments. Five of the companies are large, profit-making firms operating in four different industries. The sixth is a large private university. We conducted one lengthy, semi-structured interview with the appropriate representative of each organization. All interviewees are senior managers in corporate planning or information systems. All the interviewees have general knowledge of the IT investment of the entire organization. The interviewees were frank and open; we have made every attempt to conceal the identity of the organizations and any sensitive information that was discussed.

The primary questions were the following:

- How do you define IT for the purposes of determining the level of investment?
- How do you manage and track investments in IT?
- Where in the firm is the impetus for IT investment (i.e., top-down or bottom-up)?
- What other factors influence IT investment decisions (e.g., political considerations, competitors' strategies)?

The results of the cases are presented in two stages. This section concentrates on a brief presentation of the demographics and interesting findings in each company. In the next section, we present a brief aggregate analysis of all cases, organized by the key factors emerging from the interviews.

Organization I: Insurance

Organization 1 is a large insurance company with three major businesses: individual insurance, group insurance, and investment. The interviewee is manager of corporate technology planning. The company performs an inventory of all IT investment each year. The inventory includes all DP center expenditures, communications (excluding voice-based technology), personal computers, software, education, and consultants. The definition does not include consumables (such as paper, diskettes, etc.) and full-time computer operators. The company is decentralizing rapidly, and the management of IT investment is also becoming more decentralized. A steering committee oversees major investments.

Return on investment analysis is used to justify investment where ROI is considered appropriate; however, there have been some cases where investment was driven by a need to stay with the competition or by "gut feeling." There is a general perception that, in the industry, which is of course highly information intensive, a certain threshold investment is required to be competitive.

Historically, IT investment decisions at the firm were "less political, as there were fewer experts," but the interviewee noted a definite increase in the political nature of the process. Approximately 14% of the total budget is devoted to IT investment. The impetus to invest is both "bottom-up and top-down," and the intention is that corporate strategy should drive investment in IT. However, the competition also drives IT investment, and IT also impacts corporate strategy; thus there is a complex circular relationship between IT investment and corporate strategy.

The interviewee believes that the MIS function is becoming more complex and sophisticated, but he had not observed any attempt to increase management control of this function at a higher management level. He perceives the issue of converting IT investment into useful output as "just another example of the familiar problem of resource management" and observes that some organizations are more skilled at this task than others.

Organization 2: Banking

Organization 2 is a large, geographically dispersed commercial bank. The interviewee is director of corporate systems. The bank includes in its working definition of IT "everything related to IT." According to the interviewee, the bank is "getting carried away" with its investment in IT. Individual projects are assessed using ROI techniques, but there is no notion of an overall return on IT investment. The bank tracks IT investment at both the project and corporate levels; multiple measures are used because no simple measure is considered "comfortable" in terms of accuracy. Some divisions are considerably more dependent on intensive information technology than others, but the bank attempts to estimate company-wide IT investment.

Historically, the impetus to invest was bottom-up. According to the interviewee, the bank views information as the product of the organization; it makes a distinction between administrative IT (e.g., MIS and office automation) and IT for productive capacity (e.g., transaction processing and general ledger).

The interviewee had not observed an increase in the political nature of IT investment; however, there have been attempts to change organizational power structures by shifting responsibility for the management of the IT resource.

Organization 3: Insurance

Organization 3 is a large, diversified insurance group. The interviewee is the manager of capacity planning in corporate administration. The company defines corporate investment in IT as only those expenditures in the corporate administration information systems budget and includes all data centers and communications. In 1987, this expenditure was approximately $270 million. This does not include divisional expenditures for information systems, including personal computers, programmers, and some minicomputers. The latter expenditure was estimated to be at least as large as the corporate budget expenditure. The company has begun to decentralize its IT management and budgets, in line with perceived trends of its competition.

The interviewee was unaware of any formal tracking of the IT resource; he understood that there was a rough estimate of investment by the president's staff for the entire organization. The company justifies IT projects by traditional ROI measures but emphasizes cost containment. There is some "seed money" available for experimenting with the use of new IT, but it is very limited. The interviewee does not believe there is a common perception of a threshold investment in IT in the industry.

The nature of IT investment decisions in the company is highly political; according to the interviewee, the "technological aspects are secondary." The decision is usually a function of the "clout of the division based on revenue generation." Divisions with poor performance have a harder time obtaining resources. This could be due to political power of a division or to a "standard" guideline of percent of last year's revenues to determine this year's divisional IT budgets. The impetus for investment is purely bottom-up.

There is a long-term IT strategy, but it is not well integrated into overall corporate planning. There are separate departments for IT planning and corporate planning. There is an organization-wide attempt to "keep IT investment flat, as it is difficult to understand the return on investment phenomenon."

Organization 4: Consumer Products

Organization 4 is a large, multi-national conglomerate in the consumer products industry. The interviewee is manager of corporate information resources planning.

According to the interviewee, the company makes no attempt to aggregate or manage IT investment at the corporate level; this process is the responsibility of the individual divisions. Within the context of planning, the definition of IT includes MIS-related expenditures as well as all communications expenditures excluding voice, facsimile, and telex. IT also includes expenditures on personal computers and external consultants. All IT investment directly related to manufacturing capacity is excluded.

The interviewee believes there is an optimal level of IT investment that will maximize return but says the company makes no attempt to estimate it. He sees the impetus for investment to be primarily from "middle-down" rather than top-down. Bottom-up push for investment is rare but does occur. Investment decisions are often highly political in nature, and the hot issues change over time. The company considers itself "lean and mean"; its current strategy to extract short-term profitability from products reduces the opportunity for significant speculative IT investment, since all investment is charged against product revenues.

Organization 5: Education

Organization 5 is the business school in a large private university in New York City; the interviewee is the associate dean for administration.

According to the interviewee, investment in IT is measured and tracked in an aggregate form. He estimates IT to account for half the management and administration budget, and thus 10% of the school's total annual budget. IT is defined to include computer centers, departmental equipment budgets, telecommunications (except voice-based), and IT expenditures in administrative departments, such as the bursar's office. Word processors and maintenance are significant expenditures. More capital is invested in non-DP areas on personal computers than the school's total remaining DP investment.

IT is tracked and managed at the school level, although there is no attempt to include the "relatively small expenses." Some departmental expenditures on personal computers are not captured in this tracking. No measure of ROI for IT investment is used, although each department is assessed on an input-versus-output basis. Inputs include expenditures on people, space, and equipment; output is defined as research output, student course ratings, and estimated contribution to revenues via tuition.

The interviewee feels that a threshold investment in IT is "required to survive" and remain competitive among business schools. Strategic IT investment is linked to the overall "strategy of quality to maintain and add competitive advantage." One performance measure of quality is GPA and SAT scores of incoming students; this is a relative measure commonly used to rank business schools. As a result of this strategy, the impetus for IT investment is top-down, to either cut costs or improve quality. Some cuts in administrative staff have been achieved with IT investment.

Organization 6: Telecommunications

Organization 6 is a very large communications company that is currently restructuring its operations. The interviewee is manager of communications.

According to the interviewee, the definition of IT is "moving to include everything." This change has not yet been achieved, and many managers "would not know their departmental IT investment."

He says that, in the opinion of top management, there is no evidence of a return from IT investment encompassing the "thousands of systems" currently installed. The organization, he says, spends more on IT than its size would suggest.

The company has recently created the new position of director of information management, whose responsibilities are primarily the "stewardship and quality of data." The management of IT resources is a very political issue, and some of the recent changes have been "politically unpalatable."

A major initiative is under way for overall planning for information systems. The scope of the plan includes hardware, software, development projects, databases, telecommunications staff, and organizational design. This plan is driven by the organizational strategy, and two of the primary objectives are "to cope with rapidly changing technology and to understand data as a corporate resource."

ANALYSIS OF KEY FACTORS

This section presents an analysis of the key factors that emerged from six cases. Tables 2 to 4 show the findings in summary form.

The Definition and Tracking of IT

Table 2 shows the industry of each organization, the level of the interviewee in the organization, and its definition of IT. Each organization uses a different definition of IT investment. All include hardware, software, related personnel, overhead, and some communications expenditure. The majority does not include voice, telex, facsimile, or postal services. IT dedicated to productive capacity (e.g., computers used for process control in manufacturing or general ledger in banking) is or is not defined and treated separately for different industries. There appears to be an overall trend to broaden the definition of IT to include many more aspects of the technology.

IT Investment Management

Table 3 summarizes the methods of tracking and calculating return on IT investment. The companies track IT technology expenditure with varying degrees of rigor. One difficulty is that IT investment by departments or divisions in their own discretionary budgets is generally not captured in corporate IT tracking. As a consequence, corporate (centralized) IT investment in MIS or DP centers is adequately tracked (in all but one organization) but decentralized expenditure is not. The general trend toward decentralization of responsibility for IT investment compounds this problem. All respondents say that with the large investments in IT, senior managers had initiated (or were likely to initiate) methods of tracking the organization-wide investments in all types of IT. Unfortunately, this type of information is not collected in the usual costing and accounting databases and is rarely a line item shown on the consolidated budget.

The interviewees agree that integrated coordination of IT investment is necessary. The organizations use different methods for calculating returns, but none is completely comfortable with its approach. Major investments are often subjected to traditional ROI methods on a project basis. However, a threshold investment is often necessary to compete in the industry and is often not subject to ROI or any other type of return calculation. Most interviewees believe there to be an optimal level of IT investment, but none feels it is possible to achieve it or even be sure of what it is.

In Table 3, "centralized tracking" refers to whether the company tracks expenditures for the centralized MIS departments. "Decentralized tracking" refers to the coordinated tracking of all functional department or division IT expenditures. As indicated in the table, the organizations

TABLE 2
Demographics and IT Definition

Company	*Industry*	*Level of Interviewee*	*IT Definition*	*Define Productive Capacity?*
1	Insurance	VP	All except voice, operators, consumables	No
2	Bank	Director	All	Yes
3	Insurance	Corporate Manager	All	No
4	Consumer Products	Corporate Manager	All except facsimile, voice	Yes
5	Education	Dean	All except voice	No
6	Communications	Manager	Moving to all	No

TABLE 3
IT Investment Management and Tracking

Company	*Centralized Tracking*	*Decentralized Tracking*	*Return Calculations*	*Threshold Investment*
1	All	All	Some projects	Yes
2	All major items	All major items	All projects	Yes
3	All MIS centers	Not PCs or minis	ROI for Cost containment	No
4	No attempt	Divisional responsibility	Divisional level only	No, but cost driven
5	All except voice	Equipment budgets	No	Yes
6	Moving to	Moving to	No	No

are generally moving to corporate-wide tracking of all IT investment (i.e., total centralized and decentralized investment).

Political and Organizational Influences

Political considerations, important in most of the organizations, significantly impact investment decisions. These considerations sometimes eclipse the technical and economic considerations and are generally perceived as becoming more and more important. (See Table 4.)

The interviewees note a trend to decentralized responsibility for the management of the IT resource. This is consistent with the increased decentralized IT expenditure.

The impetus for investment comes from different levels within organizations. In some cases where lower-level management was unable to complete the required transactions with the available resources (e.g., process the required number of invoices), the impetus for investment was bottom-up. Top-down impetus for IT investment occurred where the business strategy included the objective to incorporate more IT into the organization. These initiatives usually originated at the CEO level. In one organization, the impetus for change originated middle-down and not from top management. This company was divisionalized, and IT investment decisions were left to the discretion of each division.

TABLE 4
Political and Organizational Influences

Company	*Political Considerations*	*Decentralized Management of IT*	*Impetus for Investment*	*Link to Strategy*
1	Increasing	Yes	Bottom-up and top-down	Circular link
2	Moderate	Yes	Bottom-up	Not explicit
3	High	Begun	Bottom-up	Cost containment
4	High	Yes	Middle-down	Not explicit
5	Moderate	No	Top-down	Yes, to increase quality
6	High	Unclear	Top-down	Yes

The link between IT and corporate strategy varies. In two organizations, there is no explicit link. In one organization, the link is purely for cost containment; that is, IT plays a role in the explicit strategy of cost reduction. In the university, IT is included in the strategy of improving the quality of the program. In the communications company, there is an explicit link from corporate strategy to IT support for it. In one of the insurance companies, a circular link is identified where IT is input to strategy formulation, and then the IT required to implement that strategy is identified.

Conversion Effectiveness of IT Investment

All interviewees agree that the effectiveness with which IT investment is converted to useful output was important and could vary from one organization to another. We have called this phenomenon *conversion effectiveness*. The majority of interviewees believe that this conversion is just another resource management task and that organizations that are good at general resource management would be effective converters of IT investment. They consider the implementation process, culture of the organization, and skill of the managers to be major determinants of conversion effectiveness. This result appears to reinforce the findings of Harris and Katz (1988) and the PIMS Program (1984) that high-performing companies invest more heavily in IT. It appears they are more efficient at converting resources, including IT, to organizational performance than lower-performing companies.

IMPLICATIONS OF THE FINDINGS

Three major implications emerge from these findings for practitioners responsible for IT investment. The three areas are listed here and recommendations to deal with these three areas follow in the next section.

1. **Define and Track IT.** As expenditure on IT becomes a larger part of the average firm's budget, the need to define and track the firm-wide expenditure becomes critical. Organizations have different working definitions of IT; however, there is a general trend to broaden the definition to include all aspects of IT.
2. **IT Investment and Return.** The problem of determining the return on investments in IT is difficult, and most organizations studied are not comfortable with their approach. Much investment was made with the hope of a real return, but little direct evidence was available that these returns occurred.
3. **Organizational Issues.** Investment in IT

alone is not sufficient. The investment must be converted into useful outputs. A number of important organizational issues were identified in the study, including the impetus for the investment and the link to strategy. We believe, however, that the conversion effectiveness of the investment is critical. This point has received little attention.

RECOMMENDATIONS

The three implication areas are of general concern to most firms as they grapple with the problems of managing IT investment. The following are our recommendations for beginning to deal with these problems.

Define and Track IT Investment

The first step in managing IT investment is to know exactly what that investment is. The measurement and tracking of the investment is difficult, and the following suggestions are offered:

1. Adopt a broad definition of IT, including all hardware, software, people, communications (phone, telex, facsimile, satellite, etc.), overhead, training, and maintenance.
2. Measure and track this expenditure over time against a convenient base. The base could be revenues, total expenses, or management-controlled costs (Strassmann, 1985). It is important to capture all decentralized expenditure by functional departments.

In many cases the general definition of IT investment is too broad, as different investments in IT are often made with very different objectives. Our recommended approach is to track separately the amount of IT investment of each type and treat these as a portfolio of IT investments with different aims and associated levels of risk.

The objective for which IT investment is embedded in the business strategy can be categorized into three general types: strategic, informational, and transactional (Turner and Lucas, 1985). *Strategic IT* is defined as IT that "changes a firm's product or the way a firm competes in the industry" (Ives and Learmonth, 1984; p. 1193) and can be identified as IT investment made to increase sales. An example of strategic IT is the establishment of a communications network to deliver information-based products to customers. Strategic IT is generally associated with long-term considerations, including a long lead time between investment and return. Strategic IT is temporal; that is, what is strategic today will not be strategic in 10 years.

Informational IT provides the information and communication infrastructure of the organization. Included is a range of investments from telephones and facsimile machines to computing resources for accounting and production planning purposes. The commodity of this infrastructure is information for management of the firm. Informational IT is usually concerned with a medium-term time horizon.

Transactional IT supports operational management, with short time horizons and significant repetition, and is associated with short-term ROI. Transactional IT usually involves the processing of repetitive transactions, such as accounts payable and order entry. Transactional IT can be precisely defined as IT investment made with the objective of cutting costs.

Although the resulting systems objectives are quite different for these three types of IT investment, it is possible that they are housed in the same large information system.

A fourth type of IT investment emerged in this study: a *threshold IT* investment required to compete in a certain industry. Returns on threshold investments are likely to be smaller and of a different nature than other IT investments, since this threshold investment is mandatory to compete in the industry and may not be directly associated with real, short-term returns. Thus, threshold investment can be considered a sunk cost required to participate in the industry and often may not be amenable to ROI approaches for justification.

We recommend that a firm's IT investment be categorized into these four types and that investment be tracked over time. This approach highlights the need to link IT investment with the strategy of the firm.

A few organizations have a centralized IT planning function that coordinates firm-wide IT expenditure and ensures consistency between the organization's strategy and IT investments. The tracking of IT investment would be a logical extension of this function.

Finally, there are differences in IT investment needs, as well as profiles among different industries, making comparisons across industries difficult and potentially misleading. This is due primarily to the different cost structures and stage of technical adoption, in addition to the nature of investment in productive capacity (i.e., in manufacturing versus an information-intensive organization, such as insurance). As a consequence, we suggest great care in interpreting the broad investment-level statistics often found in the popular trade press. The fact that the average level of IT investment in all manufacturing is 2% of sales is of little help to the small chocolate manufacturer trying to determine the appropriate level of IT investment in his or her firm.

IT Investment and Return

The justification of IT investment is a major stumbling block for many firms. ROI calculations and other capital budgeting techniques do not apply to many IT investments. Decisions to invest in other types of technology (e.g., an NC machine tool) are often made on a project basis using ROI methods. Investment in IT does not lend itself to ROI approaches for two reasons:

1. It is very difficult to determine the income stream, particularly for IT investment made for strategic purposes. Thus, the calculation of a meaningful ROI is not possible.
2. Investment in IT tends to be pervasive throughout the organization, and the consequences of investment are likely to be the result of interactions among a number of different project investments in IT (e.g., database, data entry, hardware, etc.).

A firm-wide approach to determining the relationship between IT investment and organizational performance is needed.

We have the following recommendations:

1. Recognize that ROI-type calculations are not relevant for many IT investments. The exception is transactional IT investment, where a common objective is cost displacement by the substitution of capital for labor.
2. Recognize that if threshold investment is required to compete in an industry, meeting that threshold may actually have negative ROI but be essential to the firm's survival.
3. Many IT investments, particularly of the strategic type, are made on "gut feelings." For these types of investments, companies should measure and track the investment and attempt to determine its effect on organizational performance.

Many contemporary business organizations contain a number of distinct businesses. To analyze IT investment at the organizational level will not provide a boundary that logically connects IT investment to performance. We choose to analyze each business as a separate unit; to achieve this we adopt the definition of Strategic Business Unit (SBU).[2] SBU has a distinct set of products or services and serves a distinct group of customers. The SBU also has a well-defined set of competitors. Analysis at this level removes many of the aggregation problems of assessing the firm level.

The performance measures used in the assessment of the impact of IT investment must be chosen to reflect the business objectives of the IT investment.

A large number of financial and social measures of firm (SBU) performance are available (Zammuto, 1982). We suggest a combination of measures in order to capture the diverse impacts of IT investment on the different aspects of performance. The aim is to choose measures that can be logically linked to the performance objective of each type of IT investment. We suggest the following matchings, inspired by the strategic intent for investing in various types of IT:

Type of IT Investment	**Performance Measure**
Strategic	Revenue Growth Rates
Informational	Return on Assets[3]
Transactional	Indirect Labor[4]

Performance measures for each type of IT investment match the objective of each type of IT investment. We based our choice of measure on discussions with IT managers,[5] as explained below.

- Strategic IT investment is generally made with long-term goals relating to competitive advantage. To capture this goal, we suggest using the sales growth rate of the firm.
- Informational IT investment is generally made with medium-term goals of improving management decision making; we suggest using the financial measure ROA to match it to performance.
- Transactional IT investment is generally made with goals of reducing the costs of doing business by substituting capital for labor. To capture this performance, we suggest assessing the change in non-production labor (adjusted for sales).

These measures are suggested as a guide and should be tailored to the industry and then the specific organization's needs. It is absolutely vital to perform any assessment of the effect of IT investment on performance over time. Cross-sectional assessment will lead to misleading pictures of the effects, as any lag between investment and performance effects will be ignored. This is one of the reasons that many of the studies reviewed earlier produced conflicting findings.

The objective is to track both IT investment (by type) and the appropriate measures of firm performance over time and look for trends and patterns. This process will assist linking IT investment to business strategy and give general indications of IT investment and performance associations. The technique would be particularly useful when applied to several SBUs in a single firm.

We do not suggest using a performance measure for threshold IT investment. This investment is required for the firm to enter, compete, or remain in a marketplace where a threshold IT exists.

Organizational Issues

One issue that emerged during this study is the importance of conversion effectiveness. Investment in IT is likely to be converted to productive outputs with different effectiveness, depending on the firm. If it is true (as suggested by the respondents) that conversion effectiveness is just another resource allocation task, then good organizational performers will have superior conversion effectiveness. We suggest that firms with good conversion effectiveness will have a stronger association between IT investment and performance. Conversion effectiveness is likely to have many components, four of which were identified by the respondents and are presented without justification below.

- Top management commitment to IT
- Previous firm experience with IT
- User satisfaction with systems
- The turbulence of the political environment of the organization

We recommend that attention be given to these four components when any significant IT investment is made. All four are intuitively appealing and consistent with implementation literature. Furthermore, an aggregation of these four components was found to significantly moderate the relationship between IT investment and firm performance in 33 manufacturing firms (Weill, 1988).

CONCLUSION

Previous studies of IT investment and its relationship to organizational performance have had conflicting findings. In the case studies reported here, designed to gain insight into the way firms define and track IT investment and its relationship to performance, we found a wide variety of definitions and methods of tracking. We also found that while firms acknowledge the importance of IT investment to organizational performance, they do not have a satisfactory way of assessing the relationship. We believe that some of the more interesting observations—the importance of political considerations, the concept of an industry-based threshold investment, and the notion of conversion effectiveness with which investment affects performance—are useful to managers who are responsible for managing and tracking IT investment.

In future research, we are seeking to refine and operationalize the appropriate definition of organizational performance, depending on the type of IT investment. We are also refining the concepts of threshold investment, investment types, and conversion effectiveness in a larger, cross-sectional survey of 33 manufacturing firms. Future research also needs to take into account the problems of comparing investment levels across industries and, as expenditures and authority for those expenditures become more and more decentralized, of measuring IT investment firmwide.

NOTES

1. The series of *Datamation* studies unfortunately cannot be analyzed for trends because different sampling techniques and measures for IT investment were used in each survey. For a detailed assessment of these results and their validity and usefulness, see Weill (1986).
2. The SBU was developed by The Strategic Planning Institute in Cambridge, Massachusetts, and was used in the PIMS (1984) study.
3. A study of the relationship between the process of corporate planning and financial performance of firms (Grinyer and Norburn, 1975) used 16 financial measures that assessed firm size, profitability, performance, and growth. This study showed statistically that the one measure of return on assets could substitute for all the others. For each firm, this measure was compared to its competitors to take account of industry differences, and this relative performance was used as the dependent variable.
4. Or fixed and variable product cost in service environments.
5. These discussions took place after a presentation by Weill to representatives of affiliate companies of the Center for Research on Information Systems (CRIS) at New York University in September 1987.

REFERENCES

Bender, D. "Financial Impact of Information Processing." *Journal of MIS* 3: 2 (Summer 1986): 232–238.

Cron, W. and Sobol, M. "The Relationship Between Computerization and Performance: A Strategy for Maximizing Economic Benefits of Computerization." *Information and Management* 6 (1983): 171–181.

Datamation. "Data Processing Budget Study." *Datamation* (New York, 1985).

Datamation. "Data Processing Budget Study." *Datamation* (New York, 1986).

Cahners Publishing Market Research Department. *Datamation Budget Study 1987/1988—Executive Summary*. New York: Cahners, 1987.

The Diebold Group. *MIS/Telecommunications Budgets and Key Indicators*. New York: The Diebold Group Inc., 1982.

The Diebold Group. *MIS/Telecommunications Budgets and Key Indicators*. New York: The Diebold Group Inc., 1984.

Grinyer, P. and Norburn, D. "Planning for Existing Markets: Perceptions of Executives and Financial Performance." *Journal of the Royal Sta-*

tistical Society, Series A 138: Part A (1975): 70–97.

Harris, S. E. and Katz, J. L. "Profitability and Information Technology Capital Intensity in the Insurance Industry." In *Proceedings of the Twenty-First Hawaii International Conference on System Sciences* Vol. IV (January 1988), pp. 124–130.

Ives, B. and Learmonth, G. "The Information System as a Competitive Weapon." *Communications of the ACM* 27: 12 (December 1984): 1193–1201.

Lucas, H. C., Jr. "The Use of an Accounting Information System, Action and Organizational Performance." *The Accounting Review* L: 4 (October 1975): 735–746.

Panko, R. "Spending on Office Systems: A Provisional Estimate." *Office Technology and People* 1: 2, 3 (September 1982): 177–194.

The PIMS Program. *Management Productivity and Information Technology*. (Cambridge, MA: The Strategic Planning Institute, 1984.)

Strassmann, P. *Information Payoff*. (New York: Free Press, 1985).

Turner, J. "Organizational Performance, Size and the Use of Data Processing Resources." Working Paper, #58, Center for Research in Information Systems, New York University, New York, 1985.

Turner, J. and Lucas, H. C., Jr. "Developing Strategic Information Systems." In *Handbook of Business Strategy*, edited by W. Guth, 21/1–21/35. Boston, MA: Warren, Gorham and Lamont, 1985.

U.S. Office of Management and Budget. *A Five Year Plan for Meeting the Automatic Data Processing and Telecommunications Needs of the Federal Government*. U.S. Government Printing Office, Washington, DC, 1985.

Weill, P. "Information Technology Investment in Utilities," Working Paper, #138, Center for Research in Information Systems, New York University, New York, 1988.

Weill, P. *The Relationship Between Investment in Information Technology and Firm Performance in the Manufacturing Sector*, unpublished Ph.D. dissertation, New York University, New York, 1988.

Zammuto, R. F. *Assessing Organizational Effectiveness: Systems Change, Adaption and Strategy*, State University of New York Press, Albany, NY, 1982.

QUESTIONS

1. Why is it so difficult to measure the relation between IT investment and performance?
2. Why is it necessary to have more than one type of measure to determine the results of IT investments?
3. Why is it important to track IT investment performance over time?

6 Implementation

THE same technology in two different firms can produce totally different outcomes depending on how it is implemented and deployed. Implementation, therefore, is more than a technology issue. It is a management issue.

Technology by itself does not have intrinsic value beyond the amortized asset value of its hardware, software, and telecommunications components. The real business value of technology depends on the benefits created by the management process.

This chapter begins with descriptions of two major movements that spur productivity: the diffusion of new innovations and the reengineering of existing work flows. The next two sections deal with the specifics of implementing change in manufacturing through organizational change and project management. The next section describes how the computer serves to integrate productivity in the office and the factory. Finally, we present case studies of corporations that have implemented elements of the virtual corporation.

6.1 Understanding Innovation Diffusion Helps Boost Acceptance Rates for New Technology

JAMES C. BRANCHEAU AND
JAMES C. WETHERBE

New products and ideas are required to create improvements in productivity. However, we know that only a few new products are hits. Most that are offered in the marketplace flash briefly but are never seen again. This observation led people in marketing to study the spread of technology innovations. Their results were quite simple. They found that innovations followed an S-curve like that followed by the spread of epidemics. Initially, some people will try anything because it is new. These "pioneers" become opinion leaders "infected" with the innovation. They tell their friends about it and, if it is a good innovation, the friends tell their friends. Initially, of course, only a few people are infected and hence the spread of the "disease" is slow. At some point, there are enough people infected that the spread of the disease increases quickly and the curve rises steeply. Eventually, however, the number of people left to infect becomes small and the curve becomes quite flat.

To grow, firms must implement new technology from time to time. The technology may be on the factory floor or in the office or in the sales force. Brancheau and Wetherbe show that managers who know how technologies diffuse can use that knowledge to speed up the implementation process.

Source: *Chief Information Officer Journal* (Fall 1989).

IF chief information officers understand how certain personality types and other underlying forces will influence the technology diffusion process, they can control end-user acceptance rates of new technology, as well as the effectiveness of new methods.

Recent research on innovation diffusion has created a valuable tool for developing a set of guidelines for managing the introduction of new technology.

INNOVATION DIFFUSION THEORY

Although a body of knowledge of technology change has not yet developed, a general theory explaining the spread of new ideas and new technologies among people has emerged from research in other fields. For example, rural sociologists have studied the spread of agricultural technology among farmers. The theory developed from the research is known as innovation diffusion theory.

Applying innovation diffusion theory to the study of end-user computing provides a basis for managing the introduction of new information technology. The major components of the theory are:

- S-shaped adopter distribution
- Innovativeness and adopter categories
- Individual adoption process
- Diffusion networks and opinion leaders

S-SHAPED ADOPTER DISTRIBUTION

One of the most salient components of the theory is the notion that the adoption of successful new ideas and products traces an S-shaped curve over time. Research in other fields has shown that the number of individuals adopting per period of time roughly follows a normal bell-shaped curve. Plotted on a cumulative frequency basis, these data describe the traditional S-shaped curve.

The S-shaped curve implies that relatively few people adopt a new idea at first. Then as word spreads and a few pioneers have shown that the idea works, more people are willing to try it. As the idea catches on, it reaches a point where it really takes off and most people will adopt it. Near the end of the cycle, the last few holdouts finally adopt the idea.

Initial research has confirmed the existence of the S-shaped curve for information technology. As shown in Figure 1, actual data on the adoption of spreadsheet software among finance/accounting professionals in 22 organizations conformed closely to the expected pattern. For example, one accounting executive in an educational products manufacturing organization installed a PC on every desk in his controller group during 1984. The department went from less than 5% to 100% adoption in a single year. Obviously, the number of adopters did not follow the S-shaped curve in this case. However, in reviewing the utilization data for his organization, he found that the number of hours the PCs were used each week did follow the predicted S-shaped curve. This suggests that even when a new technology is made

Figure 1 Adopter Distribution for Spreadsheet Software

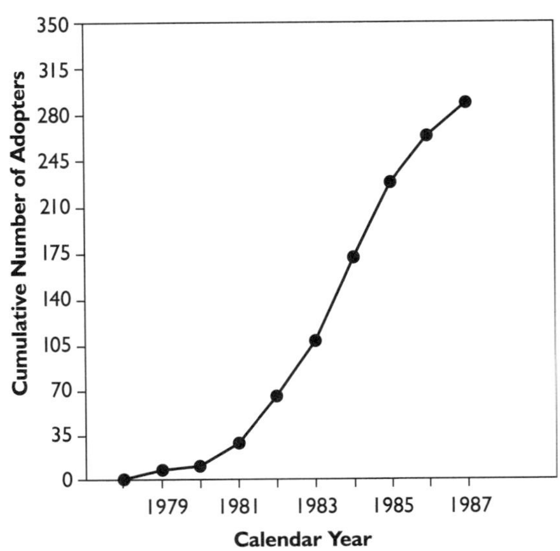

available early in its diffusion cycle, a learning curve phenomenon may still be at work.

INNOVATIVENESS AND ADOPTER CATEGORIES

Another major component of the theory is the notion that individuals vary in the time they take to adopt an innovation. One measure of an individual's innovativeness is the relative earliness or lateness that he or she adopts an innovation (Figure 2).

The rationale for the classification is that the people at the head of the curve (pioneers and early adopters) are considerably different than the people at the tail of the curve (majority and laggards) (Table 1). In discussing how these types of differences affected adoption of spreadsheet software in a specialty manufacturing organization, a group of managers referred to the pioneers and

Figure 2 Diffusion Curves and Adopter Categories

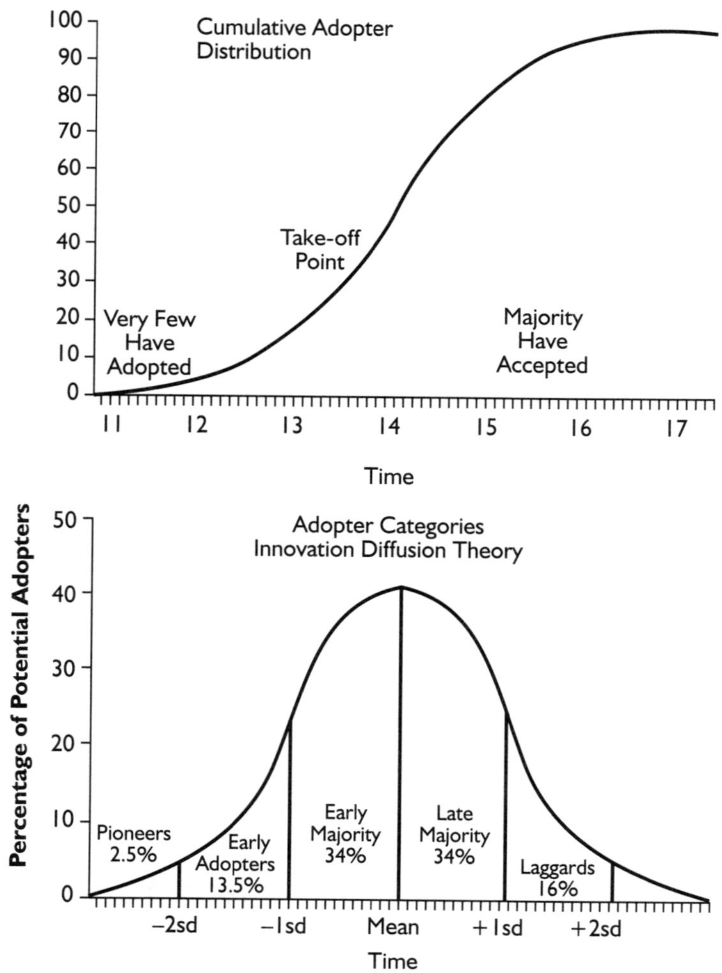

TABLE 1

Initial Findings on Key Adopter Categories for End-User Computing

Pioneers	
Findings	*Implications*
About average age.	Venturesome.
College graduate.	High tolerance for uncertainty.
Reads a lot.	Well connected (externally).
Professionally active outside company.	Often acts as "gatekeeper."
Talks to IS/IC frequently.	
Usually a computer opinion leader.	
Very interested in technology.	

Early Adopters	
Findings	*Implications*
About average age.	Reduces uncertainty for majority adopters.
College graduate.	Well connected (internally).
Reads a lot.	Often acts as "boundary spanner."
Less active outside company.	
Rarely talks to IS/IC.	
Often a business opinion leader.	
Less interested in technology.	

Laggards	
Findings	*Implications*
About ten years older than average age.	Traditional outlook.
Most likely a clerk or manager.	Suspicious of technology.
Holds associate degree.	Can slow adoption through position in hierarchy.
Reads half as much as pioneers.	
About half as much outside participation as pioneers.	
Rarely talks to IS/IC.	
Some business opinion leadership.	
No computer opinion leadership.	
Minimal interest in technology.	

Source: Brancheau, 1987.

early adopters as "self-starters . . . All you need to do is throw a manual at them . . ." These managers indicated that differences among individuals were apparent well after most of the organization had started to use the software. The self-starters, it turns out, were still more innovative in developing new applications and transferring them to others as appropriate.

INDIVIDUAL ADOPTION PROCESS

The third major component of the theory suggests that individual adoption is not an instantaneous act, but a process that occurs over time. In making a decision to adopt or reject an innovation, individuals normally go through a series of stages in their thinking; (e.g., knowledge, persuasion, decision, and implementation).

The knowledge stage occurs as an individual becomes aware of the existence of an innovation and gains an understanding of its function. Both the need-driven and the technology-driven paths to awareness can operate. The persuasion stage occurs as an individual begins to form a favorable or unfavorable attitude toward the innovation. Attitudes form as additional information is received. Evaluative information from interpersonal networks becomes particularly important. The decision stage occurs as an individual consciously decides to adopt or reject the innovation. At this stage, the innovation may be used on a trial basis. Finally, implementation occurs as an individual acquires and makes initial use of the innovation. At this point, an individual is an adopter.

Initial research on end-user computing only partially confirms the prior finding in other fields. While the expected trends in the importance of the various channels (i.e., vendor literature, company newsletters, journals) were confirmed, mass media channels were not as important to adopters as anticipated. As shown in Figure 3, work colleagues were the most influential sources of information across the stages of decision making. This appears to be an important difference between end-user computing and other contexts. However, consistent with innovation diffusion theory, the influence of

Figure 3 Importance of Communication Channels across Stages of Adoption Decision Making

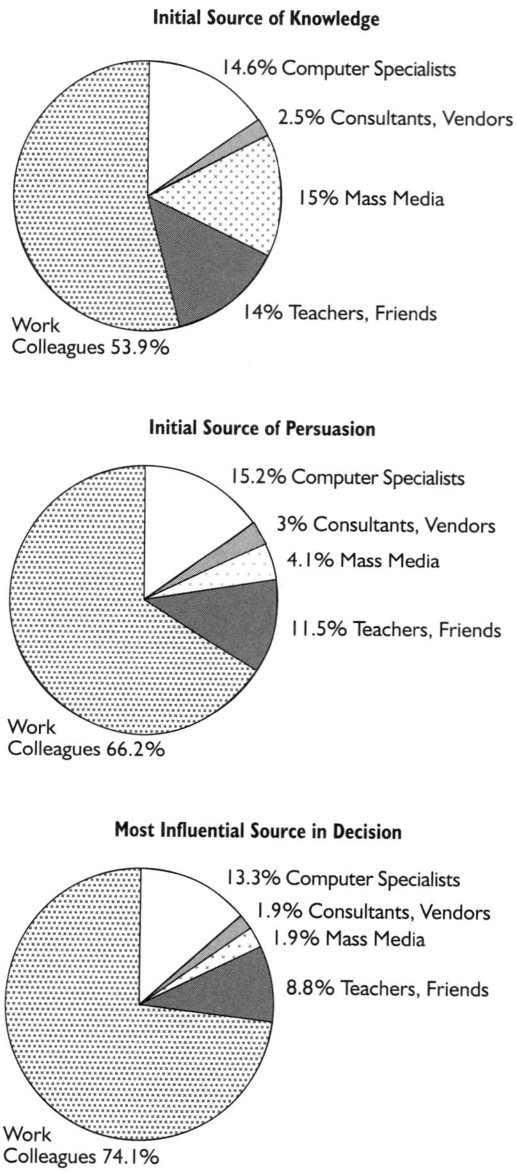

local/interpersonal channels increased as individuals' thinking progressed from knowledge, through persuasion, to the final decision.

An information center manager from a large investment products organization relayed his experience with this phenomenon. "Back in 1983 when we started up, the finance people were getting a fair number of ideas from us by way of newsletters, product demos, etc. Now there is very little we know about Lotus that is not already common knowledge in the finance group. They use the software every day and are virtual experts with it. The few novice users left in finance learn more about the technology from their local gurus than they do from us."

DIFFUSION NETWORKS AND OPINION LEADERSHIP

The fourth major component of the theory deals with the way information about innovations spreads through a company. In deciding whether to adopt, individuals depend mainly on information from others like themselves who have already adopted the innovation. The interpersonal networks these communications flow through are referred to as diffusion networks.

Closely related to diffusion networks is the concept of opinion leadership. As its name implies, opinion leadership involves informal influence over others' attitudes and behavior. Due to their extensive interpersonal networks and their willingness to experiment with new ideas, opinion leaders can activate the diffusion network in an organization. They affect diffusion to such a degree that the diffusion curve may have its usual S-shape because of the time at which opinion leaders adopt.

One of the early adopters at an energy distribution company echoed these findings. He said that use of spreadsheet software in most departments did not take off until someone inside the department acted as the "seed." Then, others could look over his shoulder and find out what the software could do. "Until enough seeds were planted, most people were very skeptical."

MANAGING THE INTRODUCTION OF NEW TECHNOLOGY

Despite strong support for innovation diffusion theory in the context of end-user computing, one research project cannot conclusively establish the validity of the theory as well as the implications that flow from it. Thus, the guidelines offered here are preliminary implications of the theory assuming it is confirmed and further developed through follow-up research.

Tactical guidelines for introducing new technology can be offered in a number of areas.

1. Change agents. One of the most effective means for introducing new information technology is to utilize internal change agents in the early stages of the diffusion cycle. Change agents act as intermediaries between the technicians responsible for creating or maintaining the new technology and the users of that technology. In the context of end-user computing, the information systems department normally assumes the change-agent role. For example, most large organizations have set up an information center to help manage the introduction of new technology. But, unless change agents intervene before 10–25% of the target population has adopted a new technology, the natural forces within the diffusion network can be very difficult to counteract.

   Furthermore, effective intervention requires a great deal of personal contact with users and empathy for their situation to establish credibility. The critical factors are for change agents to establish effective communication channels with potential users and to get involved in diffusion as early as possible. A big problem in the early days of end-user computing was that many companies waited too long before establishing their information centers. As a result, the strength of the diffusion network made it difficult to control users' perceptions of new technology.
2. Opinion leaders are more effective than change agents in influencing new users once diffusion has reached a certain point (estimated at 10–25% adoption). If opinion leaders are identified in advance in key areas, they can be instrumental in assisting efforts to manage new technology introductions. Of course, the opposite situation can also occur. If opinion leaders reject a new technology, they can stop the diffusion process in its tracks. The key factors are for change agents to recognize potential opinion leaders early and target their efforts accordingly.
3. Communication channels vary in effectiveness across the stages of adoption decision making. For example, mass media channels may be somewhat effective in making potential users aware of new technology, but they are not effective in persuading them to adopt it. In contrast, informal communication with peers who have used the new technology is very persuasive. Because potential users rely almost exclusively on interpersonal channels, change agents should be aware of this and rely heavily upon informal networks to influence adoption decision making.
4. Modeling and demonstration play an important role in spreading new ideas. New information technology is tried "vicariously" when individuals are provided with conspicuous examples of how the technology works. Modeling and demonstration are most effective when they take place in the end user's work environment and are conducted by peers. Opinion leaders are particularly effective in modeling and demonstration.
5. Trial use of a new technology also can increase the adoption rate. Information technology's portability can be exploited in this regard. Change agents can speed diffusion by loaning hardware and software to potential end users. Hands-on experience with the new technology reduces the uncertainty involved in making an adoption decision. This tactic is most effective if aimed at recognized opinion leaders within the organization. Combining trial use with per-

sonalized technical support helps to insure successful trials.

6. Local experts. Closely related to the issue of influencing individual adoption is the issue of providing support to sustain long-term utilization. Individuals tend to be influenced by others who are close to them in physical distance and relatively alike in social characteristics. Because of this, alternatives to a centralized support organization need to be examined over the long term. One alternative is to use local experts within user departments. Housing technical support personnel inside the functional office areas brings them closer to their "customers." In addition, over time, it will broaden the range of shared experience and improve the effectiveness of communication. If the local expert approach is not taken formally, it will occur informally. And while informal support should be encouraged, it may not meet the long-term needs of the organization. A formal effort to identify and recognize local experts may be required. In the long run, IS managers must work to close the communication gap and build organizational learning.

AN OVERALL STRATEGY

A review of what is known about innovation diffusion in organizations suggests an overall strategy for managing the introduction of new information technology. The key points are illustrated in Figure 4 and discussed below.

In the first stage, CIOs should encourage pioneers by providing slack resources. No organizational mechanism needs to be established, however, since pioneers work best in an undisciplined structure. The results of pioneer work should be examined for feasibility and organizational implications. Information interchange with early adopters should be facilitated by publicizing the results of the pioneers' experimentation.

In the second stage, when early adopters begin to appear (or when the organization chooses to en-

Figure 4 An Overall Strategy for Introducing New Technology

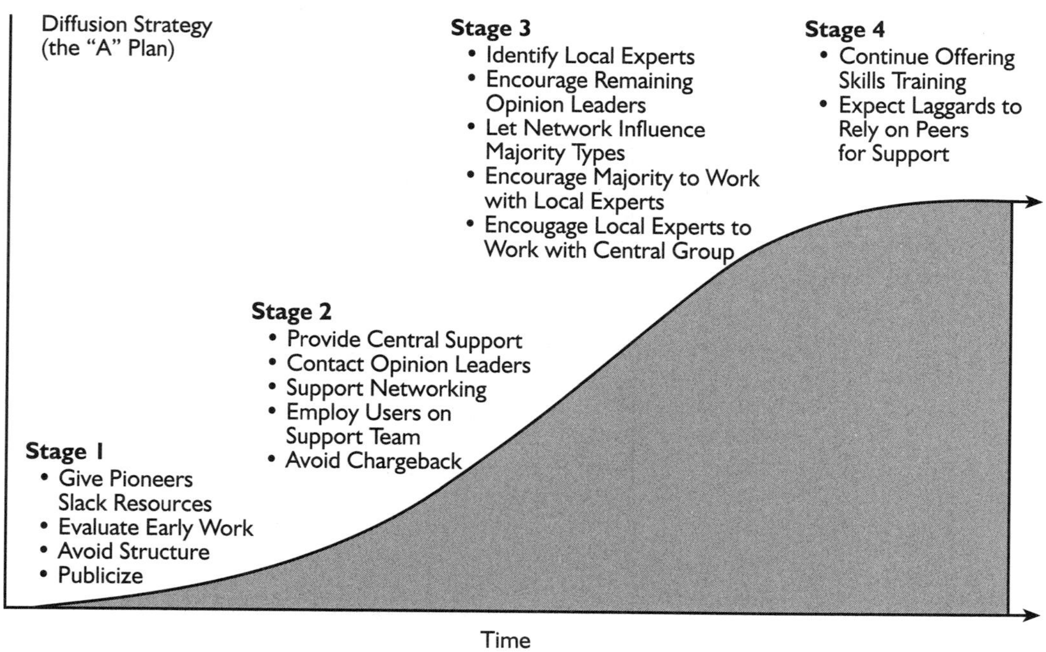

courage them), IS should establish a centralized support center. The potential for the new technology is publicized in general terms through newsletters, bulletins, etc. Where feasible, opinion leaders are contacted personally and encouraged by offers of free trial use. Networking among early adopters and pioneers is encouraged and supported through user groups and focus meetings. The support center should employ user-empathetic support personnel and actively encourage and support early adopters. Charging for support center services should be avoided if possible.

In the third stage, when early adopters have provided solid organizational evidence of the value and feasibility of the new technology, emphasis should shift toward identification and support of local experts. Opinion leaders who have not previously adopted are actively recruited and encouraged. The role of influencing others to adopt is shifted from support center personnel to opinion leaders. The support center provides initial training and consulting but turns over encouragement and on-the-job skill development to local experts. Majority-type adopters are encouraged to seek help from local experts. Local experts work with the centralized support center to solve specialized problems.

In the fourth stage, the diffusion process reaches a stable state as laggards adopt the technology. Laggards can be expected to rely heavily on support systems composed of peers within their functional areas.

This cycle of events can be repeated in a miniversion as new equipment options, new software packages, and new business applications are selected for introduction. The target population necessarily changes with each new technology since it will not be equally useful to all segments of the organization. Pioneers may be willing to try out the new technology during initial technical debugging. Indeed, they are often more adept at this than support center staff. Early adopters may be willing to work through the changes in user and organizational procedures (including training processes). This can help make the new technology operate smoothly in the organization. As the technology diffuses to the remainder of the target population, early adopters will be able to provide the local expertise needed to supplement general training provided by the support center.

The foregoing discussion assumes the managers will lead the introduction of new technology, not react to it. Unfortunately, in day-to-day operations, urgency tends to drive out importance. Even the best managers will not always be positioned appropriately at the beginning of a diffusion cycle. In such cases, an alternate strategy is needed. As might be expected, the strategy calls for staying close to the informal network. Being aware of what pioneers and opinion leaders are doing with information technology will insure that diffusion does not advance too far without an organizational review. In most cases, offering skills training and other basic support services can help get the "late" change agent back into the diffusion network. By offering value-added services, change agents can get pioneers and opinion leaders to work with them instead of against them.

CONCLUSION

The past half-decade has been a period of rapid change in the role of information systems in the organization. Many IS managers have found that traditional implementation strategies have not helped them deal with new end-user information technology. Managing the introduction of new technology among relatively autonomous knowledge workers requires a new approach. By understanding what drives the diffusion of technology, managers can work with the natural forces instead of against them.

SUGGESTED READING

Allen, Thomas J. *The World, Your Company: A Gate for Information! Who Guards the Gate? Innovation*, 8 (1969), pp. 12–19.

Bandura, Albert. *Social Learning Theory*. Englewood Cliffs, NJ: Prentice-Hall, 1977.

Brancheau, James C., Vogel, Douglas R., and

Wetherbe James C. "An Investigation of the Information Center from the User's Perspective." *Data Base*, 17: 1 (Fall 1985): 4–17.

Brancheau, James C., and Davis, Gordon B. *A Model for the Diffusion of End-User Computing*. MISRC Working Paper 87-07, MIS Research Center, Carlson School of Management, University of Minnesota, Minneapolis, MN, November 1986.

Brancheau, James C. *The Diffusion of Information Technology: Testing and Extending Innovation Diffusion Theory in the Context of End-User Computing*. Unpublished Ph.D. Dissertation, University of Minnesota, Minneapolis, MN, 1987.

Brancheau, James C., and Wetherbe, James C. "A Comparative Analysis of Higher and Lower Rated Information Centers." *Journal of Information Management* (Winter 1988).

Davis, Gordon B., and Olson, Margrethe H. *Management Information Systems Conceptual Foundations, Structure and Development*. New York, NY: McGraw-Hill, 1985.

Drucker, Peter. *The Age of Discontinuity*. New York, NY: Harper & Row, 1978.

Kanter, Rosabeth Moss. *The Change Masters*. New York, NY: Simon and Schuster, 1983.

Leitheiser, Robert L., and Wetherbe, James C. *Avoiding the Pitfalls of End-User Computing*. MISRC Working Paper 85-09, MIS Research Center, Carlson School of Management, University of Minnesota, Minneapolis, MN, October 1984.

Leitheiser, Robert L., and Wetherbe, James C. "Service Support Levels: An Organized Approach to End-User Computing." *MIS Quarterly*, 10: 4 (December 1986): 337–349.

Peters, Thomas J., and Waterman, Jr., Robert H. *In Search of Excellence: Lessons from America's Best-Run Companies*. New York, NY: Harper & Row, 1982.

Petty, R., and Cacioppo, J. *Attitudes and Persuasion: Classic and Contemporary Approaches* Dubuque, IA: Brown, 1981.

Rogers, Everett M. *The Diffusion of Innovations*, 3rd Edition. New York, NY: Free Press, 1983.

Rogers, Everett M., and Kincaid, D. Lawrence. *Communication Networks: Toward a New Paradigm for Research*. New York, NY: Free Press, 1981.

Tushman, Michael L., and Scanlan, Thomas J. "Boundary Spanning Individuals: Their Role in Information Transfer and Their Antecedents." *Academy of Management Journal*, 24: 2 (June 1981) 289–305.

Wetherbe, James C., and Leitheiser, Robert L. "Information Centers: A Survey of Services, Decisions, Problems, Successes." *Journal of Information Systems Management* (Summer 1985): 3–10.

QUESTIONS

1. How can innovation diffusion theory provide guidance for managing the implementation of an office information system?
2. Is it possible to make the S-shaped diffusion curve steeper to accelerate the diffusion process? If yes, how? If no, why not?
3. Is innovation diffusion theory adequate for analyzing all types of information systems? Explain why or why not.

6.2 The New Industrial Engineering: Information Technology and Business Process Redesign

THOMAS H. DAVENPORT AND
JAMES E. SHORT

The concept of "reengineering" the business processes of the organization has dominated the 1990s. Millions read Hammer and Champy's *Reengineering the Corporation* in which the authors call for transformations that obliterate existing ways of doing business. Yet they claim that success rates for such reengineering projects are less than 40%. Studies subsequent to their book bear out the fact that large numbers of reengineering projects fail.

To get beyond the slogans, a business process is a set of tasks performed to achieve a business purpose. It may be as mundane as booking an order or sending a bill. But even mundane tasks may not be simple. Consider what happens at a particular airplane manufacturer. An airline places an order. Marketing credits the order when the client commits. Legal credits the order when a contract is signed. Finance recognizes the order when the down payment is received. Who is right? The definition of what is a booked order affects the financial statement.

Most business processes were developed before the computer and telecommunications age. Even new companies still use traditional processes because the people involved learned that these practices are standard. The first wave of computer and communications technology merely automated the old practices. With the advances in technology and with the second wave of automation upon us, it is no longer necessary to follow obsolete processes.

Davenport and Short take a less apocalyptic view of reengineering than do Hammer and Champy. They argue that most reengineering projects involve information technology as the key enabler. Furthermore, most processes are cross-organizational and hence need good communi-

---

Source: *Sloan Management Review* (Summer 1990).

cation among the participants. The Davenport and Short article takes you through the steps needed to reengineer a process, including the managerial roles and the organizational changes.

AT the turn of the century, Frederick Taylor revolutionized the workplace with his ideas on work organization, task decomposition, and job measurement. Taylor's basic aim was to increase organizational productivity by applying to human labor the same engineering principles that had proven so successful in solving the technical problems in the work environment. The same approaches that had transformed mechanical activity could also be used to structure jobs performed by people. Taylor came to symbolize the practical realizations in industry that we now call industrial engineering (IE), or the scientific school of management.[1] In fact, though work design remains a contemporary IE concern, no subsequent concept or tool has rivaled the power of Taylor's mechanizing vision.

As we enter the 1990s, however, two newer tools are transforming organizations to the degree that Taylorism once did. These are *information technology*—the capabilities offered by computers, software applications, and telecommunications—and *business process redesign*—the analysis and design of work flows and processes within and between organizations. Working together, these tools have the potential to create a new type of industrial engineering, changing the way the discipline is practiced and the skills necessary to practice it.

This article explores the relationship between information technology (IT) and business process redesign (BPR). We report on research conducted at MIT, Harvard, and several consulting organizations on 19 companies, including detailed studies of five firms engaged in substantial process redesign. After defining business processes, we extract from the experience of the companies studied a generic five-step approach to redesigning processes with IT. We then define the major types of processes, along with the primary role of IT in each type of process. Finally, we consider management issues that arise when IT is used to redesign business processes.

IT IN BUSINESS PROCESS REDESIGN

The importance of both information technology and business process redesign is well known to industrial engineers, albeit as largely separate tools for use in specific, limited environments.[2] IT is used in industrial engineering as an analysis and modeling tool, and IEs have often taken the lead in applying information technology to manufacturing environments. Well-known uses of IT in manufacturing include process modeling, production scheduling and control, materials management information systems, and logistics. In most cases where IT has been used to redesign work, the redesign has most likely been in the manufacturing function, and industrial engineers are the most likely individuals to have carried it out.

IEs have begun to analyze work activities in nonmanufacturing environments, but their penetration into offices has been far less than in factories. IT has certainly penetrated the office and services environments—in 1987 *Business Week* reported that almost 40% of all U.S. capital spending went to information systems, some $97 billion a year—but IT has been used in most cases to hasten office work rather than to transform it.[3] With few exceptions, IT's role in the redesign of nonmanufacturing work has been disappointing; few firms have achieved major productivity gains.[4] Aggregate productivity figures for the United States have shown no increase since 1973.[5]

Given the growing dominance of service industries and office work in the Western economies, this type of work is as much in need of analysis and redesign as the manufacturing environments to which IT has already been applied. Many firms have found that this analysis requires taking a broader view of both IT and business activity, and of the relationships between them. Information technology should be viewed as more than an automating or mechanizing force; it can fundamentally reshape the way business is done. Business activities should be viewed as more than a collection of individual or even functional tasks; they should be broken down into processes that can be designed for maximum effectiveness, in both manufacturing and service environments.

Our research suggests that IT can be more than a useful tool in business process redesign. In leading edge practice, information technology and BPR have a recursive relationship, as Figure 1 illustrates. Each is the key to thinking about the other. Thinking about information technology should be in terms of how it supports new or redesigned business processes, rather than business functions or other organizational entities. And business processes and process improvements should be considered in terms of the capabilities information technology can provide. We refer to this broadened, recursive view of IT and BPR as the new industrial engineering.

Taylor could focus on workplace rationalization and individual task efficiency because he confronted a largely stable business environment; today's corporations do not have the luxury of such stability.[6] Individual tasks and jobs change faster than they can be redesigned. Today, responsibility for an outcome is more often spread over a group, rather than assigned to an individual as in the past. Companies increasingly find it necessary to develop more flexible, team-oriented, coordinative, and communication-based work capability. In short, rather than maximizing the performance of particular individuals or business functions, companies must maximize interdependent activities within and across the entire organization. Such business processes are a new approach to coordination across the firm; information technology's promise—and perhaps its ultimate impact—is to be the most powerful tool in the twentieth century for reducing the costs of this coordination.[7]

Figure 1 The Recursive Relationship Between IT Capabilities and Business Process Redesign

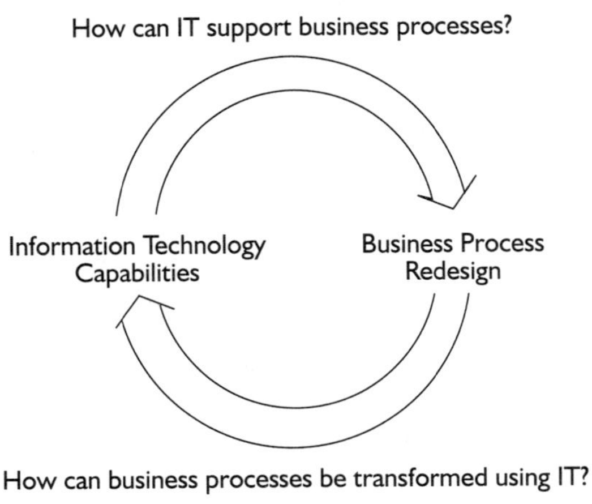

WHAT ARE BUSINESS PROCESSES?

We define business processes as a set of logically related tasks performed to achieve a defined business outcome. This definition is similar to Pall's: "The logical organization of people, materials, energy, equipment, and procedures into work activities designed to produce a specified end result (work product)."[8]

A set of processes forms a business system—the way in which a business unit, or a collection of units, carries out its business. Processes have two important characteristics:

- They have customers; that is, processes have defined business outcomes, and there are recipients of the outcomes. Customers may be either internal or external to the firm.
- They cross organizational boundaries; that is, they normally occur across or between organizational subunits. Processes are generally independent of formal organizational structure.

Common examples of processes meeting these criteria include:

- Developing a new product
- Ordering goods from a supplier
- Creating a marketing plan
- Processing and paying an insurance claim
- Writing a proposal for a government contract

Ordering goods from a supplier, for example, typically involves multiple organizations and functions. The end user, purchasing, receiving, accounts payable, etc., and the supplier organization are all participants. The user could be viewed as the process's customer. The process outcome could be either the creation of the order, or, perhaps more usefully, the actual receipt of the goods by the user.

Our examples so far are of large-scale processes that affect whole organizations or groups. There are more detailed processes that meet the definitional criteria above. These might include installing a windshield in an automobile factory, or completing a monthly departmental expense report. IT-driven process redesign can be applied to these processes, but the implications of redesigning them may be important only in the aggregate. In many of the firms studied, analyzing processes in great detail was highly appropriate for some purposes, for example, the detailed design of an information system or data model to support a specific work process. However, the firms that were truly beginning to redesign their business functions took a broader view of processes.

A BRIEF HISTORY OF PROCESS THINKING

Process thinking has become widespread in recent years, due largely to the quality movement. Industrial engineers and others who wish to improve the quality of operations are urged to look at an entire process, rather than a particular task or business function. At IBM, for example, "process management will be the principal IBM quality focus in the coming years."[9] But process discussions in the quality movement's literature rarely mention information technology. Rather, the focus is usually on improving process control systems in a manufacturing context; when IT is discussed, it is in the context of factory floor automation. Recent IE literature also borders on process thinking when advocating cross-functional analysis,[10] although, as we will discuss, cross-functional processes are only one possible type of process.

Other than quality-oriented manufacturing process redesign, most processes in major corporations have not been subject to rigorous analysis and redesign. Indeed, many of our current processes result from a series of ad hoc decisions made by functional units, with little attention to effectiveness across the entire process. Many processes have never even been measured. In one manufacturing company studied, for example, no one had ever analyzed the elapsed time from a customer's order to delivery. Each department (sales, credit checking, shipping, and so on) felt that it had optimized its own performance, but in fact the overall process was quite lengthy and unwieldy.

Even fewer business processes have been analyzed with the capabilities of IT in mind. Most business processes were developed before modern computers and communications even existed. When technology has been applied, it is usually to automate or speed up isolated components of an existing process. This creates communication problems within processes and impediments to process redesign and enhancement. For example, in a second manufacturing firm studied, the procurement process involved a vendor database, a materials management planning system, and accounts payable and receivable systems, all running on different hardware platforms with different data structures. Again, each organizational subunit within the process had optimized its own IT application, but no single subunit had looked at (or was responsible for) the entire process. We believe the problems this firm experienced are very common.

REDESIGNING BUSINESS PROCESSES WITH IT: FIVE STEPS

Assuming that a company has decided its processes are inefficient or ineffective, and therefore in need of redesign, how should it proceed? This is a straightforward activity, but five major steps are involved: develop the business vision and process objectives, identify the processes to be redesigned, understand and measure the existing process, identify IT levers, and design and build a prototype of the new process (see Figure 2). We observed most or all of these steps being performed in companies that were succeeding with BPR. Each step is described in greater detail below.

Develop Business Vision and Process Objectives

In the past, process redesign was typically intended simply to "rationalize" the process, in other words, to eliminate obvious bottlenecks and inefficiencies. It did not involve any particular business vision or context. This was the approach of the "work simplification" aspect of industrial engineering, an important legacy of Taylorism. An example of the rationalization approach appears in a 1961 "Reference Note on Work Simplification" from the Harvard Business School:

> A good manager asks himself *why* things are done as they are, extending his inquiry to every aspect of the job and the surroundings in which it is performed, from the flow of paper work to the daily functioning of his subordinates. . . . He is expected to supply the stimulus and show that job improvement or simplification of work is not only important but also is based on commonsense questioning aimed at uncovering the easiest, most economical way of performing a job.[11]

Figure 2 Five Steps in Process Redesign

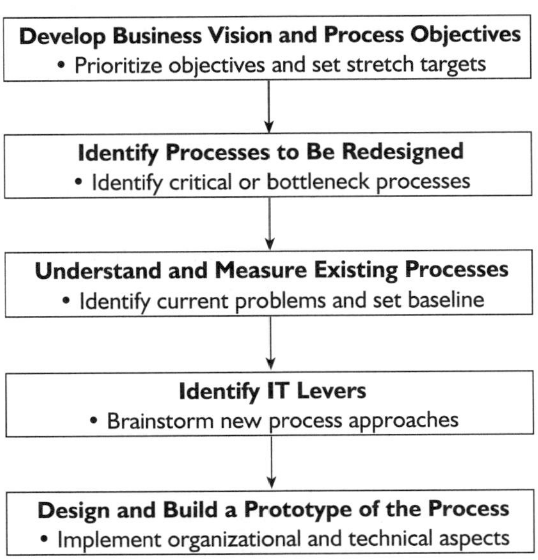

Our research suggests strongly that rationalization is not an end in itself, and is thus insufficient as a process redesign objective. Furthermore, rationalization of highly decomposed tasks may lead to a less efficient overall process. Instead of task rationalization, redesign of entire processes should be undertaken with a specific business vision and related objectives in mind.

In most successful redesign examples we studied, the company's senior management had developed a broad strategic vision into which the process redesign activity fit.[12] At Xerox, for example, this vision involved taking the perspective of the customer and developing systems rather than stand-alone products; both required cross-functional integration. At Westinghouse, the vision consisted largely of improving product quality. Ford's involved adopting the best practices of Japanese automobile manufacturers, including those of Mazda, of which it is partial owner.

Each of these visions implied specific objectives for process redesign. The most likely objectives are the following:

- Cost Reduction. This objective was implicit in the "rationalization" approach. Cost is an important redesign objective in combination with others, but insufficient in itself. Excessive attention to cost reduction results in tradeoffs that are usually unacceptable to process stakeholders. While optimizing on other objectives seems to bring costs into line, optimizing on cost rarely brings about other objectives.

- Time Reduction. Time reduction has been only a secondary objective of traditional industrial engineering. Increasing numbers of companies, however, are beginning to compete on the basis of time.[13] Processes, as we have defined them, are the ideal unit for a focused time reduction analysis. One common approach to cutting time from product design is to make the steps begin simultaneously, rather than sequentially, using IT to coordinate design directions among the various functional participants. This approach has been taken in the design of computers, telephone equipment, automobiles, and copiers (by Digital Equipment, AT&T Bell Labs, Ford, and Xerox, respectively).
- Output Quality. All processes have outputs, be they physical—such as in manufacturing a tangible product—or informational—such as in adding data to a customer file. Output quality is frequently the focus of process improvement in manufacturing environments; it is just as important in service industries. The specific measure of output quality may be uniformity, variability, or freedom from defects; this should be defined by the customer of the process.
- Quality of Worklife (QWL)/Learning/Empowerment. IT can lead either to greater empowerment of individuals, or to greater control over their output. Zuboff points out that IT-intensive processes are often simply automated, and that the "informating" or learning potential of IT in processes is often ignored.[14] Moreover, Schein notes that organizations often do not provide a supportive context for individuals to introduce or innovate with IT.[15] Of course, it is rarely possible to optimize all objectives simultaneously, and in most firms, the strongest pressures are to produce tangible benefits. Yet managers who ignore this dimension risk failure of redesigned processes for organizational and motivational factors.

Some firms have been able to achieve multiple objectives in redesigning processes with IT. American Express, for example, set out to improve the cost, time, and quality of its credit authorization process by embedding the knowledge of its best authorizers in an "Authorizer's Assistant" expert system. This successful redesign led to a $7 million annual reduction in costs due to credit losses, a 25% reduction in the average time for each authorization, and a 30% reduction in improper credit denials.

Finally, all firms found it was important to set specific objectives, even to the point of quantification. Though it is difficult to know how much improvement is possible in advance of a redesign, "reach should exceed grasp." Setting goals that will stretch the organization will also provide inspiration and stimulate creative thinking. For example, a company might decide to reduce the time to bring new products to market by 80%. In the accounts payable process at Ford, the "stretch" goal was to eliminate invoices—to pay suppliers upon receipt of their products or services. This goal has been achieved with help from an information system to confirm expected deliveries at the loading dock. As a result, Ford has eliminated three-quarters of the jobs in accounts payable.

Identify Processes to Be Redesigned

Most organizations could benefit from IT-enabled redesign of critical (if not *all*) business processes. However, the amount of effort involved creates practical limitations. Even when total redesign was the ultimate objective, the companies we studied selected a few key processes for initial efforts. Moreover, when there was insufficient commitment to total redesign, a few successful examples of IT-enhanced processes became a powerful selling tool.

The means by which processes to be redesigned are identified and prioritized is a key issue.

This is often difficult because most managers do not think about their business operations in terms of processes. There are two major approaches. The *exhaustive* approach attempts to identify all processes within an organization and then prioritize them in order of redesign urgency. The *high-impact* approach attempts to identify only the most important processes or those most in conflict with the business vision and process objectives.

The exhaustive approach is often associated with "information engineering" (developed by James Martin in the early 1980s), in which an organization's use of data dictates the processes to be redesigned.[16] For example, one information engineering method, employed at several divisions of Xerox, involves identifying business activities and the data they require using a data-activity matrix. The clusters of data activity interactions in the cells of the matrix are the organization's major business processes. Once processes are identified, Xerox managers prioritize them in the order in which new IT applications support should be provided. Although process identification in some Xerox divisions has taken as little as three months, many companies find this approach very time consuming.

The alternative is to focus quickly on high-impact processes. Most organizations have some sense of which business areas or processes are most crucial to their success, and those most "broken" or inconsistent with the business vision. If not, these can normally be identified using senior management workshops, or through extensive interviewing.[17] At IBM, the salesforce was surveyed to determine the relative importance of various customer support processes; the generation of special bids emerged as the highest priority and was the first process to be redesigned.

Companies that employed the high-impact approach generally considered it sufficient. Companies taking the exhaustive approach, on the other hand, have not had the resources to address all the identified processes; why identify them if they cannot be addressed? As a rough rule of thumb, most companies we studied were unable to redesign and support more than 10–15 major processes per year (i.e., one to three per major business unit); there was simply not enough management attention to do more. And some organizations have abandoned the exhaustive approach.[18]

Whichever approach is used, companies have found it useful to classify each redesigned process in terms of beginning and end points, interfaces, and organization units (functions or departments) involved, particularly including the customer unit. Thinking in these terms usually broadens the perceived scope of the process. For example, a sales manager may be aware that there are inefficiencies in customer order entry. A skilled process consultant might decide that the whole process—negotiating, receiving, and fulfilling orders—needs to be redesigned. Whether the problem is broken down into three processes or viewed as one is not important; expanding the *scope* of the process analysis is the key issue.

High-impact processes should also have owners.[19] In virtually all the process redesigns we studied, an important step was getting owners to buy in to both the idea and the scope of process redesign at an early stage. In several companies, managers felt that the process owner's job should be either above the level of the functions and units involved, or, if on the same level, that the owner should be willing—and able—to change the status quo. The difficulty, however, is that some processes only come together at the CEO level. In this situation, the CEO should designate a senior manager as owner and invest him or her with full authority. Processes that are fully contained within a single function or department can normally be owned by the manager of that area.

Understand and Measure Existing Processes

There are two primary reasons for understanding and measuring processes before redesigning them. First, problems must be understood so that they are not repeated. Second, accurate measurement can serve as a baseline for future improvements. If the objective is to cut time and cost, the time and cost consumed by the untouched process must be measured accurately. Westinghouse Productivity and Quality Center consultants found that simply graphing the incremental cost and time consumed by process tasks can often suggest initial areas for redesign. These graphs look like "step functions" showing the incremental contribution of each major task.

This step can easily be overemphasized, however. In several firms, the "stretch" goal was less to eliminate problems or bottlenecks than to create radical improvements. Designers should be informed by past process problems and errors, but they should work with a clean slate. Similarly, the process should not be measured for measurement's sake. Only the specific objectives of the redesign should be measured. As with the high-impact process identification approach, an 80–20 philosophy is usually appropriate.

Identify IT Levers

Until recently, even the most sophisticated industrial engineering approaches did not consider IT capabilities until after a process had been designed. The conventional wisdom in IT usage has always been to first determine the business requirements of a function, process, or other business entity, and then to develop a system. The problem is that an awareness of IT capabilities can—and should—influence process design. Knowing that product development teams can exchange computer-aided designs over large distances, for example, might affect the structure of a product development process. The role of IT in a process should be considered in the early stages of its redesign.[20]

Several firms accomplished this using brainstorming sessions, with the process redesign objectives and existing process measures in hand. It was also useful to have a list of IT's generic capabilities in improving business processes. In the broadest sense, *all* of IT's capabilities involve improving coordination and information access across organizational units, thereby allowing for more effective management of task interdependence. More specifically, however, it is useful to think about IT capabilities and their organizational impacts in eight different ways (See Table 1).

There are undoubtedly other important IT capabilities that can reshape processes. Organizations may want to develop their own lists of capabilities that are specific to the types of processes they employ. The point is twofold: IT is so powerful a tool that it deserves its own step in process redesign, and IT can actually create new process design options, rather than simply support them.

Design and Build a Prototype of the Process

For most firms, the final step is to design the process. This is usually done by the same team that performed the previous steps, getting input from constituencies and using brainstorming workshops. A key point is that the actual design

TABLE 1
IT Capabilities and Their Organizational Impacts

Capability	*Organizational Impact/Benefit*
Transactional	IT can transform unstructured processes into routinized transactions
Geographical	IT can transfer information with rapidity and ease across large distances, making processes independent of geography
Automational	IT can replace or reduce human labor in a process
Analytical	IT can bring complex analytical methods to bear on a process
Informational	IT can bring vast amounts of detailed information into a process
Sequential	IT can enable changes in the sequence of tasks in a process, often allowing multiple tasks to be worked on simultaneously
Knowledge Management	IT allows the capture and dissemination of knowledge and expertise to improve the process
Tracking	IT allows the detailed tracking of task status, inputs, and outputs
Disintermediation	IT can be used to connect two parties within a process that would otherwise communicate through an intermediary (internal or external)

is not the end of the process. Rather, it should be viewed as a prototype, with successive iterations expected and managed. Key factors and tactics to consider in process design and prototype creation include using IT as a design tool, understanding generic design criteria, and creating organizational prototypes.

- IT as a Design Tool. Designing a business process is largely a matter of diligence and creativity. Emerging IT technologies, however, are beginning to facilitate the "process" of process design. Some computer-aided systems engineering (CASE) products are designed primarily to draw process models. The ability to draw models rapidly and make changes suggested by process owners speeds redesign and facilitates owner buy-in. Some CASE products can actually generate computer code for the information systems application that will support a modeled business process.

  Several Xerox divisions, for example, are moving directly from process modeling to automated generation of computer code for high-priority processes. They report improved productivity and high user satisfaction with the resulting systems. A further benefit is that when the business process changes, the IS organization can rapidly modify the affected system. Use of code generation products generally presumes that process designers will use the exhaustive approach to process identification.
- Generic Design Criteria. Companies used various criteria for evaluating alternative designs. Most important, of course, is the likelihood that a design will satisfy the chosen design objectives. Others mentioned in interviews included the simplicity of the design, the lack of buffers or intermediaries, the degree of control by a single individual or department (or an effective, decentralized coordinative mechanism), the balance of process resources, and the generalization of process tasks (so that they can be performed by more than one person).
- Organizational Prototypes. Mutual Benefit Life's (MBL) redesign of its individual life insurance underwriting process illustrates a final, important point about process design. At MBL, underwriting a life insurance policy involved 40 steps with over 100 people in 12 functional areas and 80 separate jobs. To streamline this lengthy and complex process, MBL undertook a pilot project with the goal of improving productivity by 40%. To integrate the process, MBL created a new role, the case manager. This role was designed to perform and coordinate all underwriting tasks centrally, utilizing a workstation-based computer system capable of pulling data from all over the company. After a brief start-up period, the firm learned that two additional roles were necessary on some underwriting cases: specialists such as lawyers or medical directors in knowledge-intensive fields, and clerical assistance. With the new role and redesigned process, senior managers at MBL are confident of reaching the 40% goal in a few months. This example illustrates the value of creating organizational prototypes in IT-driven process redesign.

Creating prototypes of IT applications has already gained widespread acceptance. Advocates argue that building a prototype of an IT change usually achieves results faster than conventional "life cycle" development, and, more important, that the result is much more likely to satisfy the customer. Building prototypes of business process changes and organizational redesign initiatives can yield similar benefits.[21] The implications of this extension are that process designs, after agreement by owners and stakeholders, would be implemented on a pilot basis (perhaps in parallel with existing processes), examined regularly for problems and objective achievement, and modified as necessary. As the process approached final acceptance, it would be phased into full implementation.

DEFINING PROCESS TYPES

The five steps described above are sufficiently general to apply to most organizations and processes. Yet the specifics of redesign vary considerably according to the type of process under examination. Different types require different levels of management attention and ownership, need different forms of IT support, and have different business consequences. In this section, we present three different dimensions within which processes vary.

Understanding and classifying the different types of processes is important because an organization can appear to be a seamless web of interconnected processes. With various process *types* in mind, a manager can begin to isolate particular processes for analysis and redesign, including activities that, without process thinking, might otherwise be overlooked.

Three major dimensions can be used to define processes (see Table 2). These are the organizational entities or subunits involved in the process, the type of objects manipulated, and the type of activities taking place. We describe each dimension and resulting process type below.

TABLE 2
Types of Processes

Process Dimension and Type	*Typical Example*	*Typical IT Role*
Entities		
Interorganizational	Order from a supplier	Lower transaction costs; eliminate intermediaries
Interfunctional	Develop a new product	Work across geography; greater simultaneity
Interpersonal	Approve a bank loan	Role and task integration
Objects		
Physical	Manufacture a product	Increased outcome flexibility; process control
Informational	Create a proposal	Routinizing complex decisions
Activities		
Operational	Fill a customer order	Reduce time and costs; increase output quality
Managerial	Develop a budget	Improve analysis; increase participation

Defining Process Entities

Processes take place between types of organizational entities. Each type has different implications for IT benefits.

Interorganizational processes are those taking place between two or more business organizations. Increasingly, companies are concerned with coordinating activities that extend into the next (or previous) company along the value-added chain.[22] Several U.S. retail, apparel, and textile companies, for example, have linked their business processes to speed up reordering of apparel. When Dillard's (department store) inventory of a particular pants style falls below a specified level, Haggar (apparel manufacturer) is notified electronically. If Haggar does not have the cloth to manufacture the pants, Burlington Industries (textile manufacturer) is notified electronically. At this example of electronic data interchange (EDI) illustrates, information technology is the major vehicle by which this interorganizational linkage is executed.

For most companies, simple market relationships are the most common source of interorganizational processes. All the tasks involved in a selling-buying transaction form a critical process for sellers, and an increasingly important one for buyers seeking higher quality, cost efficiency, and responsiveness. Yet much of the focus has been on a simple transaction level, rather than on an interorganizational business process level. Again, how EDI is used illustrates this point.

Buyers and sellers have used EDI largely to speed up routine purchasing transactions, such as invoices or bills of materials. Few companies have attempted to redesign the broader procurement process—from the awareness that a product is needed, to the development of approved vendor lists, or even to the delivery and use of the purchased product. In the future, sellers will need to look at all buyer processes in which their products are involved.

Moreover, many firms will need to help the buyer improve those processes. Du Pont's concept of "effectiveness in use" as the major criterion of customer satisfaction is one leading approach to measuring the effectiveness of interorganizational processes. Du Pont is motivated not simply to sell a product, but to link its internal processes for creating value in a product, to its customer's processes for using the product. This concept led Du Pont to furnish EDI-provided Material Safety Data Sheets along with the chemicals it sells to its customers to ensure their safe use.

Westinghouse used an interorganizational process approach in dealing with Portland General Electric (PGE), a major consumer of power generation equipment. PGE managers called upon Westinghouse's Productivity and Quality Center, a national leader in process improvement, to help them implement EDI, but the Westinghouse team asked if it could analyze the entire process by which PGE procured equipment from Westinghouse and other suppliers. They found that, while implementing EDI could yield efficiencies on the order of 10%, changing the overall procurement process, including using EDI and bypassing the purchasing department altogether for most routine purchase orders, could lead to much greater savings. In one case, the time to execute a standard purchase order, for example, could be reduced from 15 days to half a day; the cost could be reduced from almost $90 to $10.

A second major type of business process is interfunctional. These processes exist within the organization, but cross several functional or divisional boundaries. Interfunctional processes achieve major operational objectives, such as new product realization, asset management, or production scheduling. Most management processes—for example, planning, budgeting, and human resource management—are interfunctional.

Many manufacturing companies that focused on quality improvement found that producing quality products and services required addressing difficult interfunctional issues. Yet most firms have never even listed their key interfunctional processes, let alone analyzed or redesigned them, with or without the aid of IT.

Two companies that recently analyzed their key interfunctional business processes are Baxter Healthcare Corporation and US Sprint Communications Company. Baxter's 1985 merger with American Hospital Supply provided the context for a major analysis of key business strategies, and the alignment of the IT infrastructure with those strategies.[23] As part of a seven-month IT planning effort, the company defined 29 major interfunctional processes and analyzed the current and future role of IT in supporting them. For example, in the distribution area, the company identified order entry, inventory, warehouse management, purchasing, transportation, and equipment tracking as key processes. The success of this IT planning effort led Baxter to incorporate the process definition approach into its annual corporate planning process.

At US Sprint, well-publicized problems with the customer billing system prompted the company's IT function to develop a model of information flows for the entire business as part of a comprehensive systems improvement program. This model defined the critical information and key interfunctional processes necessary to run the business. Sprint is now assigning ownership to key processes and continuing to identify improvements—and ways to measure them—in each process. The systems improvement program raised the IT organization's composite internal quality index by more than 50% in one year.[24]

A major problem in redesigning interfunctional processes is that most information systems of the past were built to automate specific functional areas or parts of functions. Few third-party application software packages have been developed to support a full business process. Very few organizations have modeled existing interfunctional processes or redesigned them, and companies will run into substantial problems in building interfunctional systems without such models.

Interpersonal processes involve tasks within and across small work groups, typically within a function or department. Examples include a commercial loan group approving a loan, or an airline flight crew preparing for takeoff. This type of process is becoming more important as companies shift to self-managing teams as the lowest unit of organization. Information technology is increasingly capable of supporting interpersonal processes; hardware and communications companies have developed new networking-oriented products, and software companies have begun to flesh out the concept of "groupware" (e.g., local area network-based mail, conferencing, and brainstorming tools).[25]

Several companies, including GM's Electronic Data Systems (EDS), are exploring tools to facilitate the effectiveness of meetings and small group interactions. At EDS, the primary focus is on enhancing the interpersonal processes involved in automobile product development. The company's Center for Machine Intelligence has developed a computer-supported meeting room, and is studying its implications for group decision making and cooperative work.[26]

We should point out that IT can make it possible for employees scattered around the world to work as a team. As an example, Ford now creates new car designs using teams that have members in Europe, Central America, and the United States. Because Ford has standardized computer-aided design systems and created common data structures for the design process, engineers can share complex three-dimensional designs across the Atlantic. Similarly, a small team at Digital Equipment used the company's electronic mail and conferencing capabilities to build the core of a new systems integration business. The team was scattered around the United States and Europe and only rarely met in person.

Defining Process Objects

Processes can also be categorized by the types of objects manipulated. The two primary object types are physical and informational. In physical object processes, real, tangible things are either created or manipulated; manufacturing is the obvious example. Informational object processes create or manipulate information. Processes for making a decision, preparing a marketing plan, or designing a new product are examples.

Many processes involve the combination of physical and informational objects. Indeed, adding information to a physical object as it moves through a process is a common way of adding value. Most logistical activities, for example, combine the movement of physical objects with the manipulation of information concerning their whereabouts. Success in the logistics industry is often dependent on the close integration of physical and informational outcomes; both UPS and Federal Express, for example, track package movement closely.

The potential for using IT to improve physical processes is well known. It allows greater flexibility and variety of outcomes, more precise control of the process itself, reductions in throughput time, and elimination of human labor. These benefits have been pursued for the past three decades. Still, manufacturing process flows are often the result of historical circumstance and should usually be redesigned before further automation is applied. This is particularly true in low volume, job shop manufacturing environments.[27] Redesigners of physical processes should also consider the role of IT in providing information to improve processes; Shoshana Zuboff has described this "informating" effect in detail for the paper industry.[28]

Strangely, the proportion of informational processes already transformed by IT is probably lower than that of physical processes. True, legions of clerks have become unemployed because of computers. But the majority of information processes to which IT has been applied are those involving high volume and low complexity. Now that these processes are well known even if not fully conquered, the emphasis needs to shift to processes that incorporate semistructured and unstructured tasks and are performed by high-skill knowledge workers. Relevant IT capabilities include the storage and retrieval of unstructured and multimedia information, the capturing and routinizing of decision logic, and the application of far-flung and complex data resources. A computer vendor's advertising videotape, for example, illustrates how artificial intelligence and "hyper-text," or mixed-media databases, combine to lead a manager through the process of developing a departmental budget. The IT capabilities in the video are available today, but they are rarely applied to such information-intensive yet unstructured processes.

Defining Process Activities

Our examples of business processes have involved two types of activities: operational and managerial. Operational processes involve the day-to-day carrying out of the organization's basic business purpose. Managerial processes help to control, plan, or provide resources for operational processes. Past uses of IT to improve processes, limited as they are, have been largely operational. We will therefore focus almost entirely on managerial processes in this section.[29]

Applying IT to management tasks is not a new idea. The potential of decision support systems, executive support systems, and other managerial tools has been discussed for over 20 years. We believe, however, that the benefits have not been realized because of the absence of systematic process thinking. Few companies have rigorously analyzed managerial activities as processes subject to redesign. Even the notion of managerial activities involving defined outcomes (a central aspect of our definition of business processes) is somewhat foreign. How would such managerial processes as deciding on an acquisition or developing the agenda for the quarterly board meeting be improved if they were treated as processes—in other words, measured, brainstormed, and redesigned with IT capabilities?

The generic capabilities of IT for reshaping management processes include improving analytic accuracy, enabling broader management participation across wider geographical boundaries, generating feedback on actions taken (the managerial version of "informating" a process), and steamlining the time and resources a specific process consumes. Texas Instruments and Xerox's corporate headquarters provide excellent examples.

Texas Instruments has developed an expert system to facilitate the capital budgeting process. Managers in a fast-growing and capital-intensive TI division were concerned that the time and experience necessary to prepare capital budget request packages would become an obstacle to the division's growth. The packages were very complex and time consuming, and few employees had the requisite knowledge to complete them accurately. The expert system was developed by two industrial engineers with expertise in both the technology and the budget process.

TI's system has radically improved the capital budget request process. Requests prepared with the system require far less time than the manual approach and conform better to the company's guidelines. One experienced employee reported a reduction in package preparation time from nine hours to 40 minutes; of the first 50 packages prepared with the system, only three did not conform to guidelines, compared to an average of 10 using a manual approach.[30]

At Xerox Corporation headquarters, IT has been used to improve the review of division strategic plans. Prior to the development of the company's Executive Information System (EIS), the planning process was somewhat haphazard; each division prepared its planning documents in a different format and furnished different types of information to corporate headquarters. Plans often came in too late for the corporate management committee to review them before the quarterly or annual review meeting. The EIS was developed to include standard information formats and a user friendly graphical interface enabling fast comprehension. Divisional plans are now developed on the EIS and delivered instantaneously over Xerox's network to all corporate management

IT-DRIVEN PROCESS REDESIGN AT RANK XEROX U.K.

▪ ▪ ▪ ▪ ▪

Rank Xerox U.K. (RXUK), a national operating company of Xerox Corporation, has undertaken the most comprehensive IT-driven process redesign we have studied. The process was led by David O'Brien, the division's managing director, who arrived at the company in 1985. O'Brien quickly came to two realizations: first, the company needed to focus on marketing "office systems" in addition to its traditional reprographics products; and second, the company's strong functional culture and inefficient business processes would greatly inhibit its growth. He began to see his own organization as a place to test integrated office systems that support integrated business processes; if successful, he could use RXUK as a model for customers.

The company began to redesign its business in 1987. In a series of off-site meetings, the senior management team reappraised its external environment and mission, then identified the key business processes needed if the company was to achieve its mission. The group began to restructure the organization around cross-functional processes, identifying high-level objectives and creating task forces to define information and other resource requirements for each process. It created career systems revolving around facilitation skills and cross-functional management, rather than hierarchical authority. O'Brien decided to keep a somewhat functional formal structure, because functional skills would still be needed in a process organization and because the level of organizational change might have been too great with a wholly new structure.

The level of change was still very high. Several senior managers departed because they could not or would not manage in the new environment. Two new cross-functional senior positions, called "facilitating directors," were created, one for organizational and business development, the other for process management, information systems, and quality. O'Brien took great advantage of the honeymoon period accorded to new CEOs, but managing the change still required intense personal attention:

> Of course, this new thinking was in sharp contrast to some of the skills and attitudes of the company. We were introducing a change in management philosophy in a company that, in many ways, was very skillful and effective, but in a different product-market environment. We faced all the issues of attitudinal change and retraining that any such change implies. We were moving to a much more integrated view of the world and had to encourage a major shift in many patterns of

committee members. These members can now read and discuss the plans beforehand and can move directly to decisions at the review meetings. The workstations are even used in the meetings themselves, allowing revisions to be made and agreed upon before adjournment. As one manager put it, ". . . [the system] lets us communicate at higher speed and in greater depth."[31]

MANAGEMENT ISSUES IN IT-ENABLED REDESIGN

Companies have found that once a process has been redesigned, several key issues remain. These include the management role in redesigned activity, implications for organization structure, new skill requirements, creating a function to perform the existing culture. This meant a very hard, tough program of selling the new ideas within the organization as well as an extensive and personal effort to get the new messages and thinking to our potential customers.*

As the key processes were identified and their objectives determined, the company began to think about how information technology (its own and from other providers) could enable and support the processes. The facilitating director of processes and systems, Paul Chapman, decided that the firm needed a new approach to developing information systems around processes. His organization used the information engineering approach discussed earlier and worked with an external consultant to refine and confirm process identification. They uncovered 18 "macro" business processes (e.g., logistics) and 145 "micro" processes (e.g., fleet management).

The senior management team reconvened to prioritize the identified processes and decided that seven macro processes had particular importance; customer order life cycle, customer satisfaction, installed equipment management, integrated planning, logistics, financial management, and personnel management. It selected personnel management as the first process to be redesigned because this was viewed as relatively easy to attack and because personnel systems were crucial in tracking the development of new skills. The personnel system has now been successfully redesigned, using automated code generation capabilities, in substantially less time than if normal methods had been used.

RXUK's financial situation began to improve as it redesigned its business processes. The company emerged from a long period of stagnation into a period of 20% revenue growth. Jobs not directly involved with customer contact were reduced from 1,100 to 800. Order delivery time was, on average, reduced from 33 days to six days. Though many other market factors were changing during this time, O'Brien credits the process redesign for much of the improvement.

Other Xerox divisions heard of RXUK's success with process redesign and began efforts of their own. Xerox's U.S. product development and marketing divisions now have major cross-functional teams performing process redesign. Paul Chapman has been loaned to Xerox corporate headquarters, where he is heading a cross-functional team looking at corporate business processes. Commitment to IT-driven process redesign by Xerox senior corporate management is also growing.

*David O'Brien, quoted in B. Denning and B. Taylor, "Rank Xerox U.K., Office Systems Strategy (C): Developing the Systems Strategy" (Henley on Thames, England: Henley—The Management College case study, September 1988). Other Rank Xerox U.K. information comes from personal interviews.

IT-enabled BPR, the proper direction for the IT infrastructure, and the need for continuous process improvement. We discuss each below.

Management Roles

Perhaps the greatest difficulty in IT-driven redesign is getting and keeping management commitment. Because processes cut across various parts of the organization, a process redesign effort driven by a single business function or unit will probably encounter resistance from other parts of the organization. Both high-level and broad support for change are necessary.

To perform the five redesign steps described above, several companies created a cross-functional task force headed by a senior executive. These task forces included representatives from key staff and line groups likely to be affected by the changes, including IT and human resources. It was particularly important that the customer of the process be represented on the team, even when the customer was external. The team composition was ideal if some members had some record of process or operations innovation involving IT.

As the redesign teams selected processes and developed objectives, they needed to work closely with the managers and staff of the affected units. Managing process change is similar to managing other types of change, except that its cross-functional nature increases the number of stakeholders, thereby increasing the complexity of the effort.

It was also important to have strong, visible commitment from senior management. Employees throughout the organization needed to understand that redesign was critical, that differences of opinion would be resolved in favor of the customer of a process, and that IT would play an important role. In many cases, the CEO communicated any structural implications of the redesign effort.

An example of the importance of the CEO's role is found at GUS Home Shopping, the largest home shopping company in Europe. GUS undertook a $90 million project to redesign its logistical processes with IT. Redesign objectives involved both cost and time: to be able to sell a product within five minutes of its arrival on the loading dock, and to be able to deliver a product to the customer's door at an average cost of 60 cents. The company's managing director commented on his role in meeting these objectives:

> To change our business to the degree we have [done] demands integration. How involved should the managing director get in designing computer systems? My view is totally, because he's the one who can integrate across the entire organization.[32]

Process Redesign and Organizational Structure

A second key issue is the relationship between process orientation and organizational structure. Certainly someone must be in charge of implementing a process change, and of managing the redesigned process thereafter. But process responsibilities are likely to cut across existing organizational structures. How can process organization and traditional functional organization be reconciled?

One possible solution is to create a new organization structure along process lines, in effect abandoning altogether other structural dimensions, such as function, product, or geography. This approach presents risks, however; as business needs change, new processes will be created that cut across the previous process-based organization. This does not mean that a process-based structure cannot be useful, but only that it will have to be changed frequently.

While no firm we studied has converted wholly to a process-based structure, a few organizations have moved in this direction. For example, Apple Computer recently moved away from a functional structure to what executives describe as an IT-oriented, process-based, customer satisfaction-driven structure called "New Enterprise." The company relishes its lack of formal hierarchy; Apple managers describe their roles as highly diffuse, and team and project based.

A more conservative approach would be to create a matrix of functional and process responsibilities. However, because of the cross-functional nature of most processes, the functional manager who should have responsibility for a given process is not always easy to identify. The company may also wish to avoid traditional functional thinking in assigning process responsibilities. For example, it may be wiser to give responsibility for redesigning supplies acquisition to a manager who uses those supplies (i.e., the customer of the process), rather than to the head of purchasing.

New Skill Requirements

For process management to succeed, managers must develop facilitation and influence skills. Traditional sources of authority may be of little use when process changes cut across organizational units. Managers will find themselves trying to change the behavior of employees who do not work for them. In these cases, they must learn to persuade rather than to instruct, to convince rather than to dictate. Of course, these recommendations are consistent with many other organizational maxims of the past several years; they just happen to be useful in process management as well.[33]

Several organizations that are moving toward IT-driven process management are conducting programs intended to develop facilitation skills. These programs encourage less reliance on hierarchy, more cross-functional communication and cooperation, and more decision making by middle- and lower-level managers. Such a program at American Airlines is being used to build an organizational infrastructure at the same time a new IT infrastructure is being built.

An Ongoing Organization

Organizations that redesign key processes must oversee continuing redesign and organizational "tuning," as well as ensure that information systems support process flows. In most companies, the appropriate analytical skills are most likely to be found in the IT function. However, these individuals will also require a high degree of interpersonal skills to be successful as the "new industrial engineers." The ideal group would represent multiple functional areas, for example, information systems, industrial engineering, quality, process control, finance, and human resources.

There are already some examples of such process change groups. Silicon Graphics has created a specific process consulting group for ongoing process management; it is headed by a director-level manager. At United Parcel Service, process redesign is traditionally concentrated in the industrial engineering function. The UPS group is incorporating IT skills in the IE function at a rapid rate, and creating task forces with IT and IE representation for process redesign projects. Federal Express has gone even further, renaming its IE organization the "Strategic Integrated Systems Group," placing it within the Information Systems function, and giving it responsibility for designing and implementing major IT-driven business changes.

Process Redesign and the IT Organization

Just as information technology is a powerful force in redesigning business processes, process thinking has important implications for the IT organization and for the technology infrastructure it builds. Though few IT groups have the power and influence to spearhead process redesign, they can play several important roles. First of all, the IT group may need to play a behind-the-scenes advocacy role, convincing senior management of the power offered by information technology and process redesign. Second, as demand builds for process redesign expertise, the IT group can begin to incorporate the IE-oriented skills of process measurement, analysis, and redesign, perhaps merging with the IE function if there is one. It can also develop an approach or methodology for IT-enabled redesign, perhaps using the five steps described above as a starting point.

What must the information systems function do technologically to prepare for process redesign? IT professionals must recognize that they will have to build most systems needed to support (or enable) processes, rather than buy them from software package vendors, because most application packages are designed with particular functions in mind. IT professionals will need to build robust technology platforms on which process-specific applications can be quickly constructed. This implies a standardized architecture with extensive communications capability between computing nodes, and the development of shared databases. However, like the organizational strategies for process management described above, these are appropriate technology strategies for most companies, whether or not they are redesigning processes with IT.

Continuous Process Improvement

The concept of process improvement, which developed in the quality movement, requires first that the existing process be stabilized. It then becomes predictable, and its capabilities become accessible to analysis and improvement.[34] Continuous process improvement occurs when the cycle of stabilizing, assessing, and improving a given process becomes institutionalized.

IT-enabled business process redesign must generally be dynamic. Those responsible for a process should constantly investigate whether new information technologies make it possible to carry out a process in new ways. IT is continuing to evolve, and forthcoming technologies will have a substantial impact on the processes of the next decade. The IT infrastructure must be robust enough to support the new applications appropriate to a particular process.

SUMMARY

We believe that the industrial engineers of the future, regardless of their formal title or the organizational unit that employs them, will focus increasingly on IT-enabled redesign of business processes. We have only begun to explore the implications and implementation of this concept, and only a few companies have ventured into the area. Many companies that have used IT to redesign particular business processes have done so without any conscious approach or philosophy. In short, the actual experience base with IT-enabled process redesign is limited.

Yet managing by customer-driven processes that cross organizational boundaries is an intuitively appealing idea that has worked well in the companies that have experimented with it. And few would question that information technology is a powerful tool for reshaping business processes. The individuals and companies that can master redesigning processes around IT will be well equipped to succeed in the new decade—and the new century.

REFERENCES

1. Gulick, L. "Notes on the Theory of Organization." In *Papers on the Science of Administration,* edited by L. Gulick and L. Urwick, 9. New York: Institute of Public Administration, 1937).
2. Sakamoto, S. "Process Design Concept: A New Approach to IE." *Industrial Engineering* (March 1989): 31.
3. "Office Automation: Making It Pay Off." *Business Week,* 12 (October 1987): 134–146. For an alternative perspective, see Kraut, R. E. ed. *Technology and the Transformation of White-Collar Work.* Hillsdale, NJ: Lawrence Erlbaum Associates, 1987.
4. Loveman, G. W. "An Assessment of the Productivity Impact of Information Technologies" (Cambridge, MA: MIT Sloan School of Management, Management in the 1990s, Working Paper 90s:88–054, July 1988). Loveman studied microeconomic data from manufacturing firms to estimate econometrically the productivity impact of IT in the late 1970s and early 1980s. In finding no significant positive productivity impact from IT, he argues that his findings in manufacturing raise serious questions about impacts in nonmanufacturing firms as well. Baily and Chakrabarti (1988) studied white-collar productivity and IT as one part of a broader inquiry into poor productivity growth. They found no evidence of significant productivity gain. See Baily, M. N., and Chakrabarti, A. *Innovation and the Productivity Crisis.* Washington, D.C.: Brookings Institution, 1988.
5. Loveman (1988); Baily and Chakrabarti (1988). See also Thurow, L. C. "Toward a High-Wage, High-Productivity Service Sector." Washington, D.C.: Economic Policy Institute, 1989.
6. Robert Horton, who became chairman and chief executive of British Petroleum in

March 1990, argues that his major concern in setting BP's course in the next decade is "managing surprise." Horton's belief is that the external business environment is so unpredictable that surprise, rather than managed change, is inevitable. See Horton, R. "Future Challenges to Management." *MIT Management* (Winter 1989): 3–6.

7. Malone, T. "What is Coordination Theory?" Cambridge, MA: MIT Sloan School of Management, Center for Coordination Science, Working Paper No. 2051–88, February 1988; Crowston, K. and Malone, T. "Information Technology and Work Organization" Cambridge, MA: MIT Sloan School of Management, Center for Information Systems Research, Working Paper No. 165, December 1987.
8. Pall, G. A. *Quality Process Management.* Englewood Cliffs, NJ: Prentice-Hall, 1987. Our definition also complements that of Schein, who focuses on human processes in organizations—for example, building and maintaining groups, group problem solving and decision making, leading and influencing, etc. See Schein, E. H. *Process Consultation: Its Role in Organization Development,* Vol. 1, 2d ed. Reading, MA: Addison-Wesley, 1988.
9. Kane, E. J. "IBM's Total Quality Improvement System." Purchase, NY: IBM Corporation, unpublished manuscript, p. 5.
10. See, for example, Morris, M. F. and Vining, G. W. "The IE's Future Role in Improving Knowledge Worker Productivity." *Industrial Engineering* (July 1987): 28.
11. "Reference Note on Work Simplification." Boston: Harvard Business School, HBS Case Services #9–609–0601961, 1961.
12. The relationship between business vision and IT has been explored by several researchers under the auspices of the MIT Sloan School's five-year "Management in the 1990s" research program. An overview volume is scheduled for publication by Oxford University Press in August 1990.
13. See, for example, Stalk, Jr., G. "Time—The Next Source of Strategic Advantage." *Harvard Business Review* (July–August 1988): 41–51.
14. Zuboff, S. *In the Age of the Smart Machine.* New York: Basic Books, 1988.
15. Schein, E. H. "Innovative Cultures and Organizations." Cambridge, MA: MIT Sloan School of Management, Management in the 1990s, Working Paper 90s: 88–064, November 1988.
16. Information engineering and other redesign approaches based on data modeling are necessarily limited in scope. More than data is exchanged in many process relationships. Note too that many companies have used information engineering methods *without* a specific process orientation.
17. Examples of IT planning approaches where high-impact objectives and/or goals are defined include critical success factors (CSFs) and business systems planning (BSP). See Rockart, J. F. "Chief Executives Define Their Own Data Needs." *Harvard Business Review* (March–April 1979): 81–93; and IBM. *Information Systems Planning Guide,* 3d ed. Business Systems Planning Report No. GE20–05527–2, July 1981.
18. Goodhue, D., Quillard, J., and Rockart, J. "Managing the Data Resource: A Contingency Perspective." Cambridge, MA: MIT Sloan School of Management, Center for Information Systems Research, Working Paper No. 150, January 1987.
19. Rockart, J. F. "The Line Takes the Leadership—IS Management in a Wired Society." *Sloan Management Review* (Summer 1988): 57–64.
20. Henderson, J. C., and Venkatraman, N. "Strategic Alignment: A Process Model for Integrating Information Technology and Business Strategies." Cambridge, MA: MIT Sloan School of Management, Center for Information Systems Research, Working Paper No. 196, October 1989.
21. Dorothy Leonard-Barton introduced the concept of organizational prototyping with regard

to the implementation of new information technologies. See Leonard-Barton, D. "The Case for Integrative Innovation: An Expert System at Digital." *Sloan Management Review* (Fall 1987): 7–19.

22. Johnston, R., and Lawrence, P. R. "Beyond Vertical Integration—The Rise of the Value-Adding Partnership." *Harvard Business Review* (July–August 1988): 94–101. See also Venkatraman, N. "IT-Induced Business Reconfiguration: The New Strategic Management Challenge." Cambridge, MA: Paper presented at the annual conference of the MIT Center for Information Systems Research, June 1989.
23. Main, T. J., and Short, J. E. "Managing the Merger: Building Partnership through IT Planning at the New Baxter." *Management Information Systems Quarterly* (December 1989): 469–486.
24. Hall, C. R., Friesen, M. E., and Short, J. E. "The Turnaround at US Sprint: The Role of Improved Partnership Between Business and Information Management." In progress.
25. Johansen, R. R. *Groupware: Computer Support for Business Teams.* New York: The Free Press, 1988. Also see Bullen, C. V., and Johansen, R. R. "Groupware: A Key to Managing Business Teams?" Cambridge, MA: MIT Sloan School of Management, Center for Information Systems Research, Working Paper No. 169, May 1988.
26. See Applegate, L. M. "The Center for Machine Intelligence: Computer Support for Cooperative Work." Boston: Harvard Business School Case Study No. 189–135, 1988, rev. 1989.
27. Ashton, J. E., and Cook, F. X. "Time to Reform Job Shop Manufacturing." *Harvard Business Review* (March–April 1989): 106–111.
28. See cases on "Tiger Creek," "Piney Wood," and "Cedar Bluff" in S. Zuboff (1988); other industries discussed by Zuboff primarily involve informational processes.
29. One might consider managerial processes synonymous with informational processes. Certainly the vast majority of managerial processes, such as budgeting, planning, and human resource development, involve informational objects. Yet it is important to remember that informational processes can be either operational or managerial, so we believe that this separate dimension of process types is warranted.
30. A case study describes the process and the creation of the expert system. See "Texas Instruments Capital Investment Expert System." Boston: Harvard Business School Case Study No. 188–050, 1988.
31. Some aspects of this process improvement are described in Applegate, L. M., and Osborne, C. S. "Xerox Corporation: Executive Support Systems." Boston: Harvard Business School Case Study No. 189–134, 1988, rev. 1989.
32. Pugh, R. H. C. address to McKinsey & Co. information technology practice leaders, Munich, Germany, June 1989.
33. See, for example, Cohen, A. R., and Bradford, D. L. "Influence Without Authority: The Use of Alliances, Reciprocity, and Exchange to Accomplish Work." *Organizational Dynamics* (Winter 1989): 4–17.
34. See G. A. Pall (1987).

QUESTIONS

1. What is the role of IT in business process redesign?
2. What is the difference between continuous process improvement advocated by TQM experts and business process redesign?
3. It has been reported that nearly two-thirds of business process redesign efforts fail. Why are reengineering projects so difficult to implement? What can be done to make them more successful?

6.3 Implementing Advanced Manufacturing Technologies: Rules of the Road

CAROL A. BEATTY

By this point it should be clear that merely having a new technology is not enough. Active steps must be taken to implement it in the firm. This axiom is particularly important in manufacturing, which is often a bastion of conservatism about change. Changes are counterculture. They force production lines to shut down temporarily rather than produce. They require capital investment. People have to be retrained. Implementation has many pitfalls which must be avoided. All in all, it may seem better to do nothing than to change.

Beatty, writing just before the term Information Highway came into vogue, uses a road metaphor to talk about the implementation of advanced manufacturing technology. She bases her advice about what to do on the experience of 10 companies who implemented CAD/CAM. Her results reinforce the conventional wisdom:

- You need a champion with skill and clout. The champion must have people as well as technical skill.
- Be sure you have an integrated system. Islands of automation don't work.
- Not only must the system be integrated technically, the organization must be integrated as well.

Failure to follow this wisdom leads to potholes in the implementation road.

We were a company struggling to learn to ride a tricycle, and we went out and bought a bicycle.

AFTER buying a bicycle, the aspiring cyclist must learn to ride it well enough to use it for transportation. The company purchasing an advanced manufacturing technology (AMT) must likewise learn to use it effectively. Managers often believe that choosing or adopting a new technology is the most important task, and they neglect implemen-

Source: Sloan Management Review (Summer 1992): 49–60.

tation. But getting good use from a bicycle or AMT is a far more daunting and more bruising challenge than merely acquiring it.

Along the bumpy road to successful AMT use, one needs rules and a good map in order to avoid dead ends and potholes. In this article, I report on 10 companies that tried to implement AMT. I describe the potholes they fell into and the rules of the road that helped them succeed. I also consider why, despite much published research on technology inplementation, so many companies continue to take wrong turns and run into dead ends.

CAD/CAM AT TEN COMPANIES

Over the course of three years, my researchers and I have followed the implementation of computer-aided design and manufacturing systems (CAD/CAM) in 10 companies' manufacturing operations. The manufacturing facilities ranged in size from a small startup plant of 80 employees to a 4,000-employee division of a large multinational firm. The companies represented diverse industries, including auto parts, telecommunications equipment, aerospace, agricultural and industrial equipment, mass transit vehicles, and tool and die makers (see Table 1).

We studied a range of applications from simple (a CAD system integrated with numerically controlled [NC] machines on the shop floor) to complex (a complete flexible manufacturing system [FMS] facility). All the companies had recently purchased a new CAD system or new CAM equipment, or both, and were planning to integrate CAD with CAM. The study's design called for several "snapshots" of the projects over three years; we periodically interviewed senior and middle managers in manufacturing, engineering/R&D, and other relevant departments. We also interviewed and surveyed a cross section of operators at each site. For each implementation, a case history was constructed and subjected to content analysis.

Project Goals

We don't know where their thinking is with CAD. There is no planning, at least not that we see.

—a frustrated operator

Companies have various objectives in mind when they start an AMT project: reduced costs, enhanced quality, reduced time, or increased flexibility.[1] We asked the managers for their specific objectives for the AMT system and their long-term technology investment goals. Tables 2A and 2B outline the projects' goals and results. Some of the projects that did not meet their objectives within the three-year period may yet achieve them. It is also possible that a successful project could suffer a setback if a crucial factor changes. The findings, then, are valid only within the period studied.

As Tables 2A and 2B show, most companies achieved their goals only partially, but we classified as successful those companies that achieved at least the majority of their goals. However, even some of the "successful" projects fell short of full integration between CAD and CAM. For example, one of Genco's unofficial goals was to integrate design with NC machines. Although Genco did integrate some operations and may achieve full integration eventually, progress toward this goal was not swift.

Goal achievement was difficult to assess in some cases because the goals were stated vaguely. For example, Aerocan1 wanted to become a "showcase plant." But it proposed no method of measuring achievement except in extremely subjective terms. We believe this goal was not achieved—the general manager and the AMT champion left, and the project stagnated.

Many companies make the mistake of setting only short-term financial goals for AMT; they are generally disappointed by the results. AMT usually will not generate quick returns because many projects take from one to two years to install fully and even more time to realize productivity increases or other benefits.[2] Agco, for example,

TABLE 1
Sites and Implementation Characteristics

Company	*Size of Site*	*Current Technology*	*Implementation Characteristics*	*Success*
Aerocan1	500 employees	Full CAD/CAM system.	Retrofit. Fast. Integrated system. Leaders left division before implementation complete. Fuzzy goals. Some use of teams. Resistance.	Goals not achieved. Implementation stagnated.
Aerocan2	180 employees	CIM.	Greenfield. Fast. System integrated with difficulty. Strong leadership. Clear goals. Use of teams. Extensive structural and job design changes. Orientation plan.	Goals largely achieved.
Agco	870 employees	CAD system (addition of process planning and CAM planned).	Retrofit. Slow. Integrated system. Strong leadership. Clear goals. Wide use of teams. Downsizing by attrition. Orientation plan.	Goals largely achieved.
Genco	580 employees	CAD system with tie to NC (more planned).	Retrofit. Slow. System not integrated. Strong leadership. Clear goals. Use of teams.	Goals largely achieved.
Moldco1	140 employees	NC machines integrated to computerized process planning. Some CAD.	Retrofit. Slow. Integrated system. Inconsistent support of top management. The champion left. Clear goals. No teams.	Goals not achieved. Implementation stagnated.
Moldco2	88 employees	CAD/CAM.	Greenfield. Fast. System not integrated. Weak leadership. Some clear goals. No teams. Resistance.	Goals not achieved. Implementation stagnated.
Telco1	1,250 employees	CAD system and process planning (integration to CAM planned).	Retrofit. Slow. System not integrated. Strong champion, weak top management support. Some clear goals but too many. Minimal use of teams. Some resistance.	Goals not achieved but integration continues.
Telco2	4,000 employees	FMS line.	Retrofit. Fast. Integrated system. Strong champion, weak top management support. Clear goals. Minimal use of teams. Communication plan.	Goals largely achieved.
Telco3	3,000 employees	CAD, process planning, and some CAM (more CAM planned).	Retrofit. Slow. Integrated. Strong champion, weak top management support. Clear goals. Use of teams.	Goals largely achieved.
Transco	600 employees	CAD system (addition of CAM planned).	Retrofit. System not integrated. Weak leadership. Fuzzy goals. Some use of teams. Drafting office was downsized. Resistance.	Goals not achieved. Implementation stagnated.

achieved its financially oriented goals, but it took four years.

Companies investing in AMT in order to improve their market position were likely to have clearer, more specific, more challenging, and ultimately more motivating goals than those companies that focused on short-term financial ones. For example, some of Telco3's goals were targeted at enhancing competitive advantage, such as reducing design cycle time by 25%. Such a result

TABLE 2A
Stated and Achieved Goals of the Successful Projects

Company	*Stated Goals*	*Achieved Goals*
Aerocan2	Perform 8,500 hours of work in 1988.	6,000 achieved.
	Operate within budget.	Achieved.
	40-day lead-time on 28 parts.	Achieved.
	100% on-time delivery.	Achieved.
	Reduce defects to 50% of sister plant level.	Level not measured but quality better than sister plant.
Agco	More work with the same staff.	Saved 10% staff.
	Reduce repairs and NC tape tries.	Reduced almost 100% and 4:1, respectively.
	Reduce engineering change orders and prototype changes.	Not measured.
	Increase drafting and design productivity 2:1.	Increased 4:1 to 10:1 on many tasks.
	Decrease design cycle by one year.	Reduced from average 3 years to 27 months.
Genco	Improve productivity to 1.5:1 or 2:1 in design.	Not achieved.
	Reduce product costs.	Partially achieved.
	Improve response and delivery time.	Down to 35–40 weeks from 35–50.
	Improve schedule time.	Schedule kept, one week instead of three.
	Avoid increase in indirect labor.	Achieved.
	Avoid rework costs.	Reduced from 3% to 2.5% of budget.
Telco2	Reduce WIP to 2 days.	Achieved 9 days.
	Improve yield to 95%.	Achieved 93%.
	Reduce direct labor 70%.	Achieved 53%.
	Reduce overhead 20%.	Achieved 16%.
	Impact on costs $4.5M.	Impact $2.9M.
	Learn how FMS can be applied.	Achieved.
Telco3	Integrate design and decision support system.	Achieved.
	Reduce engineering change orders by 25% within 3 years.	Not measured yet.
	Reduce design cycle time by 25%.	Some improvement but not 25% yet.
	Attract and keep good staff.	Achieved.
	Base for future product development and complexity.	Achieved.

would help get the product to market sooner, providing an important advantage in the telecommunications business.

If AMT is not directed at improving the firm's key competitive factors, the project may lose credibility, as happened at Transco. One of our interviewees there said, "CAD/CAM is generally seen as a money eater by senior management. It is something that must be contained rather than the key to the future."

In summary, then, AMT project goals should be realistic, clear, well communicated, accepted throughout the organization, and, most important, targeted not at short-term payback but at improving the company's competitive positions.

THE RULES OF THE ROAD

Three implementation rules emerged from this research: develop an effective champion, plan for a high level of systems integration, and use organizational integration techniques (see Table 3). The suc-

TABLE 2B
Stated and Achieved Goals of the Unsuccessful Projects

Company	*Stated Goals*	*Achieved Goals*
Aerocan1	Double throughput with 20% staff increase.	Not measured.
	Become "showcase" plant.	Not clear.
	Increase accuracy, machine utilization, and quality.	Not achieved.
	Decrease costs, errors, data entry, and wastage.	Some improvement.
Moldco1	Stay on "big 3" supplier list.	Not yet definite.
	Lead-time less than 20 weeks.	Not achieved.
	95% CAD/CAM production.	Not achieved.
	Reduce NC programming time from 1:1 to 5:1.	Not achieved.
Moldco2	Reduce die building time 30–40% over completion.	Not achieved.
	Utilize CAM efficiently.	Not achieved.
	Convert tool designers to CAD.	Not achieved.
	Integrate whole process.	Integration on hold.
Telco1	Lower costs, improve quality.	Not measured.
	Reduce design staff and subcontracted design work.	Saved $160,000/year (3.5 people).
	Reduce errors and changes.	Not measured.
	Design cycle of 6 months.	Can't achieve without NC integration.
Transco	Increase productivity, quality, and accuracy of designs.	Some increases but perceived as minimal by senior management.
	Improve engineering changes and standardization.	Not achieved.
	Integrate NC and BOM.	Not achieved.

cessful projects followed at least two of these rules; it was safer to go with all three (see Table 4). No company following only one rule attained its goals.

The factor that appeared absolutely necessary to project success was an effective champion. This finding confirms other research on the importance of champions in various types of innovations.[3] Some companies were able to achieve their goals despite the lack of systems or organizational integration but none without a champion.

RULE OF THE ROAD ONE: AMT REQUIRES A SKILLED CHAMPION

He is the only manager I know who could do this. Our success is due to him because he listens, supports, and is creative.

—manager at Aerocan2

Companies without an effective technology champion found that implementation proceeded more slowly, remained restricted to a small corner of the organization, or, more important, did not achieve the objectives. At each site we asked who was the driving force behind the implementation effort. All but one of the companies had a clearly identifiable technology champion, and we were invariably directed to this individual by senior managers or other informants. Of the nine champions, seven were engineers, one was a draftsman, and one was a former machinist and CAD/CAM consultant. However, only six of these people fit the criteria of an *effective* champion.

Leavitt has described three leadership roles: pathfinder, problem solver, and implementator.[4] The pathfinder determinedly promotes a mission or vision, based on a set of values. This descrip-

TABLE 3
Project Outcomes Correlated to Implementation Factors

Use of Implementation Factors		*Goals Achieved*	
		Yes	*No*
Effective champion	Yes	Telco2 Telco3 Genco Agco Aerocan2	Telco1
	No		Transco Moldco1 Moldco2 Aerocan1
Systems integration	Yes	Telco2 Telco3 Agco Aerocan2	Aerocan1
	No	Genco	Transco Moldco1 Moldco2 Telco1
Cross-functional team structures (both steering and implementation teams)	Yes	Telco2 Agco Aerocan2 Genco	Aerocan1
	No	Telco2[1]	Moldco1 Moldco2 Telco1 Transco[2]

1. A corporate steering committee planned the project, but cross-functional implementation teams were not used.
2. The corporate steering committee and implementation team was disbanded after equipment selection. The team was not cross-functional.

tion fits the effective champions in this study, some of whom displayed a love of the new technology bordering on passion. Some of them saw AMT as the solution to a number of problems, such as high error rates, inefficiency, wastage, long lead times, and poor quality. They persisted in selling this vision, sometimes over a period of years, until they achieved their objectives.

The problem solver has technical expertise and analytic skills. Our effective champions were fa-

TABLE 4
Project Outcomes Correlated to Number of Implementation Factors Present

Number of Implementation Factors	*Goals Achieved*	
	Yes	*No*
3	Telco3 Agco Aerocan2	
2	Telco2 Genco	Aerocan1
1		Telco1
0		Transco Moldco1 Moldco2

miliar with computers and possessed technical expertise in design or manufacturing in their particular industries. If they had knowledge gaps, they hired subordinates to make up for them. They also possessed analytic skills that enabled them to plan and execute the project—budgeting, planning, controlling, and other project management skills.

The implementor is good at persuading, selling, gaining influence, being politically sensitive, and motivating subordinates. These skills were crucial in translating each champion's vision into reality, especially as the idea had to be sold to peers in other departments, over whom the champion lacked authority. As Table 5 shows, this was the most problematic role for the "rationally" educated engineers who made up the majority of the champions.

The majority of the champions played the pathfinding role effectively. Indeed the reason they were in charge of the implementation appeared to be their strong and enthusiastic belief in AMT and their vision of what it could do for their company or department:

> *I'd like to make us the best manufacturing plant in the world for this type of product. I'm very competitive, and I'd like to see manufacturing remain in Canada. I want my kids to make*

TABLE 5
Number of Roles Played by Champion Compared to Goal Achievement

	Leadership Roles Played Effectively		
Company	*Path-finding*	*Problem Solving*	*Implementing*
Successful implementors			
Aerocan2	√	√	√
Agco	√	√	√
Genco	√	√	√
Telco2	√	√	√
Telco3	√	√	√
Unsuccessful implementors			
Aerocan1	√		
Moldco1	√	√	
Moldco2			
Telco1	√	√	√
Transco		√	

something, not work in a service organization. My personal goals in life are very close to my day-to-day goals.
—director of advanced technology at Telco2

My vision is to create a medium-sized, five-hundred-employee center of manufacturing excellence. I want to take care of that last 10% of quality problems, like the Japanese do. I don't want to let them beat us!
—general manager of Aerocan2

But a vision is not enough. Patience and persistence are also necessary to back the vision. The champions wanted to direct the project and see it through to completion. One summarized his five-year effort to convince his company to adopt an integrated CAD/CAM system:

The conversion process was not like Saul on the road to Damascus. It was more a case of slow yet persistent effort over a number of years.
—computer-aided engineering (CAE) manager at Telco3

In the problem-solving role, champions need technical and analytical expertise in order to gain and maintain credibility and to help solve the inevitable technical problems that plague an implementation effort. They also need project management skills to design and execute the plan. One comment was typical:

We decided that both the organization and its people needed preparation for CAD/CAM. A number of papers and outlines were written, an orientation was given, a training course for upper management was planned, and close attention was paid to communications during implementation of the system.
—engineering manager of Agco

Differences among the champions were most obvious in the implementor role. Some champions were more effective at motivating others; at persuading subordinates, superiors, and peers in other departments; and at charting the politically sensitive waters surrounding implementation. Comments from these champions illustrate how they worked:

I'm able to make about one good conversion a year. Who you convert matters, and often the right people put themselves forward by volunteering. The movers do the moving, and the rest follow along.
—CAE manager at Telco3

Some people are afraid of technology, but I've found if you put in a PC and some software, people fool around with it and come around. They start to see all sorts of applications.
—manufacturing vice-president at Agco

The experiences of these managers showed that the three leadership roles—pathfinder, problem solver, and implementor—were essential to AMT implementation. Some experts argue that the necessity of a champion indicates failure of the organization's decision-making and people management systems.[5] But AMT is a radical change, and even in a perfectly managed company, implementation would require good leadership. Sound organizational systems, structures, and policies are the stable foundation upon which the champion can build. After a technology has been tested, customized, and routinized, it may work effectively without exceptional management effort. But in each of these companies, a champion was necessary, even in addition to a corporate-level sponsor, who protected and encouraged the champion's efforts.

The role of the champion, then, is to facilitate others' conversion to the new methods. The champion's vision, problem-solving, and implementation skills help build trust, acceptance, and commitment in those who are the essential operators of the new system. Without these ingredients, operators are unlikely to embark upon the intense effort required to learn the new tasks. The champion serves as indefatigable cheerleader and role model.

POTHOLE ONE: CHAMPION FAILURE

It will come as no surprise to most managers that an effective champion is recommended for AMT implementation. Numerous studies from such diverse fields as product innovation, organizational change, and technology diffusion have surely proven this rule. But still companies often fail to follow it. Why?

One of the most common reasons for champion failure was a lack of implementation skills, that is, persuasion, motivation, networking, and communications skills, and political sensitivity. Most of the champions had technical backgrounds, and many had never been in line management positions. Unless they had natural abilities in these areas, they were likely to stumble. In many cases, they did not even realize these skills were necessary. We observed many an earnest technocrat explaining the features of AMT but failing to put the benefits in terms meaningful to other departments. Instead of persuading resisters, some harshly criticized their opponents or tried to coerce them into cooperating. As has been previously described, many engineers with high technical competence tend to reject the managerial role.[6] The unsuccessful champions in this study tended to be just such engineers.

The problem for companies is twofold. First, managers who do not realize the necessity of certain skills may not select the right champion. In fact, the majority of companies we investigated allowed champions to self-select and so had no control over their skill mix. Second, such broadly skilled champions are in short supply in most companies implementing AMT because of their employees' predominantly technical orientation. Indeed, the wide range of championing skills necessary to promote change would be rare in most companies, but none of the companies offered training in these skills. Nor did they appear to value them.

Another reason for champion failure is the lack of sufficient reward or support. Many companies do not encourage champions to take risks, and some champions lie low or leave the company.[7] What happened to the champion at Moldcol, "George," is illustrative. George had previously worked with Moldcol as a CAD/CAM consultant, and the president offered him carte blanche to implement a system. In short order, he installed a bidirectional, interactive system integrating the shop-floor machines with six CAD/CAM workstations—no mean feat in such a small facility. But soon after this, the president began to overrule many of George's decisions, including a crucial one that would have converted some of the design work from manual methods to CAD. George also incurred displeasure for taking part in a trade mission to promote AMT. Ultimately he resigned when he discovered that the president was reading his mail and taking the advice of a new employee, a recent engineering graduate, in preference to his own.

Those who do not support and encourage champions can lose them and subsequently find their AMT projects at a dead end. Why do companies not support their champions? In Moldcol's case, and in others we observed, senior management appeared threatened by the champion's superior technical knowledge. Dependent on this knowledge, but unable to trust, senior managers can easily sabotage the champion, perhaps without being conscious of the fact, and hence jeopardize the project.[8]

Conversely, champions can also suffer from too much support from the senior manager if others in the organization do not support them as well. For example, the president of Aerocan1 brought in "Cal" to implement AMT and ruled in his favor in several disputes Cal had with the engineering manager. Sure of the president's support, Cal did not develop lateral relationships but pushed hard for quick and full implementation of CAD/CAM. When the president received a promotion and left, Cal fell into a deep pothole of resistance and left the company himself shortly thereafter.

Champions from outside the organization appeared to be especially vulnerable to these sorts of hazards, being ignorant of company politics and lacking a support network that can take years to build. The champion performs a precarious balancing act between senior management and peers in other departments. Jealousy and mistrust are never far from the surface during technological change, when status and the future are often at stake.

In summary, companies that want successful AMT projects need to identify, encourage, train, support, and reward their potential champions. They must also help them maintain the balance between senior and lateral support. This balance can be facilitated by educating all levels about AMT before embarking on the project. The company needs a common vision, a sort of "holographic" picture of the future company, including the AMT strategy, which the champion can refer to.[9] Without such a vision, the champion fights a losing battle.

RULE OF THE ROAD TWO: PLAN FOR SYSTEMS INTEGRATION

We go CAD, CAD, CAD, paper, paper, paper. And the minute paper enters, you've lost the advantage of CAD.

—CAD/CAM manager at Telco1

The second important rule on the road to AMT success is to ensure systems integration. The ultimate goal of factory automation is to integrate the entire database, including its subsystems, such as CAD, CAM, bill of materials (BOM), and computer-aided process planning (CAPP).[10] At a higher level of integration, these data could be shared automatically with financial and accounting systems. However, everything in each subsystem need not be included in an integrated database as long as there is a means of accessing nonshared data. This requires the ability to convert the data from one subsystem to another. Users must also be able to file, store, back up, and retrieve the data efficiently and effectively. Only five of the companies we studied were able to achieve a high level of integration and not without great effort. Of these five companies, four met their AMT goals. Agco attributed its achievement of systems integration to a mainframe strategy. One interviewee said:

Mainframe systems promote integration as they allow more centralized control over design standards. Procedures can be built into the system itself. Now everything is more public and visible. You can't hide things in piles of paper.

Nonetheless, Agco had difficulty integrating CAD with CAM. The CAD software was inadequate for this use, and, until the vendor revised it, integration was impeded. Even after the revision, Agco still had to assemble a team to customize the software. The team, consisting of an experienced machinist, who had also become an NC programmer, and a new computer graduate, took an entire year to forge the CAD/CAM link.

Other companies, such as Telco3, pursued networking instead of mainframe strategies and also achieved database integration. Telco3 decided to scrap its old systems and replace everything with networked workstations for both CAD and CAM users. This did not come cheaply, but Telco3 thought it was crucial:

The most critical issue is a common database. Every time you enter a "human channel" in the information process, you open up the possibility of problems. As for errors, on occasion we may do a run of five hundred before realizing the wrong part is being used, and it takes awhile for R&D to hear about it through manual channels. This means we run into either costly rework or junk. Both are costly. A fully integrated CAD/CAM will totally eradicate this problem.

—CAE manager at Telco3

Telco2's FMS system achieved a high degree of integration between design and manufacturing despite initial resistance. Manufacturing wanted a separate database; engineering wanted it integrated. In order to meet manufacturing's needs, the system was kept separate at that level and reintegrated at a higher organizational level. This took considerable effort and ingenuity. The design is now run through a translator that generates information in the form desired by manufacturing. Both engineers and manufacturing personnel can access information in the form they prefer without hampering each other. Gathering input from all concerned made designing the translation system a long process—an entire year. And Telco2 has a large number of communications software experts, more resources than smaller firms. But the champion credited this deliberate attempt to accommodate differences with making the project successful.

The other highly integrated AMT was at Aerocan2, although integration was achieved only after extreme difficulty. Aerocan2 purchased its FMS system from several vendors, and, although they promised compatibility, the delivered systems performed unevenly. The company's FMS planning committee had not looked for the best system for the entire company but had chosen the best tool for each task, intending to link them. The general manager said, "Virtually everyone underestimated the difficulty in introducing the complex software that a totally integrated manufacturing cell requires."

Because of this approach, Aerocan2 suffered an 18-month delay and extra costs. In retrospect, the general manager said that, rather than purchasing the machining cells from one vendor and a state-of-the-art controller (still on the drawing board) from a rival vendor, a better decision would have been to buy a compatible controller from the machining cell vendor.

Another manager explained this dilemma of choosing between compatibility and performance:

If you can get the best CAD system and the best CAM system and then integrate them, that's ideal. The problem is trying to integrate totally different systems. It's as if you tried to put together the ideal organization regarding expertise. The shop-floor level would be Japanese, the engineering level West German, the marketing level American, and so forth. Then you have to try to have all these different languages communicate with one another.

—CAD/CAM manager of Aerocan1

In summary, companies must plan for integration or risk development of incompatible islands of automation, which reduces the technology's benefits. Again, this is not astonishing news to most companies implementing AMT. Experts have been warning against the roadblocks posed by system incompatabilities for years.[11] Then why did only half of the companies we investigated heed this advice?

POTHOLE TWO: INCOMPATIBLE SYSTEMS

Over the past three years, the CAD systems in [two cities] have grown and developed separately. As a result, very little standardization between the two has occurred, and even the hardware is of different types and flavors.

—management information system (MIS) manager at Transco

The major reason that companies fail to plan for system integration at present is that the state of the technology makes it difficult if not impossible to plan, especially if systems are purchased from different vendors.[12] Yet committing to a single supplier is also risky—the supplier may go out of business, raise its prices dramatically, or stop supporting the software. Often a vendor has an excellent solution for one component, say, a good CAD system, but interfaces it to another component that is poorly developed or not suited to the needed application. In this case, departments may push hard for systems that suit their own needs, sacrificing interdepartmental compatibility. Or a supplier invents a new product that makes your present solution obsolete. For these reasons, companies are reluctant to throw in their lot with one vendor. Until vendors agree on more standardization, this roadblock will be difficult to overcome.

Integration problems also occur when companies make incremental investments in AMT. Economically, such a policy makes sense; the systems are expensive, and capital costs must be rigidly justified. But it can result in a maze of imcompatible systems in which the company has "sunk costs." The company then must hire personnel to integrate the systems, a route that may well be more expensive than originally investing in integrated systems.

For example, although Telco2's FMS system achieved a high degree of integration between its design and manufacturing subsystems, the same cannot be said for all of the company's systems. Telco2 finally decided it had to make systems integration a corporate priority. A task force found over 271 different software systems running on five different mainframe computers within the division; building the necessary bridges among them would mean writing thousands of custom software interfaces. The 14 full-time people employed to write these interfaces could not keep up. One manager said, "We've locked ourselves into a complex architecture that is rapidly becoming obsolete." Such is the price of not planning for integration before acquiring systems.

Even with the roadblocks posed by incremental investment strategies and lack of standards, companies could plan for a certain degree of present and future compatibility among AMT subsystems. But many do not. Why? Often, upper-level managers neglect to plan, and lower-level managers have little incentive to do so. Senior managers may lack the technical expertise to address compatibility, or the company may have no overall design and manufacturing strategy.

In conclusion, companies considering AMT investment need to develop an overall strategy that addresses compatibility issues, both present and future, and they need to plan at a high enough level to overcome parochial departmental orientations. To do so, they must either develop a degree of technological sophistication at the strategic level or obtain expert advice from inside or outside the organization. To a certain extent, cooperating with vendors and ensuring that they have some stake in the outcome will help, but their advice is not always disinterested.

RULE OF THE ROAD THREE: MORE ORGANIZATIONAL INTEGRATION

> *The key is to integrate the entire organization. We want to use the CIM [computer-integrated manufacturing] committee to provide leadership. The committee will consist of members from manufacturing, engineering, and information systems.*

In implementing AMT such as CAD/CAM, the key interface is between design and manufacturing, and true integration depends on their cooperation.[13] These organizations often have trouble coordinating:

> *The manufacturing people don't want to try out CAM unless it's polished and working because they're under the pressure of production and it would hold up the line. Much more planning is necessary. They won't release their people for a one-week course and a two- to three-week, up-to-speed period. They're too busy.*
>
> —engineering manager at Agco

We observed three effective strategies for overcoming these coordination roadblocks. At Agco, both departments reported to the same vice-president, and many attributed the success of the initial CAD implementation to his leadership. (According to some, when the vice-president left the company, the departments were separated and put under two vice-presidents. This made integration with CAM more complex.) This type of strategy, however, is rare except in small companies or new ventures without formal departmental structures.

Genco appeared to have few coordination problems. Its managers attributed this to "cross-fertilization." That is, most of the managers working on the CAD/CAM project had experience in manufacturing and design, as well as other departments, so they understood and respected each other's priorities and concerns. This was the only company in the study that used generalized managers.

The most common integration device was cross-functional teams. It is widely recommended that an interdisciplinary corporate steering committee should plan the AMT implementation.[14] Four of the 10 companies we studied had such a committee at the corporate level, but only two of these met their goals. Three had such committees at the divisional level, and two of these met their objectives. It appears, then, that such committees are not sufficient to ensure success, but most felt they were useful:

> *The MMP [manufacturing modernization group] became the original project team for the initiative. This was very positive as members of the MMP all understood their own fields as they related to FMS, and they were believers.*
>
> —general manager of Aerocan2

Properly constituted and managed, corporate or divisional steering committees can iron out conflict, anticipate problems, ensure compatibility of the various systems, work toward systems integration, set targets and objectives, and most of all promote an AMT strategy that will enhance competitive advantage. It was at the corporate or divisional level that we often found a high-ranking corporate sponsor or "godfather" of the AMT project, usually an active member of the steering committee. In nine of the ten cases, the original impetus for AMT came from this sponsor, not the champion, but only one of these sponsors was active in implementation. We concluded that top management support was important to get the project going, but not necessarily to energize it during implementation. Here periodic expressions of interest and support plus protection from other executives were all that appeared necessary.

A steering committee with a sponsor was useful, but it seemed to work best in combination with an interdisciplinary implementation team. Of the seven companies that used either corporate- or divisional-level steering committees, five also delegated implementation to lower-level committees composed of middle managers and technical experts (and sometimes even operating-level personnel and union representatives). Four of these five projects met their objectives. Telco3 found that momentum increased greatly when implementation teams were formed:

> *We had a CAD/CAM steering committee at the director level. It worked well and was necessary, but we were talking a lot—there was not too much action. We proposed the formation of project teams that would have a mandate to do something. Each team would have short-term targets. We are trying to create change in chunks that are easily digestible.*
>
> —manufacturing engineering manager at Telco3

The effect of decisions cascading from a steering committee to implementation teams also worked well at Aerocan2. An interviewee recounted:

> *The team approach was patterned in our heads from day one. Each committee consists of members of the management, production, and sup-*

port teams. Each committee was given a mandate and developed milestone plans that were given to the design committee [divisional steering committee].

Cross-functional implementation teams are worth considering in an AMT project. They can act as a reality check on the decisions of the corporate steering committee or the project champion. They can also generate enthusiasm in various areas of the company, which is helpful in overcoming resistance.

POTHOLE THREE: LACK OF TEAM STRUCTURE

Many have pointed to the necessity of overcoming fragmentation and using team structures to successfully implement innovations.[15] Yet once again we found that only half of our companies used the two-tier team structure to implement AMT. Why should this be so? The major fault appeared to be in the functional structures present in most of the companies, structures that seemed so natural they were carried over into AMT implementation without thought for more integration.

Most of the companies or divisions we investigated were structured along traditional functional lines; manufacturing was a separate department from engineering (or technology or R&D). These departments were separate from marketing, sales, finance, and so forth. In these structures, disputes arose often and were difficult to resolve:

A territorial dispute appeared to develop between engineering and the MIS department over ownership of the CAD/CAM hardware. Cooperation between the two departments was at a low level for some time afterward, and few improvements were made to the system.

—MIS manager at Transco

Whatever the organization structure, the implementation of an integrated AMT system depends on the cooperation of specialized departments and people who have not necessarily come into regular contact with each other in the past. These specialists usually have different orientations, priorities, and ways of solving problems.[16] The CAE manager of Telco3 put it this way:

I used to think there was not much difference between manufacturing and R&D. We're all engineers. Now I realize manufacturing's perspective is stability. They want to crank out millions of the same product. Change is anathema, whereas R&D makes one of something and says, "Great! Let's make something new!" So no wonder manufacturing resists CAD. It represents change on a grand scale, whereas R&D says, "Whoopee! Something new!"

Even when companies used teams or task forces, these were often viewed as temporary aberrations. At Transco, for example, the steering committee and the task force disbanded once they had chosen the type of system to purchase.

At Telco2, a manager reflected on the difficulty of using teams to integrate systems:

It goes against the previous organizational structure. Also, it crosses turf and two corporations. We have tried to formalize implementation teams. There are 65 people involved now in projects strictly for process and structural change.

Teams pose big political problems for the organization. When AMT is introduced, power bases shift, and resistance builds. Conflicts surface earlier with teams, although they can often be resolved better with teams as well. Managers may think they can avoid direct confrontation if they avoid teams.[17]

Furthermore, teams do not work well within an autocratic system or in the absence of team experience or training. They encounter difficulties when team members are widely separated by geographic or psychological distances, when all relevant interest groups are not competently represented, or when planning techniques and skills are inadequate. Conflict between the champion and the teams may also cause problems. A champion who claims ownership of the project may resist giving up decision-making power to an implementation team. The steering committee needs to give clear direction to avoid this problem.

Obviously, the use of teams is even more problematic and complex than developing an effective champion or integrating systems, but we would argue that teams are just as necessary to ensure widespread acceptance and skillful use of AMT. To gain maximum benefit from team structures, companies sorely need an infusion of team-building expertise and more emphasis on organizational development. Expertise in these areas, such as in sociotechnical design, is widely available, but only one of the companies, Aerocan2, saw its value.

CONCLUSION

Among the plethora of suggestions for AMT implementation, three rules stand out as the most necessary and powerful: develop an effective champion, plan for a high level of systems integration, and use organization integration techniques. Failing to follow even one of these rules can jeopardize a project, especially if the problem is champion failure. However, it is not enough to know that these rules are important; companies must consciously plan to overcome the barriers that commonly impede implementation.

Successful AMT implementation does not automatically flow from sound manufacturing strategy or a wise equipment choice. It has its own process and logic, which have more to do with management than with the technology. Until the predominantly technocratic approach to AMT implementation is supplemented by an equal concern with structure, planning, conflict resolution, team functioning, and champion skill development, we may expect to continue to see what this study has shown. Only half of the companies embarking on an AMT voyage will reach their destination. The other half will be impeded by dead ends and potholes.

REFERENCES

1. These benefits have been discussed in the literature. See: Adler, P. "Managing Flexible Automation." *California Management Review* (Spring 1988): 34–56; Meredith, J. R. "The Strategic Advantages of the Factory of the Future." *California Management Review* (Spring 1987): 27–41; Stalk G. "Time—The Next Source of Competitive Advantage." *Harvard Business Review* (July–August 1988): 41–51; and Winter, P., and Gilbert, J. "Flexible Manufacturing Systems: Resistance and Trends." *Production and Inventory Management* 28 (1987): 10–15.
2. Beatty, C. A. "The Implementation of Technological Change," Industrial Relations Centre, Queen's University, Kingston, Ontario, 1987; and Meredith (1987).
3. Leonard-Barton, D., and Kraus, W. "Implementing New Technology." *Harvard Business Review* (November–December 1985): 102–110; Schon, D. A. "Champions for Radical New Inventions." *Harvard Business Review* (March–April 1963): 77–86; Souder, W. E. "Encouraging Entrepreneurship in the Large Corporations." *Research Management* 3 (1981): 18–22; Maidique, M. A. "Entrepreneurs, Champions, and Technological Innovation." *Sloan Management Review* (Winter 1980): 59–76; Galbraith, J. R. "Designing the Innovating Organization." *Organizational Dynamics* 10 (1982): 5–25; Bass, B. M. *Leadership and Performance Beyond Expectations.* New York: Free Press, 1985; Burns, J. M. *Leadership.* New York: Harper & Row, 1978; Kanter, R. M. *The Change Masters.* New York: Simon & Schuster, 1983; Maidique (1980); and Schon (1963).
4. Leavitt, H. L. *Corporate Pathfinders.* New York: Penguin, 1986.
5. Fischer, W. A., Hamilton, W., McLaughlin, C. P., and Zmud, R. W. "Conference Report: The Role of the Project Champion." *R&D Management* 15 (1985): 71–72.
6. Slusher, A., Van Dyke, J., and Rose, G.

"Technical Competence of Group Leaders, Managerial Role, and Productivity in Engineering Design Groups." *Academy of Management Journal* 15 (1972): 197–204.
7. Frohman, A. L. "The Performance of Innovation: Managerial Roles." *California Management Review* (Spring 1978): 5–13; and Shapero, A. "Managing Creative Professionals." *Research Management* 2 (1985): 23–28.
8. Blumberg, M., and Gerwin, D. "Coping with Advanced Manufacturing Technology." *Journal of Occupational Behavior* 5 (1984): 113–130.
9. Morgan, G., and Ramirez, R. "Action Learning: A Holographic Metaphor for Guiding Social Change." *Human Relations* 37 (1983).
10. Johnson, R. H. "Assignment: Orchestrating the CIM Database." *CIM Review* (Summer 1985): 8–12; and Williamson, J. "Factory Integration Still at a Rudimentary Level." *Computing Canada* 11 (1985): 22.
11. Steele, L. "Managers' Misconceptions About Technology." *Harvard Business Review* (November–December 1983): 133–140; Meredith, J. R. *Managerial Lessons in Factory Automation: Three Case Studies in Flexible Manufacturing Systems* (Monograph 4, Operations Management Association, 1989); Bessant, J., and Haywood, B. "Experience with FMS in the UK." *International Journal of Operations and Production Management* 6 (1986): 44–56; and Cotter, S., and Skinner, C. "True CIM Highest on Information Link Between Systems." *Computerworld* (September 1985): 79–81.
12. Winter and Gilbert (1987).
13. Ettlie, J. E., and Reifeis, S. A. "Integrating Design and Manufacturing to Deploy Advanced Manufacturing Technology." *Interfaces* 17 (1987): 63–74.
14. Majchrzak, A. *The Human Side of Factory Automation.* San Francisco: Jossey-Bass, 1988.
15. Burns, T., and Stalker, G. M. *The Management of Innovation.* London: Tavistock Publications, 1961; Kanter (1983); and Beatty, C. A. and Gordon, J. R. M. "Barriers to the Implementation of CAD/CAM Systems." *Sloan Management Review* (Summer 1988): 25–33.
16. Lorsch, J. W., and Lawrence, P. R. "Organizing for Product Innovation." *Harvard Business Review* (January–February 1965): 109–120.
17. The difficulties of starting up and working in teams have been well documented. See: Kanter (1983).

QUESTIONS

1. What are the key factors for implementing manufacturing technologies successfully?
2. Can the same factors lead to success in implementing other types of systems (e.g., office information systems)? Explain why or why not.
3. This section stresses the importance of organizational integration. What is meant by organizational integration and why is it so important?

6.4 Managing Projects in the Computer Industry

GEORGE GLASER

A large portion of the work we do involves projects. Unlike ongoing production, where the same product is produced over and over, a project typically has a larger single output that must be ready at a given time. For example, a project may involve building a bridge or completing a study. Many organizations create a matrix organization where people are assigned to projects from their professional organization. Like production, projects must be managed properly if they are to achieve their goals.

In his classic review paper, Glaser looks at the critical success factors that lead to successful project management. Not surprisingly, they turn out to be a mixture of good organization, scheduling, people sense, and tool application. Given the complexity of most projects, the tools are computer based. They look at the issues of:

- Time (will we get done in time?)
- Cost (will we do it within budget?)
- Resources (do we have the wherewithal to accomplish the job?)
- Risk (what uncertainties do we face?)
- Reporting (where do we stand?)

However, no matter how good the tools, project planning will only succeed if the planning is realistic and matched to the task.

COMPUTER professionals work on projects throughout their careers. Whatever the individual job—from the most junior trainee to the most senior manager—projects are a way of life.

Projects are as varied in their characteristics—duration, staff requirements, scope, and complexity—as in their objectives.

Unfortunately, project success varies as well. In the computer industry the track record has been spotty, at best. Objectives are missed and

Source: Computer (December 1984): 45–53.

schedule and cost targets overrun with distressing regularity and, at times, with equally distressing results. Once—in the infancy of the industry—these occurrences may have been excusable and even acceptable, but today we cannot afford frequent project failures.

The focus of this article is the management of large and complex projects. To understand what constitutes a successful project, we need to examine current project management practices (including a look at why projects often go awry) and what exactly is involved in the project management task. To this end, the article presents recommendations for project planning, for the use of project management tools, and for planning and conducting project review meetings. Specific reference is made to two large-scale sample projects: the design, construction, and start-up of a semiconductor wafer fabrication plant and the introduction of a family of high-performance peripheral devices.

AN OVERVIEW OF PROJECT MANAGEMENT PRACTICES

Not everyone has the motivation, inclination, or disposition to be a project manager any more than everyone is well suited to be a technical specialist. But, just as competent project managers must have some level of understanding and appreciation for the technology they manage, so must competent technical specialists have some level of understanding and appreciation for the management principles that apply to their work as computer professionals.

Examples of projects the computer professional is likely to encounter include:

- Researching the frontiers of computer technology.
- Developing and implementing new application systems.
- Designing, manufacturing, and marketing new hardware and software products.
- Planning, staffing, and equipping new or expanded computing facilities.

Knowing how to manage such projects is a critical skill for the computer professional. Yet project management skills must be scarce, since so many projects in the computer industry fail to achieve their objectives.

Why is this so? First, because computer professionals tend to place a higher value on technical skills than on management skills—an understandable attitude in a profession rooted in technology, but a short-sighted view for those who want to see their work bear fruit. The *real* economic value of computer technology stems from its effectiveness in meeting the needs of others. Both technical and managerial skills are needed to make that happen.

The second reason for the shortage of project management skills is that few individuals have the background and experience to teach such skills. In the academic community particularly, those dedicated to teaching the hard sciences are often openly disdainful of such "soft" subjects.

As a result, because few, if any, project managers receive formal training in the subject, most must acquire their skills on the job. All too often, this training proves to be very expensive indeed, both for hapless project managers who get into trouble and for the enterprises that must bail them out.

Fortunately, certain proven principles of general management apply to project management as well, and these principles *can* be taught. So there is help available if we can learn how to use it. First, though, we need to understand what distinguishes a project from on-going management. A project typically is distinguished by the following:

- A specific objective, such as to implement a new sales order entry system.
- A specific time period, such as to be operational by May 1, 1985.
- A budget for capital expenditures and operating expenses, such as a cost not to exceed $235,000.
- An ad-hoc team, a team of individuals—some of whom are assigned part-time because of their skills and/or particular orga-

nizational relation to the task at hand—for example, two analysts, three programmers, and the manager of sales order entry (half-time).

These elements, in and of themselves, are not enough to make project management particularly difficult. Quite the contrary: a project with a straightforward objective, a realistic due date, an adequate budget, and proper staffing should be relatively routine for a competent manager.

Unfortunately, such ideal circumstances are rare in practice; more often, one or more of the following occur.

- Project objectives are poorly defined and/or understood, even by members of the project team.
- Project deadlines are dictated by external events or imposed arbitrarily by administrative fiat.
- Project budgets are based on naive estimates by inexperienced managers.
- Project staffing is determined more by availability than ability.

The outcome of projects launched under such circumstances is easily predicted. Managing a well-planned and well-staffed project is challenging enough; with fuzzy objectives, an unrealistic schedule, an inadequate budget, and weak staffing, project managers would need a miracle to succeed.

Yet projects with these weaknesses are launched every day. In fact, projects are routinely launched under circumstances that even *further* increase the chance of failure. For example:

- *Overly ambitious objectives.* Setting ambitious project objectives is often effective in motivating the project team to do their best. However, when project objectives force team members to stretch *beyond* their capabilities, motivation drops. Often this approach is taken by companies trying to catch up with their competition. But trying to regain lost ground by embarking on a naively ambitious recovery course invites disaster.
- *Crash schedules.* Crash schedules, like ambitious objectives, can motivate a competent, well-managed project team to achieve truly outstanding performance. But when schedules are so tight that deadlines are missed consistently, the team may become so frustrated that they will "cheat" to create the impression that they are on schedule, perhaps by overstating progress or skipping important tasks altogether—both causing the project to fall even further behind schedule and even shorter of its objectives.
- *Cross-functional scope.* When multiple functional organizations must work together to reach a shared project objective, exceptional demands are placed on project managers, especially when they do not have formal authority over "outsider" members of the project team. The manager needs a high level of management skill *and* the political clout to go with it. Occasionally, a particularly capable manager will successfully substitute personal charm and diplomacy for political clout, but more often success depends on clout rather than charm.
- *Technological complexity.* Projects that depend on technological breakthroughs for their success present special management problems. Because such projects explore "uncharted territory," the manager may not be able to identify the detailed steps to be taken and, therefore, cannot realistically assess progress. Despite the high risk of failure, such projects are common in the computer industry. A few are successful, largely because survival is a strong motivator, but typically casualties are high.

No single one of these presents an insurmountable obstacle. However, they are often found in combination, sorely testing the skills of the project manager.

THE PROJECT MANAGEMENT TASK

Given that the management of projects is a demanding undertaking, what can be done to enhance the prospect for a successful outcome?

There is no short answer to this question, but the likelihood of any project's success is strongly influenced by the management philosophy (corporate culture, if you will) of the enterprise. If an organization is characterized by a high level of professionalism—a "can-do" attitude, careful planning, disciplined adherence to schedules and budgets, a sense of teamwork among organizational sub-units, and a cadre of energetic and conscientious managers—it is well equipped to take on demanding projects. Organizations that fall significantly short in one or more of these basic strengths are playing in a high stakes game with poor cards.

Launching a Project

The most opportune time for planning and organizing a project is, of course, at the time it is started. Enthusiasm is high, the outlook for success seems bright, the need for a plan of action is generally acknowledged, and the project team is eager to go to work. Moreover, at this time, the cost per day of planning is low because the number of people involved is relatively small and the atmosphere is relatively crisis-free. Later, the cost per day of slippage is substantially higher because the team and/or the operating facility is fully staffed, and the capital investment is high.

Each case has unique details, but the basic steps for launching a project are similar for a wide variety of circumstances. Guidelines are summarized in the box on page 357.

Project Management Philosophy

Every organization has a management philosophy—explicit or implicit—that governs the way it carries out day-to-day affairs. Ideally, all philosophies take advantage of management strengths and compensate for management weaknesses, although some philosophies are more effective than others. So it is with project management. Every organization has a project management philosophy, whether explicit or not, whether effective or not. To be effective, such a philosophy must be tailored to the organizational, political, and economic realities of the organization and must be consistent with its general management philosophy.

No project management philosophy, of course, can guarantee the success of any project, no matter how noble its objectives or how diligently it is applied. It can, however, materially improve the prospects for success, *provided* all project participants accept the philosophy and it is administered in a consistent and disciplined manner.

When is an explicit statement of philosophy warranted? A project that involves a large number of tasks and complex organizational interrelationships would be a likely candidate. Consider the following case based on the recently completed start-up of a semiconductor-wafer fabrication plant:

- The project cost over $100 million.
- A number of organizational units were involved, including corporate facility planners and designers, R&D staff, outside architects and building contractors, and the line management that was to operate the new plant.
- The project manager, although a well-known and respected individual within the company, was managing a project of this size for the first time; furthermore, the key individuals reporting to the project manager had not worked together as a team before. In fact, several of them were new employees, hired specifically for this project.
- The site for the new plant was approximately 1000 miles away from the corporate offices where funding decisions were made and where most of the detailed product and process planning took place.

Together, these facts argued for a formal management approach and led to the adoption of the project management philosophy shown in the box

GUIDELINES FOR LAUNCHING A PROJECT

Every project is unique in its management requirements, but certain steps can be taken at the time the project is launched to improve the prospects for a successful outcome. To that end, the following guidelines are offered:

1. Establish a realistic project objective, setting forth in detail what will be accomplished if the project is successful.
2. Appoint a competent project manager whose administrative, technical, and political skills are commensurate with the task.
3. Set up a project organization at the appropriate level and establish the appropriate communications links among all elements of the organization that must play a role in the project's success.
4. Staff the project with the proper mix of technical and administrative skills. Avoid, whenever possible, part-time assignments so that the individuals who are working on the project can devote their full attention to it.
5. Identify key project milestones that, when achieved, will demonstrate definitive progress toward the ultimate project objective.

   Note: This step, plus Steps 6–11 below, may require several iterations before a satisfactory plan, schedule, and budget can be developed and approved.

6. Plan the project in detail, identifying all tasks that must be completed to reach each milestone.
7. Assign each task to an individual or to a specific organization so that responsibility for its completion is unambiguous.
8. Estimate the *time* required to complete each task. It is essential that the time estimate for each task be made by the individual or organization that bears the responsibility for completing it.
9. Estimate the *cost* of completing each task (or groups of tasks); again, these estimates should be made by the responsible party.
10. Produce a project schedule and time-phased budget (using critical-path or similar network techniques when the size of the project warrants).
11. Distribute the plan, schedule, and budget to all responsible parties and confirm their "ownership" of the tasks assigned to them.
12. Review the project schedule and budget regularly. At each review meeting, ask for reaffirmation of plans and schedules for the forthcoming period. On large, complex projects, make project reviews formal; take minutes to document key decisions and follow-up assignments.
13. Update project plans and schedules after each review meeting and distribute them as noted previously.
14. Manage the project!

on page 358. The facility was completed, equipped, and staffed on time.

Project Plans and Schedules

Seasoned project managers, well aware of the risks they face, look for ways to reduce their exposure to delays and cost overruns. Their most valuable ally is a detailed project plan to use in tracking progress.

A detailed project plan specifies what tasks need to be completed and in what order, what end-products or deliverables result, which indi-

A SAMPLE PROJECT MANAGEMENT PHILOSOPHY

A semiconductor plant start-up

1. Make plans comprehensive and detailed. Gross plans only lead to gross blunders and gross overruns.

2. Have one—and only one—project plan and supporting schedule. This does not rule out subsidiary plans that are maintained in greater detail than what the project manager needs, but plans that are independent of the project manager's plans are inherently dangerous.

3. Report progress in detail against the plan. For each task, be prepared to report time to completion; avoid the use of "percentage complete," a particularly beguiling, but more often than not misleading figure.

4. Consider time estimates as commitments of the person who makes them. On the other hand, those responsible for tasks are encouraged to revise time estimates (upward, if necessary) whenever they encounter difficulties so that schedules are realistic at all times.

5. Urge forthright reporting of bad news: To slip a schedule is a sin, but to conceal a slippage—or to fail to reveal an anticipated slippage—is a grievous sin.

6. To avoid getting bogged down in detail, focus project reviews on activities that are now critical or that soon will be.

7. Make project review meetings formal, with a written agenda, careful preparation beforehand—and diligent follow-up afterwards—and select someone with a strong hand as chairman.

8. Hold project review meetings on a regular basis; adjust their frequency as the tempo of the project demands.

9. Make attendance at project review meetings mandatory for organizations designated as "critical" at each stage of the project. Furthermore, ensure that every organization is represented by an individual who can commit it to deadlines or revised cost objectives. Messengers bearing bad news but not responsible for it are not acceptable replacements for those in charge.

10. Report progress on all tasks scheduled for completion in the period just ended. Look ahead to the forthcoming period to reaffirm that scheduled commitments will be met. When it is known in advance that an activity scheduled for completion in the upcoming period will be behind schedule, develop and present alternative plans. These plans should, to the extent possible, minimize or offset any adverse effect on other activities by, for example, reallocating resources. At a minimum, warn others affected by the slippage in advance.

11. Ensure that capital expenditure tracking includes amounts already spent, amounts committed, and amounts necessary "to complete." Accounting should anticipate—not merely document—expenditures.

viduals or organizations are responsible, and the estimated "time-to-complete." Using this detailed plan, managers can develop a similarly detailed schedule to indicate the date each task can start and the date it must be completed.

Identifying the tasks that make up a project is demanding and time consuming. For large, complex projects, such as those that span several years and that have a company-wide scope, developing an initial plan may take the project team several months of concentrated effort.

A good way to get started is to define major

project milestones that, when reached, signify important progress toward the project's objective (see "Typical Project Milestones"). The milestones, in turn, can be used to identify tasks that precede and follow them. Note that the list in the box includes milestones for development engineering, manufacturing engineering, test engineering, customer engineering, production management, and marketing—all of which influence the development of project plans.

For projects of the size just cited, critical-path (or similar) networking techniques are essential in developing and maintaining plans and schedules.

Accountability for results is critical to any management process. Accountability, however, comes only after responsible managers have endorsed the project plans and schedule. This endorsement is key, since it carries with it an implied commitment to the tasks and dates contained within. Until such commitments are obtained, the plan is little more than a project manager's "wish list" and can hardly be expected to be binding on other parties. Once the plan is formally adopted, however, individuals (or organizations) can be held accountable for completing specific tasks and meeting approved schedule targets.

Plans and schedules drawn up at the project's outset can quickly become obsolete as events unfold. To keep them current, they must be regularly reviewed and, when necessary, revised. Such changes should always be approved by the project manager, and such approval should not be lightly given, since casual and frequent schedule slippages obviously should be avoided. Nonetheless, it is better to slip a schedule and meet the revised dates than to cling to an unrealistic one, knowing that the team will not be able to meet it. Certain characteristics that distinguish a sound project plan from others are set forth in the box on page 361.

Project Reviews

The enthusiasm for planning that accompanies a new project is often short-lived. Typically, there is an initial flurry of activity: task lists are created, time estimates are made, critical-path networks are drawn, Gantt charts are prepared for management, and budget requests are energetically presented. Once the project is approved and the funds authorized, however, the "plans" serve little purpose other than to decorate the walls of the project manager's office. No further planning takes place until the project (and, presumably, the project manager) is in hot water, at which time the exercise is repeated with no greater conviction or commitment than when the project was initiated.

The best way to keep plans (and planning) alive is through a disciplined, that is, regular and rigorous, review process. While format and content will differ just as project managers' management styles differ, project reviews are critically important in the management of any project.

Review meetings satisfy several objectives:

- To interject general management awareness, concern, and support.
- To reaffirm the correctness and completeness of current plans and schedules.
- To anticipate potential problems and to assign responsibility and deadlines for their resolution.
- To foster cooperation and communication among the individuals and organizations involved.
- To create a sense of enthusiasm, cohesiveness, and importance.

Note that "reporting on accomplishments of the previous period" is omitted from these objectives. Often, such recitations are not only a waste of time but also (worse yet) a smokescreen for avoiding discussion about what should have happened but did not.

Also absent is any suggestion that review meetings should be used to admonish those who missed deadlines. There is a proper time and place for confronting the nonperformer, but most would agree that public floggings have not proven effective in behavior modification. When schedules slip, the first order of business is to identify the implication of the slippage on the project

TYPICAL PROJECT MILESTONES
The introduction of a new product

▪ ▪ ▪ ▪ ▪

Marketing requirements defined
Functional specifications released
Performance specifications released
Design review(s) completed
Engineering prototypes built
Design validation test completed
Engineering drawings released
Beta-site test completed
Product pricing finalized
Product announced
Critical-part vendors qualified
Long-lead parts ordered
Hard tooling ordered
Test plans developed
Spare parts ordered
Manufacturing process instructions issued
Manufacturing prototypes built
Field-support plans developed
Customer engineering documentation released
Customer engineering training initiated
Product validation test complete
First customer shipment
Manufacturing cost review(s) completed

completion date, then to focus on what can be done to recover the time lost.

How often review meetings should be held depends on the nature and length of the project. For a project of six months or more, formal reviews at the top management level should be held at least monthly. Reviews at the project-team level should be held at least biweekly. A decision-level spokesman for every organizational unit should attend so that revisions to plans and schedules can be adopted on the spot. A formal agenda should be prepared and distributed in advance, and formal minutes listing issues identified and actions taken should be recorded and circulated as a confirming document to those directly responsible and as an information document to others.

To repeat, review meetings are vital to the success of any project. They are the best—and often the only—mechanism for maintaining credibility of plans and schedules and for instilling a sense of teamwork and commitment vital to success.

Project Management Tools

A number of software tools are available to aid the project manager and members of the project team in carrying out their responsibilities.

Such tools have been developed for both large mainframes and for personal computers. They vary substantially in cost (from $200 for a small system for a personal computer to $100,000 for a stand-alone system with extensive graphics capability), the size of project network they can process (from a few dozen tasks to many thousands), their architecture (batch/remote-batch/timesharing/on-line interactive), and the flexibility of their progress reporting and report generation.

These tools typically provide one or more of the following complementary capabilities: time analysis, resource analysis, and cost analysis.

Time Analysis

Using the critical path (or PERT) method, time-analysis tools calculate a time schedule for a network of project tasks. The process usually consists of the following steps:

- Individual tasks, each with its estimated time to completion, are interconnected in a network to reflect the predecessor/successor relation among them.
- For each task, a "can start" date is calculated, indicating the earliest possible start date according to the scheduled start dates for each of the task's predecessors.

CHARACTERISTICS OF A SOUND PROJECT PLAN

1. Comprehensiveness. Its scope covers *all* aspects of the project, including those that are the responsibility of support organizations.

2. Uniqueness. It is the only approved overall plan for the project. Subplans for lower level managers can, of course, be useful (even necessary) in certain situations, but independent plans, that is, plans that are developed and executed independently of the official project plan, are a common cause of schedule slippage and other difficulties that result from poor communication and coordination.

3. Attention to detail. Every significant task in the project is identified, and its logical sequence and relation to other tasks is explicitly stated.

4. Clarity. Every task is assigned to one and only one individual or organizational unit, and its deliverables (or other conditions for satisfactory completion) are crisply defined so that progress can be assessed realistically; "fuzzy" claims of completion are often used to imply progress when, in fact, little real progress has been made.

5. Authority. It has been endorsed by the line management of each of the individuals or organizational units participating in the project. Time estimates, as a result, represent commitments by those responsible for completing individual tasks.

6. Currency. It is reviewed and updated on a regular and disciplined basis (and revised when necessary) so that it represents *the* project schedule at all times.

- A "must complete" date for each task is then calculated according to the scheduled start dates for each of the task's successors and the target completion date for the overall project.
- The difference between the calculated time *available* to complete a task and the estimated time *required* to complete it—its *float* or slack time—is then calculated. If the float is zero, the task is said to be "critical," since any delay in completing it will cause a delay in completing the final task of the project.

Knowing the scheduled start and completion dates for each task in a project is fundamentally important in managing it. Knowing the float time for each task is fundamentally important in distinguishing between tasks that literally pace a project and those that allow greater scheduling flexibility.

Most time-analysis tools also provide for a variety of selecting, sorting, and reporting options so that customized schedules can be prepared, for example, for responsible organizations, levels of management, types of resources, and time periods.

RESOURCE ANALYSIS

The number and type of available resources—such as programmers, final test stations, and bulldozers—often determine when an individual task can start and the time needed to complete it. Project management software with resource-analysis capabilities can be used to produce a schedule that takes these constraints into account.

COST ANALYSIS

Tools with cost-analysis capabilities are used to compute direct costs, on the basis of resource requirements and resource rates, and to summarize costs in various ways, by time period and organization, for example.

In addition to these analytical capabilities, certain project management tools include graphics capabilities for drawing network diagrams, plotting Gantt (bar) charts, and graphing other types of project data, such as period costs. Almost all of them can be used to produce Gantt charts and relatively crude network diagrams on a line printer; still others can be used to produce both bar charts and network diagrams on a dot-matrix printer, although for projects with more than 100 tasks, printing delays become excessive; the most powerful tools can readily produce a wide variety of graphics output for very large projects (thousands of activities) on high-speed flat-bed or drum plotters.

Project management tools make the job of the project manager easier, and for projects comprising more than a few dozen tasks, they are essential. But neither one alone nor all of them together are effective in the hands of an inexperienced project manager or an undisciplined organization. Such tools are often acquired with the naive assumption that powerful tools can compensate for weak management skills—an assumption equivalent to believing that a mediocre artist can create classic sculpture if the chisel is sharp enough. Management tools can, in fact, do more harm than good if they are applied without adequate understanding of their limitations or without proper concern for the organizational and managerial issues that seem inevitably to accompany large projects.

What project management tools *will* do is relieve project managers of the more onerous clerical tasks and allow them to focus on the more challenging—administrative, technical, and political—aspects of the job.

For a project to succeed, the right people must complete the right tasks at the right time. This doesn't happen by accident. It happens because objectives have been stated crisply, obstacles to achieving them (including time and resource constraints) have been identified and realistically evaluated, detailed plans have been developed, and realistic schedules have been adopted.

Planning and managing a project in this way is hard work—and at times, even painful. Yet it is far less painful to plan at the outset of a project—and to manage carefully as it progresses—than it is to explain to our superiors why we are behind schedule and over budget, and to our team members why their hard work on a foundering project hasn't advanced their careers or gotten them a raise.

Managing a successful project, on the other hand, can be an exhilarating and rewarding experience. Those who make the investment in time and effort required to plan and manage their projects well can expect to realize a handsome payoff from that investment.

ANNOTATED BIBLIOGRAPHY

Argelo, S. M. "Pitfalls in Automation Projects." In *Beyond Productivity: Information Systems Development for Organizational Effectiveness*, edited by M. A. Bemelmans, 241–252. North-Holland, Amsterdam, 1983.

Proceedings of the IFIP WG 8.2 Working Conference on the title subject.

Brooks, F. P., Jr. *The Mythical Man Month: Essays on Software Engineering*, Menlo Park, CA: Addison-Wesley, 1975.

The *classic* book on the management of software engineering by the man who led the development of IBM's OS/360. Described by the author as "a belated answer to Tom Watson's probing questions as to why programming is hard to manage." Required reading for all project managers and those who aspire to the role.

Gall, J. *Systematics*. New York: Quadrangle Press, 1975.

A "valuable contribution to our understanding of organized failure," this book examines the pitfalls of systems analysis projects. Full of wise and wry comments on projects that got out of hand.

Glass, R. L. *The Universal Elixir and Other Computing Projects which Failed* Seattle, WA: Computing Trends, 1977.

A collection of short anecdotes about computer projects that failed and why they did so.

Kidder, T. *The Soul of a New Machine* New York: Avon Books, 1982.

A Pulitzer-Prize-winning story about the development of Data General's Eclipse MV8000 computer (then called Eagle). Focus of the book is largely on the challenges faced by the project team and on the demands imposed on them by themselves and by the corporate culture in which they worked.

Metzger, P. W. *Managing a Programming Project* Englewood Cliffs, NJ: Prentice Hall, 1973.

According to the author, a book that is "short on pure theory and long on practical 'how-to-do-it' advice." "The Definition Phase," chapter 2, contains a number of useful suggestions on project planning and management.

Quist, A. B. *Excuse Me . . . What Was That?* Beaverton, OR: Dilithium Press, 1982.

A book "full of anecdotes about things that should've worked but didn't and things that did work but for all the wrong reasons." The author notes that "someone may actually derive wisdom from what is offered here, probably by accident." Humorous and insightful.

Reynolds, C. H. "What's Wrong with Computer Programming Management?" In *On the Management of Computer Programming*, edited by G. F. Weinwurm, 35–42. Princeton, NJ: Auerbach Publishers, 1970.

Shelly, G. B., and Cashman, T. J. *Business Systems Analysis and Design* Fullerton, CA: Anaheim Publishing, 1981.

Chapter 10, "Project Management, Scheduling and Control," describes the use of Gantt charts and Pert (critical-path method) networks for systems analysis and design projects. Includes case studies.

Weinwurm, G. F. "The Challenge to the Management Community." In *On the Management of Computer Programming* edited by G. F. Weinwurm. Princeton, NJ: Auerbach Publishers, 1970.

A pragmatic book on techniques for the management of computer programming projects by a series of respected and seasoned practitioners.

QUESTIONS

1. How does project management differ from on-going management?
2. What is project management philosophy and why is it necessary to have one?
3. What are the major features of computer-based project management tools and how can they be used to manage projects?

6.5 Computers of the World, Unite!

JEREMY MAIN

The basic message of this book is that the integration of computer and communications technology changes productivity in the office and the factory. Main, writing in *Fortune*, makes this point with numerous case examples. He points out that businesses with strong communications links among their computers make the process of producing and selling goods and services much faster. The competitor who has this capability has an advantage.

Note that the message is integration—putting the pieces together into a single whole. Integration is an implementation problem. It involves all functional units of the company and it involves putting the company together with the customer to provide improved service. Thus conglomerates with truly separate product lines (e.g., turbine engines and electric light bulbs) do not (and probably cannot) integrate across those product lines.

Integration is not primarily a technical issue, although you must make sure that you have solved the technical problems. The key obstacles are managerial and organizational.

"I don't know what you're talking about, but keep talking," said Edson Gaylord, chairman of Ingersoll Milling Machine Co. of Rockford, Illinois. George Hess, his vice president for systems and planning, kept talking. When Hess finished, he had persuaded Gaylord to launch Ingersoll, an extremely competitive machine tool producer, on a complex, risky venture: turning itself into what is becoming known as a computer-integrated business. That was in 1979. Today a lot of CEOs want to do the same.

The computer-integrated business is a hot concept. The phrase describes an enterprise whose major functions—for example, sales, finance, distribution, manufacturing—exchange operating information quickly and constantly via computer. Product designers can send specifications straight to machines on the factory floor. Salesmen—or even customers—can find out which products are in stock and when they can be delivered, and can place orders, which automatically cause new units to be manufactured. Accounting receives on-line all information about sales, purchases, and prices. And high executives can get any of this information, and much more, immediately. The computer system parallels the whole process of producing and selling goods or services and makes it move faster—much faster.

Source: Fortune (September 24, 1990): 122.

Speed is the compelling reason for computer integration. Old methods of deploying a lot of different computer systems that couldn't talk to each other, or could talk only through expensive translation programs, aren't good enough anymore. Says John Rockart, head of the Center for Information Systems Research at MIT's Sloan School: "The buffers of space, time, people, and inventory are gone, so you have to have the lubrication of information to get the flow going." To react fast enough to customers' demands, corporations need a fast, seamless information network throughout the company. For example, Sony aims to use integration to cut the time it takes to make and distribute products from 50 days to 20.

It isn't easy. No large company is yet fully computer integrated. One company's network started in manufacturing and then grew to encompass management and sales. Another began on the sales side and is just now reaching into the factories. But even if a system falls short of the scale its planners envisaged, it can still dramatically improve performance.

Ingersoll's system, put in place by 1982 at a cost of $5 million and much elaborated since, has helped considerably. While half of America's machine tool companies have folded since the 1970s, Ingersoll's shipments of large tools multiplied tenfold to nearly $500 million last year. The company is just finishing what it believes to be the largest custom-built machine tool ever made, a monster the size of a three-story house to make large turbines and other parts for hydroelectric generators. The customer is Impsa, a maker of hydroelectric equipment in Argentina.

Companies much bigger than Ingersoll are following its example. Frito-Lay, PepsiCo's most profitable division, has a new network that joins the hand-held computers used by every one of its 10,000 route salespeople to the office of President Robert Beeby, with connections to area and division offices and to company plants. Saturn Corp., the new General Motors small-car subsidiary, starts off fully computer integrated. Du Pont is committed to tying all its 80 businesses in 50 countries into a uniform information network. The effort will take at least five years, swallowing up about $200 million of Du Pont's $900 million annual spending on information systems.

If you're thinking of computer-integrating your operation, brace yourself: It won't be easy. No computer company sells a ready-made integrated system. Because of any large enterprise's vast complexity, each system must be specially created, a fact that provides fertile ground for consultants, business professors, and software and hardware firms.

Digital Equipment Corp. not only is turning itself into a computer-integrated enterprise, but is also showing others how to do it. The large Tokyo office of Andersen Consulting (part of Arthur Andersen) works almost entirely on helping clients such as Sony create networks. The job is big enough that pessimistic experts, such as Brandt Allen of the University of Virginia business school, argue that big companies can never integrate themselves completely. Says he: "The effort is so gargantuan and takes so long that by the time you have finished, everything has changed and the champions of the project have long gone."

Giant conglomerates probably don't need to be fully integrated: The lawn equipment division doesn't require intimate communication with the life insurance division. In any case, argues MIT's Rockart, "total integration is not the issue—integration with the customer is the issue." He says that companies with distinct sets of clients might use a separate network for each, rather than try to tie the whole corporation into one giant network. Johnson & Johnson might have a network for hospitals, another for pharmacies, and a third for other retail stores. Competition to give the customer better and faster service will keep up the pressure to put more business online.

In achieving integration on any scale, the key obstacles are no longer technological; they are in management and organization. "Integration was inhibited in the past by cost and lack of technology, but that has changed in the past two or three years," says David Mengden, director of Du Pont's computing and networking services. Unit computing costs have dropped by a factor of 100,000 in 20 years, and telecommunications costs are falling by 10% to 15% a year. Fiber-optic cables can furnish the vast capacity that companies need when transmitting engineering drawings and other graphic material. Two such cables span the Atlantic, another crosses the Pacific, and two more will reach the Orient in the next two years.

So the machines are up to the job. The trouble comes from complicated, sometimes irrational, corporate organizations and procedures. For example, each division in a company often counts sales or profits differently. A computer network can't cope until they are counted the same way. Every process, whether in accounting or design or the CEO's office, needs to be rethought and simplified.

Small companies are easier to integrate than big ones, new companies easier than old ones. Lynda Applegate, a Harvard business school professor, says the companies having the hardest time integrating are the large ones that set up massive back-room computer systems in the 1960s and 1970s, such as banks and insurance companies. The systems are obsolete, but replacing them is difficult and costly. It's easier to computerize the operations of a brand-new company or of an existing company that wasn't heavily computerized, like Frito-Lay.

Ex-CEO Michael Jordan relates that Frito-Lay decided to reorganize its sales system in the mid-1980s because processing 100,000 sales documents a week by hand had become too cumbersome. Besides that, the business was getting more complex and fragmented. As a national company competing mostly with strong regional companies that are close to their markets, Frito-Lay was handicapped because news of competitive incursions took months to drift up to headquarters in Dallas. Says Jordan: "We needed instant actuals."

Frito-Lay anchored its new systems to the hand-held Fujitsu computers issued to all its salespeople in 1987 and 1988. The computers are about as big and tough as a long brick. A salesman carries one into the store, punching in the code numbers and quantity of Fritos, Cheetos, Tostitos, and other snacks that need replacing, and the number of "stales" he removes because they have reached the end of their 35-day shelf life. When he attaches his hand-held, as Frito-Lay people call it, to a printer in his truck, it spits out an invoice for the day's deliveries to that store, which he hands over with the snacks.

At day's end, the 10,000 Frito-Lay salespeople hook their hand-helds to telephones and the sales information pours into the company's IBM 3090 mainframes in Dallas. They pull it all together and then redistribute it in appropriate chunks to the area and division offices; to marketing, purchasing, and transportation offices, and to top management.

For salesmen, the hand-helds eliminate four to five hours of paperwork a week. To a division sales manager like Paul Davis in Dallas they mean that every Monday he gets a summary of sales, crisp and clear, on his computer screen. He can break down the sales any way he wants—by product, type of store, or district—and he can get the results daily if he wants to follow a critical campaign closely. Bad news shows up in red. Davis recalls that he used to get sales figures six weeks after the fact, in a hard-to-analyze two-foot pile of computer printouts.

When Frito-Lay and Von's, a Los Angeles supermarket chain, recently ran a joint promotion, the daily report showed that sales were up, but more important, that some stores in the chain were doing a lot better than others. A quick trip through the chain revealed a big variance in the displays. Then a call to the chain's headquarters—which didn't know that not all stores were cooperating—got the laggards revved up. The reaction time was two days. The old Frito-Lay would have noticed a slight increase in Los Angeles sales weeks later, and probably would never have known why the promotion didn't do better.

Frito-Lay keeps a product-by-product, store-by-store watch on competitors through its integrated system, although the information comes more slowly. Monthly reports from a market research firm, Information Resources, go into the database with the internal information. Competitors can buy the same information about Frito-Lay, but they can't pass it around the company in the same accessible and friendly graphic fashion. Frito-Lay finds that this use of competitive information has helped persuade store owners to give its products more shelf space. If you can show them that Frito-Lay's snacks move faster or produce a higher margin than the snacks of another brand that is getting bigger displays, says Jordan, you have a powerful argument for winning more space.

Integration certainly has produced results. Since 1988 Frito-Lay has added 400 routes without increasing its sales force of 10,000 and pushed revenues up by almost $1 billion, to $4.2 billion. Jordan, now CEO of PepsiCo World Wide Foods, says, "We couldn't manage the company today without this system." PepsiCo's soft-drink division is going to hand-held computers, and the restaurant division is considering a similar network tied to the cash registers.

HOW fast can a computer-integrated company move? Take a look at Mayday, a sophisticated $5-million-a-year machine tool shop in Lewisville, Texas. Mayday manufactures bushings, the metal sleeves that protect some moving parts on aircraft, such as the wheel struts. They are often made of exotic metals to fine tolerances.

Jim Nelson, Mayday's president, says every company in the business uses the same type of production equipment. "What we need is time management," he says. "I'm selling time on the machines." He began by buying big automated Japanese lathes for the plant. But as the plan became more productive, he realized he needed to computerize the office to process more orders faster to keep up with the machines. Nelson bought some elements of the integrated system, like the bookkeeping software, off the shelf, which helped keep costs down. Others had to be designed by his own programmer.

Today, Nelson has an automated system for quoting prices when a customer calls in for an order. When the customer gives his specifications, Nelson's computers can figure the costs of materials and machine time and quote a delivery date and suggested prices within 30 seconds, even if the part has never been made before. If the client places the order, then the system itself sends him a confirmation by fax or mail, or both, and takes other steps to make the product, such as ordering materials and scheduling machine time. As the order flows through the plant, Nelson's computers can report its progress. At the end, the system produces the shipping labels and invoices.

MIGHTY Du Pont is far behind Mayday, but the imperatives are the same. Du Pont needs to develop, sell, and deliver products faster to remain competitive. If a Du Pont salesman in West Germany wants to sell O rings to an auto company today, he can look up the parts available in Germany in a catalogue. But when Du Pont becomes computer integrated in the mid-1990s, he will be able to look into a worldwide database and find not only what O rings are available, but also what products are being developed and when they should be ready for delivery.

Du Pont has a head start on building its system. Ten years ago the company told all divisions to standardize equipment, using IBM mainframes with Hewlett-Packard and DEC minicomputers. Du Pont also has a worldwide electronic mail network linking 80,000 of its 146,000 employees. And last year it installed an executive information system that supplies some 300 top executives with key numbers and charts that can be updated daily.

Over the next five or six years, each of these Du Pont systems will be meshed into a much expanded network with the same basic hardware and software throughout. The system will grow and bend to accommodate new technologies and business needs, says Ray Cairns, vice president for information technology. The effort to simplify and rationalize the corporation to give integration a chance to work has already caused major changes at Du Pont. For example, there used to be more than 40 kinds of distributed control systems corporatewide. They are the computer brains that run continuous processes like refining oil by manipulating valves, sensing pressures, and so forth. Du Pont cut the 40 types to seven and then to two.

In contrast to Du Pont, GM's Saturn started with a clean sheet since the division is brand new and was conceived as a computer-integrated enterprise. EDS, the data-processing company that GM acquired from Ross Perot, supplies much of the expertise and equipment to tie all of Saturn into a single database. Michael Reed, the EDS group manager for Saturn, uses the slogan "from art to part," meaning that the pieces that make up a Saturn car care designed on computers, which tell purchasing, manufacturing, and other departments what they must do for that part to be made and then schedule production.

Saturn's system actually goes well beyond art to part. Top suppliers will link electronically with Saturn, so that the orders they get from GM, bills they send, and payments they receive will all move electronically. Once a car is built, a computer record will follow it to the dealer and through its life until it is scrapped, so long as the owners keep going to Saturn dealers anywhere for service. Each service visit or repair will go into the car's record. With this system plus good service at reasonable prices, Saturn hopes to hang on to two-thirds of its customers for regular service and repairs after the warranty period instead of the usual one-third.

By tracking sales and customer preferences closely and fine-tuning production, Saturn also hopes to run on slimmer inventories. Says Saturn vice president Donald Hudler: "Conventional wisdom holds that everybody should have a 60-day supply of cars. We want to operate on a 30- to 45-day supply. Nothing good happens to a car sitting in inventory." How well all this will work, of course, remains to be seen.

CREATING the computer-integrated corporation remains chancy. How can you put into a computer the subtleties and intricacies of relationships in a big corporation? Even if you succeed, won't you create an overbearing centralized management at a time when business theorists are calling for more decision-making power down the line? Supporters of integration argue that if the new networks give CEOs better knowledge of what even distant managers are doing, the CEOs will feel comfortable allowing the managers a freer hand—and the managers will have the knowledge to make good decisions.

At least, that is the goal. Reaching it is sure to be difficult. But like a number of other elusive goals, just pursuing it can make a company a lot stronger.

QUESTIONS

1. What is meant by computer-integrated business? What are the potential benefits of computer integrated business?
2. What are the major barriers to firmwide integration of information systems?
3. Given that a firm runs a large number of separate functional information systems that are quite old (called legacy systems), what approach would you recommend for converting them to a firmwide integrated system?

6.6 Virtual Managers

MONTIETH M. ILLINGWORTH

"Work is something you do, not a place to go." This quotation sums up the essence of virtual management. The conventional approach of everyone showing up at the office at 9AM and leaving at 5PM is changing for many people. The changes are due in part to the changing nature of work, with more workers being mobile. It is also due in part to enabling inventions such as portable computers, modems, and cellular phones, that allow workers to keep in touch and to access the corporate data base no matter where they are. Finally, it is due to the realization by companies that many employee offices are empty a large part of the time and real estate does not come for free.

Illingworth presents several cases of leaders in the virtual office with special emphasis on Chiat/Day, an advertising agency in Venice, California. The basic idea is to implement several concepts simultaneously. One is telecommuting (discussed in Sections 2.1 and 3.3) Another is hoteling. Hoteling is the idea that when you arrive at work your space is assigned to you just as it is in a hotel. You can even reserve space ahead of time. A third is the concept of a team room where people working on the same project get together in both real and virtual mode.

In implementing changing work patterns, the company must consider the human effects as well as the technical effects. Such issues as worker satisfaction, individual identity, and corporate community have to be resolved.

IT'S 10 A.M. Do you know where your workers are? For a growing number of companies building the "virtual office," the answer is irrelevant. They have rid themselves of scores of private offices and instead equipped their employees with portable computers, cellular phones, pagers, personal digital assistants, and other mobile computing and communications devices.

"Work is something you do, not a place to go," says Laurie Coots, senior VP of administration and business development at Chiat/Day Inc., the Venice, California, advertising agency that lets its employees work from any location they choose. "We can no longer institutionalize things just to be able to conveniently manage them. We're in the business of knowledge—managing it, broker-

Source: Information Week (June 13, 1994): 42–58.

ing it, leading it—and that means you have to innovate."

Chiat/Day and other first-generation virtual corporations—including IBM, Ernst & Young, and Arthur Andersen—have made radical changes in the way their work is done. Now, with implementation of their virtual offices almost complete, these companies are asking what effect the transition is having on individual identity, worker satisfaction, and corporate community.

"Many managers know that the non-territorial office saves them money and increases the amount of time they can spend with customers," says Professor Franklin Becker, director of international workplace studies at Cornell University and an adviser to many companies experimenting with the virtual office. "Now they want to find out whether it's truly a better way of working."

Although the mobile-worker concept has been around since the early 1970s and was hyped all through the 1980s, it's only in recent years that mobile working achieved trend status. Today roughly seven million people do at least some of their work on the road with the help of portable PCs, phones, and other devices, according to market research firm Link Resources in New York. By 2000, predicts Link, the ranks of these telecommuters will swell to 25 million people.

MOBILE MARCH

The march of mobile workers is due in part to advances in cellular communications, networking, and laptop computer technology. Declining prices for such mobility tools also have accelerated the process, particularly among the nearly 80% of mobile workers who, according to Link, work at cost-conscious businesses with 100 or fewer employees.

Cornell's Becker believes that comprehensive, systemic change in work structures produces the best results—as long as management helps it evolve and carefully monitors how well employees adapt. His studies show that while mobile-work programs driven by cost concerns produce slight improvements in productivity, similar programs driven by business concerns report that productivity has increased significantly.

IBM's motivation for making workers mobile combines the two approaches. All seven of IBM's regional U.S. markets are at some stage of converting to the non-territorial office. Although each region is evolving at its own pace, corporate headquarters decided to standardize the process and share information and experiences.

The most comprehensive program at IBM is in the Midwest, says John Frank, project manager for mobility implementation in that region. Beginning last November, IBM converted seven offices in five cities in the Midwest into "productivity centers." IBM Midwest sent into the field some 700 executives—slightly more than a third of the 2,000 employees targeted to become mobile by this September. All 30 IBM locations in the Midwest will soon be converted, including the Chicago office.

Gone from the productivity centers are the private offices of customary corporate working environments. Replacing them are small "team rooms" for use by the mobile executives—when they're around. Eight executives are assigned to each room on a rotating basis. Inside each room is a round conference table, plugs for local area network (LAN) communications, and, of course, telephones. "The executives are supposed to consider these rooms 'work pods,' where they come to solve specific problems," says Frank. "After the problem is solved, the executives go on their way back to the customer."

A few months after initial implementation, the psychological impact of the changes began to bubble up. "It forced people to ask themselves, 'Why was I sitting in the office?' " says Frank. "Now they have to get out and do something. We've found that most employees become very successful very fast, while others—a small number—wonder if they're in the right job or if they belong at IBM."

TEAM SPIRIT

A one-day seminar conducted for IBMers in the program helped ease the transition. Frank knew that taking away private offices and pushing people out the door could erode a sense of identity and curb motivation—the linchpins of any healthy corporate culture. "We needed to keep up the team atmosphere," he says. "So we talked about how to stay in touch with peers and managers, and how to make sure the customer could always reach you."

They also discussed the equipment needed to stay in touch. Every mobile IBM executive now has a text pager, cellular phone, and an IBM ThinkPad laptop—and they're learning to use the technology effectively. "We expect executives to spend 70% of their time with customers and 30% either traveling, at the office, or at home," says Frank. "But we pointed out that this was not a work-at-home program. They just have the option to do more at home than before, if that's what it takes to get the job done."

Giving executives the right tools and addressing issues of corporate culture have apparently paid off. Frank recently surveyed some 300 Indiana-based executives and found that almost three-quarters had become more productive, with estimates reaching improvements of 20%.

IBM recently completed a survey aimed at finding out where the greatest productivity gains occurred. At press time the results were still being interpreted by Becker and his staff at Cornell. IBM's Frank suspects the biggest gains have come in customer-related activities, with smaller gains resulting from employees reading their electronic mail at home and spending less time commuting. "While sitting in a customer's office, our executives can dial into the server to get information and change a proposal right there," he says. "There's no more going back to the office first."

THE HUMAN FACTOR

The IBM survey also tried to measure the so-called soft issues of the work environment. "We asked executives how they feel about going mobile, how it affects their perceptions about the company and their personal lives," says Frank. That question is vital, argues Andrea Saveri, a director at the Institute for the Future in Menlo Park, California. "Electronically mediated work questions the extent to which humans and human contact are needed," she says. "It also threatens to fragment an organization. People begin to wonder what's holding them together. Work, after all, is basically social."

For his part, Michael Wiley, IBM Midwest general manager of customer service, is happier being a telecommuter. Before going mobile, Wiley says he was usually off to work before his wife and children were up for breakfast. Now, Wiley reads E-mail at home early in the morning, eats with his family, then leaves the house. Wiley also has the flexibility to attend his son's afternoon Little League games. If that means working at night, after the kids have gone to bed, he doesn't mind. "I'm probably working a lot more," he admits, "but I also enjoy it more than I did before."

Some mobile workers at Ernst & Young say they're happier, too. The consulting and audit firm inaugurated what it calls "hoteling" in its Chicago office in June 1992. About 450 of that office's 1,050 accountants and consultants—including senior managers on the consulting side—share 100 new office spaces.

The key to Ernst & Young's program is its reservations system. When a consultant or accountant wants to use an office space, that person must reserve it at least a day in advance. Everything necessary to do the work—including personal belongings—will be set up beforehand.

The hoteling concept carries some career advantages for Ernst & Young employees like consultant Jeremy Stendeback, who works in the firm's information technology practice. Stendeback says the arrangement lets him organize his time better, which helps him get more work done. "Before, I sat with people in my own practice," he says. "If I wanted to talk to people in other areas, I had to hunt around and find them. Now, I sit with a dozen people working on various kinds of engagements. I'm finding out about new technologies all the time."

Stendeback also has more contact with his

boss. Before, he rarely saw her, and they communicated mostly by phone. Now his cube is always placed near her office. "We interact a lot more," Stendeback says.

That's music to the ears of Barbara Rose, Ernst & Young's director of administration. In focus groups held prior to implementation of the hoteling scheme, Ernst & Young employees stressed how important it was to feel just the same about themselves and their jobs as their colleagues who weren't hoteling. Now that they're not stuck in the office, it seems they may even feel better.

Just how much better was recently determined in a company survey. Ernst & Young sought to measure employee satisfaction with the service clusters. The approval rating for 1994 went up to 99%, compared with 97% last year. Ernst & Young is now considering expanding the number of hotelers to 700—perhaps including senior managers.

Still, there were problems early on. Rose says some Ernst & Young employees would reserve office space, then not show up at the specified time. Others showed up on time, but stayed too long. "We quickly nipped those problems in the bud," she says. "We generated reports comparing reservations with fulfillments and presented it to the employees. They didn't think we had that capability."

There remains, however, that 1% of satisfaction to be earned. Even true-believers like Stendeback envision potential snags. "What happens when the office space gets full?" he asks. "That might mean you would have to go to another floor, away from friends and peers." As with IBM, mobile employees at Ernst & Young worry about alienation from the group. Whether a valid concern or not, it seems a basic human reaction and as such isn't being ignored.

Rose believes that the solution is largely technological. Increase the power and capability of the hardware and software tools mobile executives use, she says, and their concerns will be allayed. "They see that they're still connected to the office," says Rose. "Now they want to go the next step. They want fuller use of the applications from the file server, like our networkable CD-ROM products. We're prepared to make the investment to give them that."

CULTURE SHOCK

Psychological issues seem even more acute at companies that include senior executives in hoteling. At Arthur Andersen, for example, high-level managers and partners lost offices that, in some cases, they spent years trying to acquire—and they're not afraid to speak up about it.

John Nendick, a partner at Andersen's Los Angeles office, admits that at first it was a cultural shock for the partners. His office started its pilot program 12 months ago for 40 of its 200 executives. "Its the thought that you're taking something away from someone," he says.

The best way to deal with those concerns, Nendick advises, is to get into a planning group as many of the people affected as possible. Then, pose a conceptual challenge. "Take away the paradigm of the existing office, and give them a clean sheet of paper to figure out the best way to do things," he says. "People will get excited coming up with solutions."

Many other firms have "failed miserably" at hoteling, says James Nixon, Andersen's managing director of real-estate management. He wouldn't reveal the companies' names but says they neglected to get early feedback from employees. And that feedback, he stresses, has to be an ongoing process. Andersen has formed groups that meet periodically to suggest ways to improve the hoteling program. One time, the group decided to replace rectangular desks with round tables to create more of a sense of interaction among people.

Andersen also wants to introduce more information technology. Nendick, for instance, would like to buy a bank of scanners so more information can be put into the network—a step designed to move Andersen closer to the paperless office. Real-estate director Nixon wishes the firm's phone technology was more powerful. "It's holding us back," he says. "It's not as mobile as we need it to be."

For ad agency Chiat/Day, technology wasn't as important as speed. The ad company, housed in a Frank Gehry-designed building with a three-story-high Claus Oldenberg sculpture of a pair of binoculars at the entrance, has already received attention for its leap toward the office of the future. And for good reason. When the Chiat/Day office opened on January 3, the agency made it clear that it was attempting something far more radical than anything tried by any other U.S. company. It also wanted results in the shortest possible time—implementation was scheduled to take six months.

A DAY NEAR THE BEACH

With the exception of support staff, all 350 employees in Chiat/Day's Venice, California, office are given the option, and the technology, to work at home or from any other location they please. Four file servers provide access to applications, calendars, file management, and E-mail. When staffers opt to work in the agency's office near the beach—a choice made by roughly 60% at any given time—there are no private offices, not even cubicles. Almost all work spaces, with the exception of a few study carrels in the library, have been converted to common areas.

These common areas vary in design. On the first floor, they're called "living rooms" and include couches. There are also "project rooms" where account teams meet. The second floor also has a "student union clubhouse," a pub-like setting where employees can meet, eat, drink coffee, and watch TV. Each employee also has a small locker.

The technology used at Chiat/Day varies, too. Executives in the building are issued pagers and radio phones to keep in touch with clients and co-workers. It's common to see Chiat/Day execs walking the halls while engaged in animated conversation on their phones. Outside, they use cellular phones. Access to data is provided by jacks, sprinkled throughout the building, that tie directly to the server or a home computer. Hard copy is stored in two ways: Data that needs to be kept handy is scanned and cataloged using an imaging system; archival material is cataloged, bar-coded, and shipped off-site. Also, the first floor has a bank of five Macintoshes, or what the agency calls "data drinking fountains," where employees can pull up and glance at E-mail messages or files. Three more banks, each with five Macs, are being added throughout the building.

Even with the changes, Chiat/Day has tried to retain the agency culture. Senior VP Coots, a key player in the implementation, set up a special research panel to preserve a sense of community at the company. There are also "team reconstruction" seminars that bring together account team members with senior management to discuss how to make the implementation work.

Steven Alburty, Chiat/Day's New York-based director of information systems, helped design the systems, participated in their implementation, and leads one of the reconstruction seminars. The typical cycle for some employees, he says, is first a sense of betrayal (as in, "Why are they doing this to me?"), then denial ("This is never going to work"), and then identity crisis ("Do I really want to work here?"). "A lot of people were unhappy at first," says Alburty. "The unhappiest are the people who don't learn how to use the new technologies. They just keep doing things the old way." The solution: Chiat/Day now makes computer training a part of every employee review.

The changes at Chiat/Day were so dramatic, and done so quickly, that achieving general worker satisfaction—at least on the scale already realized at IBM, Ernst & Young, and Andersen—appears to be some time away. Although that clearly adds work to the task of conversion, both Coots and Alburty believe that the undertaking at Chiat/Day is not only necessary but imperative. "We used to have T-shirts that said 'Innovate Or Die, And Death Is Not An Option,'" quips Coots. "Simply put, this is the way the world is going, and we see ourselves as leaders."

The changes at Chiat/Day, then, are about removing all the distractions to work. The focus is now starkly on results. Says Alburty: "More of the dialogue in meetings is about what you contributed that day or that week. Previously, you just assumed someone was contributing because you saw that person sitting at a desk."

Coots realizes that not everyone will make the change equally well. "It can be a big challenge if they can't figure out how to draw the boundaries between when to work and when to relax," she says. And senior executives are also challenged. "If you're the type of manager who always has to save the day and be the hero, it's going to be very difficult for you to make the transition to the new way. You're now going to face employees who are more empowered to make their own decisions and be responsible for them."

Coots's research panel studies show that the staff is adapting slowly but surely. For example, they used to go out in twos or threes for lunch meetings. Now it's in groups of four or larger, suggesting that there is more collaboration. "We also know that people produce more work in shorter bursts of time," she says. "But we don't know what that means yet. We're still trying to figure it out."

FINISHING TOUCHES

Adjustments are being made. The agency is adding extra architectural features, such as a giant fedora with seats inside that will allow staffers to seclude themselves and find inspiration. Chiat/Day will also unveil a graphically driven groupware interface for the computer network that, Alburty claims, will more accurately reflect the virtual elements of the agency's evolving corporate culture. The interface, designed and developed in a joint venture with the Art Technology Group of Cambridge, Massachusetts, was to be rolled out in Chiat/Day's New York office on June 1.

The interface will show on every computer screen an image of the project rooms where teams will work on specific accounts. When a team member "enters" the room, an image of other team members will appear on screen. So will the various files that hold text and artwork relating to the specific account campaign. "It reflects the personalities of the people who work here and the new electronic corporate community of Chiat/Day," says Alburty.

Chiat/Day is so convinced that the new interface is better than anything else available in the groupware market that it is now discussing a plan with software marketers and distributors to make the program widely available.

If the idea takes off, it may not matter where workers are at 10 AM—or at any other time, either.

QUESTIONS

1. What are the potential problems of implementing a virtual office?
2. What steps can managers take to avoid or overcome these problems?
3. Do you think the virtual office concept is appropriate for all types of businesses? Explain why or why not. Give examples to support your position.

7 Future Directions

7.1
Corporations of the Future

7.2
Technology of the Future

PRODUCTIVITY gains from the computer and from communications beyond those already achieved are certain to occur. What will these gains be? Although we cannot be certain about all innovations, the shape of these gains can be inferred from developments going on in the laboratory and from the concepts about the way business will be done in the future.

The two sections that follow approach the future from the business and the technology viewpoint. The first section chronicles how we can expect the way markets and organizations will change because of the new technologies. The second article focuses on how human knowledge will be augmented through the intelligent application of technology.

7.1 Computers, Networks, and the Corporation

THOMAS W. MALONE AND JOHN F. ROCKART

Computers and the communication networks connecting them have been key themes throughout this book and are the key theme of Malone and Rockart's article. They examine these technologies in terms of their implications for the way global business will be done in future years. The results are surprising.

Malone, a professor at MIT, and Rockart, head of the Center for Information Systems Research at MIT, are very bright people. Talking with them for an hour in their respective offices is an intellectual experience. You come out with new ideas and new ways of looking at the world. To use the advertising cliché, when they talk, you listen.

What you will hear in this article is that technology changes the way business is done. Classical hierarchical structures are being replaced with electronic markets. Over time, organizations reduce their production costs by buying in electronically mediated markets. Although this approach increases the cost of coordination, the result is a net saving.

Changes do not occur in isolation. Replacing hierarchies with electronic markets result in empowering individuals, in making competition in time possible, and much more.

Is Malone and Rockart's view of the future correct? The decade of the '90s will tell us whether they are or not.

ABOUT 150 years ago the economy in the U.S. and Europe began to undergo a period of change more profound than any experienced since the end of the Middle Ages. We call that change the Industrial Revolution. The industrial economies are now in the early stages of another transformation that may ultimately be at least as significant.

There is a critical difference this time, however. Changes in the economies of production and transportation drove the revolution of the last

Source: Scientific American (September 1991).

century. The revolution under way today will be driven not by changes in production but by changes in coordination. Whenever people work together, they must somehow communicate, make decisions, allocate resources, and get products and services to the right place at the right time. Managers, clerks, salespeople, buyers, brokers, accountants—in fact, almost everyone who works—must perform coordination activities.

It is in these heavily information-based activities that information technologies have some of their most important uses, and it is here that they will have their most profound effects. By dramatically reducing the costs of coordination and increasing its speed and quality, these new technologies will enable people to coordinate more effectively, to do much more coordination and to form new, coordination-intensive business structures.

The core of the new technologies is the networked computer. The very name "computer" suggests how one usually thinks of the device—as a machine for computing, that is, for taking in information, performing calculations, and then presenting the results. But this image of computing does not capture the essence of how computers are used and now they will be used even more in the future. Many of the most important uses of computers today are for coordination tasks, such as keeping track of orders, inventory, and accounts. Furthermore, as computers become increasingly connected to one another, people will find many more ways to coordinate their work. In short, computers and computer networks may well be remembered not as technology used primarily to compute but as coordination technology.

TO understand what is likely to happen as information technology improves and its costs decline, consider an analogy with a different technology: transportation. A first-order effect of transportation technology was simply the substitution of new transportation technologies for the old. People began to ride in trains and automobiles rather than on horses and in horse-drawn carriages.

As transportation technology continued to improve, people did not use it just to substitute for the transportation they had been using all along. Instead a second-order effect emerged: people began to travel more. They commuted farther to work each day. They were more likely to attend distant business meetings and to visit faraway friends and relatives.

Then, as people used more and more transportation, a third-order effect eventually occurred: the emergence of new "transportation-intensive" social and economic structures. These structures, such as suburbs and shopping malls, would not have been possible without the wide availability of cheap and convenient transportation.

Improved coordination technology has analogous effects. A first-order effect of reducing coordination costs is the substitution of information technology for human coordination. For example, data-processing systems helped to eliminate thousands of clerks from the back offices of insurance companies and banks. Similarly, computer-based systems have replaced scores of factory "expediters." Today computers track the priority of each job in the factory and indicate the most critical ones at each workstation. More generally, the long-standing prediction that computers will lead to the demise of middle management finally seems to be coming true. In the 1980s many companies flattened their managerial hierarchies by eliminating layers of middle managers.

A second-order effect of reducing coordination costs is an increase in the overall amount of coordination used. For instance, contemporary airline reservation systems enable travel agents to consider more flight possibilities for a given customer more easily. These systems have led to an explosion of special fares and price adjustments. American Airlines and United Air Lines, which provide the largest systems, have benefited significantly from the fees they charge for this service. For instance, in 1988 American made about $134 million from its reservation system—almost 15% of its total income. In addition, access to up-to-the-minute information about ticket sales on all airlines enables American and United to adjust their fare schedules to maximize profits.

Otis Elevator Company also increased the amount of its coordination—primarily to improve maintenance service for its customers. With its Otisline system, highly trained multilingual operators receive trouble calls through a national toll-free number. The operators record the problems in a computer data base and then electronically dispatch local repair people.

This real-time availability of data has vastly improved the management of repair activities. For instance, if a particular type of part has failed during the past week on eight of 100 elevators, Otis can preemptively replace that part on the other 92 elevators. Although this kind of nationwide correlation of data was possible before, the degree of communication and coordination required was impractical. These capabilities have played a major role in reducing maintenance calls by nearly 20%.

In some instances, the second-order effect of an increase in demand may overwhelm the first-order effect of substitution. For example, in one case we studied, a computer conferencing system helped to remove a layer of middle managers. Several years later, however, almost the same number of new positions (for different people at the same grade level) had been created. According to people in the company, the new specialists took on projects not considered before. Evidently, managerial resources no longer needed for simple communication could now be focused on more complex coordination tasks.

A third-order effect of reducing coordination costs is a shift toward the use of more coordination-intensive structures. A prime example is Frito-Lay, Inc., studied by Lynda M. Applegate of Harvard Business School and others. At Frito-Lay, some 10,000 route salespeople record all sales of each of 200 grocery products on hand-held computers as they deliver goods to customers on their route. Each night, the stored information is transmitted to a central computer. In turn, the central computer sends information on changes in pricing and product promotions to the hand-held computers for use the next day. Each week, the main computer summarizes the centrally stored information and combines it with external data about the sales of competitive brands. Some 40 senior executives and others can then gain access to this information through an executive information system (EIS).

THE availability of these data has enabled Frito-Lay to push key decisions down from corporate headquarters to four area heads and several dozen district managers. The managers can use the data not only to compare actual sales to sales targets but also to recommend changes in sales strategy to top management. This entire coordination-intensive structure has become possible only in the past few years because of the improved capability and reduced costs of hand-held computers, EIS software, computer cycles, and telecommunications equipment.

Coordination-intensive structures do not just link different people in the same companies. Many of the most interesting new structures involve links among different companies. For example, the U.S. textile industry has begun implementing a series of electronic connections among companies as part of the Quick Response program. As described by Janice H. Hammond of Harvard Business School and others, these electronic connections link companies all along the production chain, from suppliers of fibers (such as wool and cotton) to the mills that weave these fibers into fabric, to the factories that sew garments and, ultimately, to the stores that sell the garments to consumers.

When such networks are fully implemented, they will help companies respond quickly to demand. For instance, when a sweater is sold in New York City, a scanner reading the bar-coded label may automatically trigger ordering, shipping, and production activities all the way back to the wool warehouse in South Carolina. This new, multiorganizational structure will reduce inventory costs throughout the value chain. The textile-apparel retail industry spends about $25 billion in inventory costs every year; the Quick Response approach may save half that amount.

Wal-Mart has already established parts of a

similar system that links the retailer to Procter & Gamble Company and several of its other major suppliers. In doing so, Wal-Mart has eliminated significant parts of its own purchasing groups and contracted with its suppliers to replace products as they are sold. In one such experiment, both unit sales and inventory turnover increased by about 30%.

Sometimes technology helps to create interorganizational networks—not just among buyers and suppliers but also among potential competitors. For example, Eric K. Clemons of the University of Pennsylvania has studied the Rosenbluth International Alliance, a consortium of travel agencies around the world that share customer records, services, and software. The alliance also provides clients with toll-free English-language help lines in every major country. This consortium of independent agencies, led by Rosenbluth Travel in Philadelphia, can therefore manage all travel arrangements for international trips and for meetings of people from many parts of the globe.

The textile firms near Prato, Italy, illustrate a related kind of interorganizational alliance. As described by Michael J. Piore and Charles F. Sabel of the Massachusetts Institute of Technology, the operation of a few large textile mills was broken into many small firms, coordinated in part by electronic connections among them. This network can flexibly adjust to changes in demand, sometimes shifting orders from an overloaded mill to one with spare capacity. The structure also takes advantage of the entrepreneurial motivation of the owners: in small mills, the owners' rewards are more closely linked to their own efforts than is the case in large ones.

AS these examples show, information technology is already facilitating the emergence of new, coordination-intensive structures. What do these changes mean for the organizations of the near future?

A surprising result of our research is a prediction that information technology should lead to an overall shift from internal decisions within firms toward the use of markets to coordinate economic activity. To see why, consider that all organizations must choose between making the goods or services they need and buying them from outside suppliers. For instance, General Motors Corporation must decide whether to make tires internally or purchase them from a tire manufacturer.

Each of these two forms of coordination—internal and external—has advantages and disadvantages. As Oliver Williamson of the University of California at Berkeley and others have argued, buying things from an outside supplier often requires more coordination than making them internally. To buy tires, General Motors may need to compare many potential suppliers, negotiate contracts, and do formal accounting for the money that changes hands. Coordinating the production of tires internally, on the other hand, can often be done less formally and at lower cost, with telephone calls and meetings.

But improved information technology should reduce the costs of both internal and external coordination, much as transportation technology lowered the expense of traveling. When trains and automobiles reduced the difficulty of traveling, more people chose to live in the suburbs rather than in the cities to reap such benefits as extra living space. Similarly, when information technology reduces the costs of a given amount of coordination, companies will choose to buy more and make less. The additional coordination required in buying will no longer be as expensive, and buying has certain advantages. For instance, when General Motors buys tires, it can take advantage of the supplier's economies of scale and pick the best tires currently available from any supplier whenever its needs change. Thus, we expect networks to lead to less vertical integration—more buying rather than making—and to the proliferation of smaller firms. More electronically mediated alliances (such as the Rosenbluth International Alliance) and an increased use of electronic markets to pick suppliers (such as the airline reservation system) will result.

This argument implies that information technology will help make markets more efficient.

Buyers will no longer have to exert great effort to compare products and prices from many different suppliers. Instead an electronic market can easily and inexpensively collect and distribute such information.

These more efficient markets threaten firms whose strategic advantages rest on market inefficiencies. For instance, as Clemons described, when the London International Stock Exchange installed an electronic trading system, the trading floor became virtually deserted within weeks. Trading moved to electronic terminals around the world. The system greatly reduced the costs of matching buyers and sellers. This change, in turn, dramatically reduced the profits of brokers and trading specialists, who previously had had a monopoly on performing this function. The potential decline in profit may explain why many other exchanges still resist electronic trading.

Many other kinds of intermediaries, such as distributors and retailers, are becoming vulnerable as well. For example, consumers can now bypass retail stores entirely by using computer-based systems such as Comp-U-Card and Comp-u-store to buy household goods and services at substantial savings. Electronic markets can also make evaluating product quality easier; we expect that it is only a matter of time before networks contain extensive comments and evaluations from previous buyers, becoming a kind of instantaneous, on-line *Consumer Reports*.

Increasing market efficiency also implies that firms should focus more carefully on the few core competencies that give them strategic advantages in the marketplace. They should buy the additional, more peripheral products and services they need instead of making them. For instance, in the past few years, Ford Motor Company and Chrysler Corporation have significantly increased their proportion of externally purchased components, such as tires and batteries.

EVEN though information technology can be strategically important, single innovations in information technology are seldom in themselves a source of continuing competitive advantage. For example, American Hospital Supply (now Baxter Healthcare Corporation) won high praise for its early system that let customers place orders electronically without requiring a salesperson. This system made ordering from American Hospital easier than doing so from competitors and reduced the time salespeople had to spend on the clerical aspects of taking an order. But contrary to original expectations, systems like these do not "lock in" customers in the long run. Instead customers eventually seem to prefer electronic systems that provide a choice among several vendors. Similarly, an automatic teller machine system that may once have been a competitive advantage for a bank is now largely a competitive necessity.

One way to maintain an upper hand is to keep innovating so rapidly that other firms always lag a step behind. Another way, as Clemons has noted, is to use information technology to leverage some other structural advantage. For instance, Barclay deZoete Wedd, a British brokerage firm, continues to benefit from an electronic stock-trading system because the company already controlled the trading of far more stocks than did any of their competitors.

In addition to markets, another coordination-intensive organizational structure likely to become much more common is what some management theorists call a networked organization, or, more picturesquely, an "adhocracy," a term Alvin Toffler popularized in his book *Future Shock*. This

Figure I Relative Costs for Hierarchies and Markets

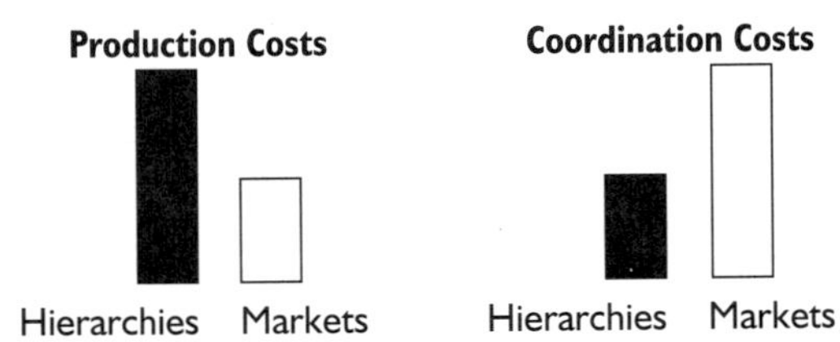

Note: Making products in vertically integrated hierarchies usually involves higher production costs than buying them in the market. Buying from an outside supplier allows a company to exploit economies of scale and other production cost advantages. But buying usually requires higher coordination costs: a firm must find a supplier, negotiate contracts, and account for payments.

form is already common in organizations such as law firms, consulting companies, and research universities. Such organizations and institutions must continually readjust to a changing array of projects, each requiring a somewhat different combination of skills and other resources. These organizations depend on many rapidly shifting project teams and much lateral communication among these relatively autonomous, entrepreneurial groups.

The adhocracy contrasts with the conventional business organization of today: the hierarchy. Hierarchies are common partly because they provide a very economical way of coordinating large numbers of people. In principle, decision makers in a hierarchy can consider all the information known to anyone in the group with much less communication than would be needed if each person communicated with everyone else.

In practice, however, hierarchies have severe limitations. Central decision makers can become overloaded and therefore unable to cope effectively with rapidly changing environments or to consider enough information about complex issues. Furthermore, people at the bottom may feel left out of the decision making and as a result be less motivated to contribute their efforts.

As information technology reduces communication costs, the nonhierarchical structures (such as markets and adhocracies) may help overcome the limitations of hierarchies. For example, because of the large amount of unpredictable lateral communication, the adhocracy is extremely coordination intensive. New media, such as electronic mail, computer conferencing, and electronic bulletin boards, can make the coordination easier and, therefore, enable the adhocracy to work much more effectively. Computer networks can help find and coordinate people with diverse knowledge and skills from many parts of an organization.

Moreover, computer-based technologies can transfer information not only faster and more cheaply but also more selectively. These capabilities help to mitigate information overload. Systems now exist to help people find, filter, and sort

Figure 2 Corporate Structures

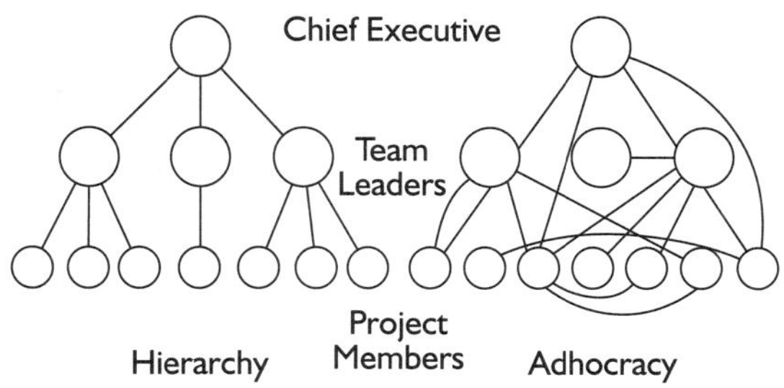

their electronic mail based on topic, sender, and other characteristics. Together these new coordination technologies can speed up the "information metabolism" of organizations—the rate at which firms can take in, move, digest, and respond to data. Abundant information poses two potential difficulties for organizational power. Some people worry that managers may become "Big Brothers" who use the information to exert stronger centralized control over those who work for them. Others fear that if power is decentralized throughout the organization, workers might use their newfound power to serve their own narrow interests, leading to organizational chaos.

In fact, neither dark vision has been realized. Instead what appears to be happening is a paradoxical combination of centralization and decentralization. Because information can be distributed more easily, people lower in the organization can now become well enough informed to make more decisions more effectively. At the same time, upper-level managers can more easily review decisions made at lower levels. Thus, because lower-level decision makers know they are subject to spot-checking, senior managers can retain or even increase their central control over decisions.

The changes at Phillips Petroleum Company illustrate this process. Previously, senior managers decided what price to set for petroleum products. These critical decisions depended on the recommendations of staff analysts several levels down. When Phillips Petroleum developed an executive information system, senior managers began to make some of these decisions more directly based on the global information provided by the system. The senior executives soon realized, however, that they could pass on this global information directly to local terminal managers, who could take into account information such as competitors' prices. By decentralizing the pricing decision in this way, the company made sounder, more profitable pricing strategies in each area of the country.

Another way of understanding this paradoxical effect is to realize that new technology does not just redistribute power. It can provide a sense of more power for everyone. For example, the agents of several insurance companies currently carry laptop computers when they visit the homes of customers. The agents use the computers to fill out applications and to project premiums and benefits. But typically, underwriters at the corporate headquarters require several weeks to review the applications and to issue new policies.

Soon the underwriting rules for certain routine policies will be included in the laptop computer itself. The agent will be able to issue these policies immediately in the customers' homes.

These systems will thus "empower" the agents, who will control the time and place of the policy-acceptance decisions and can make sales immediately. The authority of the central underwriters will increase as well, because the rules they have created will be applied consistently. The underwriters will also be able to devote more time to analyzing interesting and potentially more profitable non-routine cases.

Information technology not only changes power; it also changes time. On the one hand, time has expanded. Electronic mail, voice mail, and facsimile transmissions can be sent or received at any time of day or night, almost anywhere around the globe. Similarly, customers of automatic teller machines and some stock markets can make transactions around the clock. The "work day" has much less meaning, and companies can compete by expanding the time their services are available.

On the other hand, time has contracted. Companies can also compete on speed. For instance, effective coordination can reduce the time needed to develop new products, deliver orders, or react to customer requests. Management teams, such as the one at Phillips Petroleum, have information available throughout the management hierarchy, which enables them to react to market conditions much more quickly. Decisions that might have taken days in the past can now be made within hours or minutes.

THE changes discussed so far require no great predictive leaps; they are already happening. What will happen as information technology improves even more? What other kinds of organizations might emerge in the globally interconnected world that the technologies make feasible?

One possibility is the increasing importance of "answer networks," networks of experts available to answer questions in different areas. One might go to these services with questions such as "How many bars of soap were sold in Guatemala last year?" or "What are the prospects of room-temperature superconductivity in consumer products by 1995?" The services would include massive data bases and layers of human experts in many different topic areas. Some questions will be easily answerable from information in a data base. Others will be referred to progressively more knowledgeable human experts. Depending on how much one is willing to spend and how quickly one wants the answer, the response might range from a newspaper clipping to a personal reply from a Nobel laureate scientist. Similar but limited services exist today—product hot lines and library reference desks are examples—but computer networks and data bases will make such services much less expensive, much more valuable, and, therefore, much more widely used.

Electronically mediated markets can also assemble armies of "intellectual mercenaries" virtually overnight. For instance, there may be a large number of consultants who make their living doing short-term projects over the network. If a manager has a job to be done, such as evaluating a loan or designing a lawnmower, he or she could quickly assemble a team by advertising electronically or by consulting a data base of available people. The data base might contain not only the skills and billing rates of prospective workers but also unedited comments from others who had used their services before. Although consulting firms and advertising agencies sometimes work like this now, pervasive networks will allow teams to be assembled much more quickly, for shorter projects, and from many different organizations.

This kind of market for services might be used inside an organization as well. Instead of always relying on supervisors to allocate the time of people who work for them, extensive internal markets for the services of people and groups may exist. Murray Turoff of the New Jersey Institute of Technology has suggested how such a system might work. Someone with a short programming project to be done, for instance, might advertise internally for a programmer. Bids and payments for this internal market could be in real dollars or some kind of point system. The bids from programmers would indicate their skill and availability. The payments that programmers receive would reflect how valuable they had been to other parts of the organization.

Improved technology can also help create decision-making structures that integrate qualitative input from many people. For instance, in making complex decisions, such as where to locate a new plant, the amassing of many facts and opinions is critical. Today companies often make such decisions after incomplete discussions with only a few of the people whose knowledge or point of view might be valuable. In the future, companies may use computer networks to organize and record the issues, alternatives, arguments, and counterarguments in graphical form. Then many different people can review and critique the parts of the argument about which they know or care.

For instance, someone in a remote part of the firm might know about plans for a new highway that completely change the desirability of a proposed plant location. As such information accumulates, people can vote on the plausibility of different claims. Then, using all the information displayed in the system, a single person or group can ultimately make the decision.

WHAT will happen as the globally networked society leads to a world in which vast amounts of information are freely available or easily purchased? Clearly, this world will require services, both automated and human, to filter the tremendous amount of information available. In general, as the amount of information increases, people who can creatively analyze, edit, and act on information in ways that cannot be automated will become even more valuable.

But what else people will do will depend on the values that are important to them. When trains and automobiles reduced the constraints of travel time, other values became more significant in determining working and living patterns. As Kenneth T. Jackson of Columbia University has documented, for example, American values about the importance of owning one's home and the moral superiority of rural life played a large role in determining the nature of suburbs in the United States.

Similarly, when the costs of information and coordination are not a barrier to fulfilling people's needs and wants, other values may emerge to shape the workplace and society. The new information technologies will almost certainly help gratify some obvious wants, such as the desire for money. Some of the emerging corporate structures may be especially good at satisfying nonmaterial needs, such as those for challenge and autonomy.

But perhaps these desires are themselves manifestations of some still deeper needs. Psychologists sometimes refer to a need for self-actualization. Others might call this a desire for spiritual fulfillment. To use the new technologies wisely, we will need to think more carefully about what we truly value and how the technology can help us reach our deeper goals.

FURTHER READING

Malone, Thomas W., Yates, Joanne, and Benjamin, Robert I. "Electronic Markets and Elec-

tronic Hierarchies." In *Communications of the ACM*, Vol. 30, No. 6, (June 1987): 484–497.
Johnston, Russell and Lawrence, Paul R. "Beyond Vertical Integration: the Rise of the Value-Adding Partnership." In *Harvard Business Review*, Vol. 66, No. 4, (July–August 1988): 94–104.
Clemons, Erik K. "Evaluation of Strategic Investments in Information Technology." In *Communications of the ACM*, Vol. 34, No. 1, (January 1991): 22–36.
Global Competition and Telecommunications: Proceedings of a Symposium held at Harvard Business School, May 2–3, 1991. Edited by Stephen Bradley, Richard Noland, and Jerry Hausman (in press).

QUESTIONS

1. How does information technology make markets more efficient? What effect will efficient markets have on businesses?
2. The article contrasts two different forms of organization: hierarchy and adhocracy. Since information technology can reduce coordination costs in both types of organizations, why do the authors expect adhocracies to become more common? Do you agree with the authors' conclusions? Explain why or why not.
3. What are the potential problems with abundant information? What are the organizational impacts?

7.2 Applications of the Future

RAYMOND KURZWEIL

We conclude this book with a scenario written by Raymond Kurzweil. A scenario is a history prepared as if it was written at some time in the future. It purports to tell you what the future looks like and how we got there from here. Scenarios must meet three criteria: they should be possible, plausible, and internally consistent. That is, it should be physically possible for the scenario to occur; the scenario must be believable to the reader; and the scenario must not contain contradictions within it. Scenarios are a useful tool for communicating in business since it is a lot easier to understand a story than it is to try to absorb a set of isolated facts.

Kurzweil's scenario deals with computers and what they will be able to do. His scenario is strongly influenced by the ideas of artificial intelligence. His view of the future includes telephones that translate your language to the language of the listener and vice versa, an intelligent assistant who does your routine work for you, a world chess champion, and credit cards and keys that recognize you as the owner.

Kurzweil himself is a very interesting person. An entrepreneur and an inventor, he built a major company (Kurzweil Computer Products) and then sold it to Xerox. His specialties are in reading machines that convert printed text to computer form or into voice, automated voice recognition, and music synthesizers; that is, he works to improve the computer's human interface. Much of his work was motivated by his interest in the disabled, who need help to cope in the world. Thus, for example, his reading machines have been used by people like Stevie Wonder.

SCENARIOS

Since the founding of the computer industry almost half a century ago, one of its most salient and consistent features has been change. Functionality per unit cost has been increasing exponentially for decades, a trend that shows no sign of abating. When I attended MIT in the late 1960s, thousands of students and professors shared a *single* computer, an IBM 7094 with

Source: Raymond Kurzweil, *Intelligent Machines* (Cambridge, MA: MIT Press, 1991) pp. 401–416.

294,912 bytes of core storage (organized as 65,536 words of 36 bits each) and a speed of about 250,000 instructions per second. One needed considerable influence to obtain more than a few seconds of computer time per day. Today, one can buy a personal computer with 10 times the speed and memory for a few thousand dollars. In *Metamagical Themas*, Doug Hofstadter cites an actual job that took 10 people with electromechanical calculators 10 months to perform in the early 1940s, was redone on an IBM 704 in the early 1960s in 20 minutes, and now would take only a few seconds on a personal computer.[1] David Waltz points out that memory today, after adjustment for inflation, costs only one one-hundred millionth of what it did in 1950.[2] If the automotive industry made as much progress in the past two decades, a typical automobile today would cost about two dollars (the doubling of price performance every 22 months on average has resulted in an improvement factor of about 2,000 in 20 years; this is comparable to the difference between the 7094 of the late 1960s and a personal computer with an Intel 80386 chip today).[3] If we go back to the first relay-based computers, a personal computer today is nearly a million times faster at a tiny fraction of the cost. Many other examples of such progress abound.[4]

In addition to the basic power of computation as measured by speed and memory capacity, new hardware and software technologies have greatly improved our ability to interact with computer devices. Through the 1940s and 1950s most communication with computers was through boards with plug-in cables; through the 1960s and 1970s, with reels of punched paper tape, stacks of punched paper cards, and print-outs from line printers. Today the advent of high resolution graphic displays, the mouse, graphics tablets, laser printers, optical cameras, scanners, voice recognition, and other technologies have provided a myriad of ways for humans and machines to communicate.

Advances in software have harnessed these increasingly potent hardware resources to expand the productivity of most professions. Twenty years ago computers were used primarily by large corporations for transaction processing and by scientists (occasionally by computer scientists to explore the power of computing). Today most workers—professionals, office workers, factory workers, farmers—have many occasions to use methods that rely on the computer. I can recall that 15 years ago even thinking about changing my company's business projections was regarded as a very serious endeavor; it would take the finance department days to grind through the numbers to examine a single scenario. Today with spreadsheet programs it is possible to consider a dozen alternative plans and determine their implications in less than an hour. Twenty years ago the only people interacting with computers were computer experts and a small cadre of students learning the somewhat arcane new field of computation. Today computers appear ubiquitously on office desks, in kitchens, in play rooms, in grocery stores, and in elementary schools.

Will these trends continue? Some observers have pointed out that an exponential trend cannot continue forever. If a species, for example, happens upon a hospitable new environmental niche, it may multiply and expand its population exponentially for a period of time, but eventually its own numbers exhaust the available food supply or other resources and the expansion halts or even reverses. On this basis, some feel that after four decades, exponential improvement in the power of computing cannot go on for much longer. Predicting the end of this trend is, in my view, highly premature. It is, of course, possible that we will eventually reach a time when the rate of improvement slows down, but it does not appear that we are anywhere close to reaching that point. There are more than enough new computing technologies being developed to assure a continuation of the doubling of price performance (the level of performance per unit cost) every 18 to 24 months for many years.

With just conventional materials and methodologies, progress in the next 10 years, at least in terms of computing speeds and memory densities, seems relatively assured. Indeed, chips with

64 million bits of RAM (random access memory) and processor chips sporting speeds of 100 million instructions per second are on the drawing board now and likely to be available in the early 1990s. Parallel-processing architectures, some including the use of analog computation, are an additional means of expanding the power of computers. Beyond the conventional methods, a broad variety of experimental techniques could further accelerate these trends. Superconducting, for example, while challenging to implement in a practical way, has the potential to break the thermal barrier that currently constrains chip geometries. As I mentioned earlier, the resulting combination of smaller component geometries with the effective utilization of the third dimension could provide a *millionfold* improvement in computer power. A variety of new materials, such as gallium arsenide, also have the potential to substantially improve the speed and density of electronic circuits.[5] And optical circuits—computing with light rather than electricity—may multiply computing speeds by factors of many thousands.[6]

Will software keep up? It is often said that the pace of advances in software engineering and applications lags behind that of the startling advance of hardware technology. Advances in software are perhaps more evolutionary than revolutionary, but in many instances software techniques are already available that are just waiting for sufficiently powerful hardware to make them practical. For example, techniques for large-vocabulary speech recognition can be adapted to recognize continuous speech but require substantially greater computational speed. Vision is another application with the same requirement. There are many techniques and algorithms that are already understood but are waiting for more powerful computers to make them economically feasible.[7] In the meantime, our understanding of AI methods, the sophistication of our knowledge bases, the power of our pattern-recognition technologies, and many other facets of AI software continue to grow.

Where is all this taking us? People in the computer field are accustomed to hearing about the rapidly improving speed and density of semiconductors. People in other professions inevitably hear reports of the same progress. Numbing are the extremely small numbers used to measure computer timings and the enormous numbers used for memory capacity, time measured in trillionths of a second and memory in billions of characters. What impact are these developments going to have? How will society change? How will our daily lives change? What problems will be solved or created?

One can take several approaches in attempting to answer these questions. Perhaps most instructive is to consider specific examples of devices and scenarios that have the potential to profoundly change the way we communicate, learn, live, and work. These concrete examples represent only a few of the ways that computer and other advanced technologies will shape our future world. These examples are based on trends that are already apparent. In my view, it is virtually certain (barring a world calamity) that all of these scenarios will take place. The only uncertainty is precisely *when*. I will attempt to project current trends into the future and estimate when we are likely to see each example.

Obviously, the further into the future we look, the more uncertain the timing of these projections becomes. The history of AI is replete with examples of problems that were either underestimated or (less often) overestimated. A great irony in early AI history is that many of the problems thought most difficult—proving original theorems, playing chess—turned out to be easy, while the "easy" problems—pattern-recognition tasks that even a child can perform—turned out to be the most challenging.[8] Nonetheless, I believe that we now have a more sophisticated appreciation of the difficulty of many of these problems, and so I will attempt the thankless task of making specific projections. Of course, by the time you discover that my predictions were altogether wrong, it will be too late to obtain a refund for the purchase price of this book.

As I mentioned, these projections are based on trends that are already evident. What is most difficult to anticipate are *breakthroughs*. Any attempts to have predicted the future at the beginning of this century would have almost certainly overlooked the computer, as well as atomic energy, television, the laser, and indeed, most of electronics.

THE TRANSLATING TELEPHONE

Koji Kobayashi, chairman of the powerful Japanese corporation NEC, has a dream. Someday, according to Kobayashi, people will be able to call anyone in the world and talk, regardless of the language that they speak. The words will be translated from language to language in real time as we speak.[9]

Three technologies are required to achieve Kobayashi's dream: automatic speech recognition (ASR), language translation (LT), and speech synthesis (SS). All three exist today, but not nearly in sufficiently advanced form. Let us consider each of these requirements.

ASR would have to be at the "holy grail" level, that is, combining large (relatively unrestricted) vocabulary, accepting continuous speech input, and providing speaker independence (no training of the system on each voice). Conceivably, the last requirement could be eased in early versions of this system. Users of this capability may be willing to spend 15 minutes or so training the system on their voice. Such enrollment would be required only once. Combining the first two elements—large vocabulary and continuous speech—will take us to the early to mid 1990s. Adding speaker independence will take us another several years.

LT requires only the ability to translate *text*, not speech, since ASR technology would be translating speech input into written language. The LT capability would not require literary-quality translations, but it would have to perform unassisted. LT systems today require human assistance. Completely automatic LT of sufficient quality will probably become available around the same time that the requisite ASR is available. It should be noted that every pair of languages requires different software (going from French to English is a different problem from going from English to French). While many aspects of translation will be similar from one set of languages to another, LT technology will vary in quality and availability according to the languages involved.

SS is the easiest of the three technologies Kobayashi requires. In fact, it is available today. While not entirely natural, synthetic speech generated by the better synthesizers is quite comprehensible without training. Naturalness is improving, and SS systems should be entirely adequate by the time the necessary ASR and LT systems are available.

Thus, we can expect translating telephones with reasonable levels of performance for at least the more popular languages early in the first decade of the next century. With continuing improvements in performance and reductions in cost, such services could become widespread by the end of that decade. The impact will be another major step in achieving the "global village" envisioned by Marshall McLuhan (1911–1980) over two decades ago.[10] Overcoming the language barrier will result in a more tightly integrated world economy and society. We shall be able to talk more easily to more people, but our ability to misunderstand each other will remain undisturbed.

THE INTELLIGENT ASSISTANT

You are considering purchasing an expensive item of capital equipment and wish to analyze the different means of financing available. Issues to consider are your company's current balance sheet, other anticipated cash flow requirements, the state of various financial markets, and future financial projections. You ask your intelligent assistant to write a report that proposes the most reasonable methods to finance the purchase and analyzes the impact of each. The computer engages you in sufficient spoken dialogue to clarify the request and then proceeds to conduct its study. In the course of its research, it accesses the balance sheet and financial projections from the data bases of your company. It contacts the Dow Jones data base by telephone to obtain information on current financial markets. It makes several calls to other computers to obtain the most recent financing charges for different financial instruments. In one case, it speaks to a human to clarify certain details on a lease-repurchase plan. It speaks with your company's vice president of marketing to obtain her level of confidence in achieving the sales projections for the following two years. It then organizes and presents the information in a written report complete with color charts. The report is presented to you the following day, since it took that long to reach the two humans involved. (Had it been able to conduct the research through communication only with other computers, it would have required only a few minutes.) Other services provided by your computerized assistant include keeping careful track of your schedule, including planning your travel from one appointment to another. The system plans your work for you, doing as much of it itself as it is capable of and understanding what portion of it you need to do yourself.

When we shall see the above system depends on how intelligent an assistant we would like to have. Crude forerunners exist today. Large-vocabulary ASR has been integrated with natural-language understanding and data-base-management programs to provide systems that can respond to such commands (posed by voice) as, "Compare the sales of our western region to our three largest competitors." Such systems are, of course, highly limited in their problem-solving ability, but efforts to integrate ASR with data-base access have already begun.

The most challenging aspect of the vision is problem solving, having sufficient commonsense knowledge and reasoning ability to understand what information is required to solve a particular problem. Required are expert systems in many areas of endeavor that are less narrowly focused than the expert systems of today. One of the first intelligent assistants is likely to be one that helps get information from data bases through telecommunications.[11] It has become clear to a number of software developers that a need exists to improve substantially the ease of accessing information from such data-base systems as CompuServe, Delphi, The Source, Dialog, Dow Jones, Lexis, Nexis, and others. Such data-base systems are greatly expanding the volume and diversity of information available, but most persons do not know where to find the appropriate information they need. The first generation of office-assistant programs are now being developed that know how to obtain a broad variety of information without requiring precisely stated requests. I expect that within several years such systems will be generally available, and some of them will take ASR for input.

Thus, in the early to mid 1990s we shall see at least part of the above vision in use: flexible access to information from increasingly varied information services around the world, accessed by systems that understand human speech as well as the syntax and (at least to some extent) the semantics of natural language. They will support their own data bases and be able to access organization-specific knowledge. You will be able to obtain information in a flexible manner without having to know which data-base service has what information or how to use any particular information utility. As the 1990s progress, these systems will be integrated with problem-solving expert systems in many areas of endeavor. The level of intelligence implied in the above scenario describing a capital-equipment purchase will probably be seen during the first decade of the next century.

THE WORLD CHESS CHAMPIONSHIP

As noted earlier, the best machine chess players are now competing successfully at the national senior-master level, regularly defeating all but about 700 players.[12] All chess machines use some variant of the recursive algorithm called minimax, a strategy whose computational requirements are multiplied by some constant for each additional move ahead that is analyzed. Without a totally new approach, we thus need to make exponential progress in computational power to make linear gains in game-playing performance (though we are indeed making exponential progress in hardware). The analysis I gave before estimated that the requisite computer power to achieve world-championship chess playing should become available between nine and 54 years from now. This estimate was based on the continuing gains anticipated in the speeds of individual microprocessors. If we factor in the increasing popularity of parallel-processing architectures, the result will be much closer to the short end of this range. Some of the other scenarios in this section require significant advances in both hardware power and software sophistication. In my view, the ability of a machine to play championship chess is primarily a function of the former. Some of the possible breakthroughs in electronic hardware discussed below will be directly applicable to the chess issue. For example, if we are successful in harnessing the third dimension in chip fabrication (that is, building integrated circuits with hundreds or thousands of layers of active circuitry rather than just one), we will see a major improvement in parallel processing: hundreds or thousands of processors on a single chip. Taking into consideration only anticipated progress in conventional circuit-fabrication methodologies and continued development of parallel-processing architectures, I feel that a computer world chess champion is a reasonable expectation by the end of the century.

What will be the impact of such a development? For many, such as myself, it will simply be the passing of a long anticipated milestone. Yes, chess is an intelligent game (that is, it requires intelligence to play well), but it represents a type of intelligence that is particularly well suited to the strengths of early machine intelligence, what I call level-2 intelligence. While level-3 intelligence will certainly benefit from the increasing power of computer hardware, it will also require substantial improvements in the ability of computers to manipulate abstract concepts.

Defenders of human chess playing often say that though computers may eventually defeat all human players, computers are not able to use the more abstract and intuitive methods that humans use.[13] For example, people can eliminate from consideration certain pieces that obviously have no bearing on the current strategic situation and thus do not need to consider sequences of moves involving those pieces. Humans are also able to draw upon a wealth of experience of previous similar situations. However, neither of these abilities is inconsistent with the recursive algorithm. The ability to eliminate from consideration branches of the expanding tree of move-countermove possibilities not worth pursuing is an important part, called pruning, of any minimax program. Drawing upon a data base of previous board positions is also a common strategy in the more advanced chess programs (particularly in the early game). It is estimated that human chess masters have memorized between 20,000 and 50,000 chess boards.[14] While impressive, it is clear that this is again an area where machines have a distinct edge. There is little problem in a computer mastering millions of board positions (each of which can have been analyzed in great depth in advance). Moreover, it is feasible for computers to modify such previously stored board positions to use them even if they do not precisely match a current position.

It may very well be that human players deploy methods of abstraction other than recalling previous board positions, pruning, and move expansion. There is little evidence, however, that for the game of chess such heuristic strategies are inherently superior to a simple recursive strategy combined with massive computational power.[15] Chess, in my view, is a good example of a type of intelligent problem solving well suited to the strengths of the first half century of machine intelligence. For other types of problem solving (level-3 problems), the situation is different.

Not everyone will cheerfully accept the advent of a computer chess champion. Human chess champions have been widely regarded as cultural heroes, especially in the Soviet Union; we regard the world chess championship as a high intellec-

Figure 1 The March of Machine Intelligence Toward the World Chess Championship

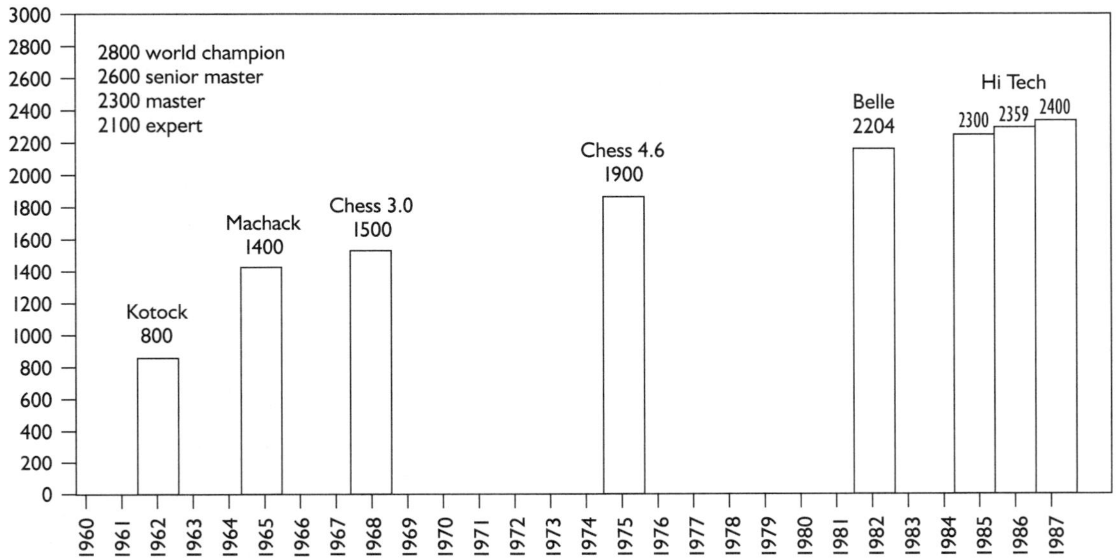

tual achievement. If someone could compute spreadsheets in his head as quickly as (or faster than) a computer, we would undoubtedly regard him as an amazing prodigy, but not as a great intellect (in fact, he would actually be an idiot savant). A computer chess championship is likely to cause a watershed change in how many observers view machine intelligence (though perhaps for the wrong reasons). More constructively, it may also cause a keener appreciation for the unique and *different* strengths (at least for the near future) of machine and human intelligence.

THE INTELLIGENT TELEPHONE-ANSWERING MACHINE

The intelligent telephone assistant answers your phone, converses with the calling party to determine the nature and importance of the call (according to your instructions), interrupts you if necessary, and finds you in an emergency. The latter may become relatively easy once cellular-phone technology is fully developed. If cellular phones become compact and inexpensive enough, most people would be able to carry them, perhaps in their wristwatches. This raises some interesting issues of privacy. Many people like the ability to be away from their phones; we do not necessarily want to be accessible to phone calls at all times. On the other hand, it could be considered irresponsible to be completely unavailable for contact in the event of an emergency. But addressing this issue will be more a matter of evolving social custom than artificial intelligence.

The other aspects of the above scenario require the same machine skills as the intelligent assistant: two-way voice communication, natural-language understanding, and automated problem solving. In some ways this application may be more challenging than the office assistant. For a cybernetic assistant that we interact with, many people would be willing to spend time learning how to use such technology if it really helped them to accomplish their work. We might not mind if it failed to handle every interaction gracefully, so long as it provided an overall productivity gain. On the other hand, we might set a higher standard for a machine intended to interact with our friends and associates.

THE CYBERNETIC CHAUFFEUR

When will computers drive our cars? Without major changes in the methods of traffic control, the advent of the self-driving car is not likely for a long time. Unlike character recognition, factory vision, and similar tasks that involve limited visual environments, driving on existing highways requires the full range of human vision and pattern-recognition skills. Furthermore, because the consequences of failure are so serious, we would demand a high level of performance from such a system before relying on it. (On the other hand, with 50,000 traffic deaths each year on American highways, the human standard of performance is far from perfect.)

Yet there is a simpler form of the cybernetic chauffeur that is easier to build and could still accomplish a sharp reduction in driving fatalities as well as substantially reduce the tedium of driving. If specially designed sensors and communication devices were placed in major thoroughfares, cars could communicate with the road as well as with other cars. They could then be placed on automatic pilot. Highways would essentially become electronic railways that our cars could join and leave. The communication system required would be similar to the complex one already in place for cellular phones. Algorithms built into each car and into the roads would maintain safe distances between cars and would handle a variety of situations.[16] Although no specific system has been developed, the underlying technology exists today to deal with the steady-state situation of cars driving on a highway and maintaining proper speed and distance. But the situation gets more complicated when several other contingencies are considered. First, we need to be able to get our car *onto* a computer-controlled road. Presumably, we would drive to an access road, where our computer driver would lock onto the road's communication system, which would take over (in conjunction with the intelligence of our own car computer). Getting *off* the road is perhaps trickier. The system has to consider the possibility that the driver has fallen asleep or is otherwise not ready to resume manual control. The best protocol might be for the system to bring the car to a halt at an exit position at which point the human driver would be expected to take over. Machine intelligence would also have to deal with possibilities of hardware failure—not just of the computer itself, but also of the engine, tires, and other parts of the vehicle. Furthermore, the designers of such systems will also have to consider the possibility of people or animals straying onto the road.

Even with all of these complications, from a technical standpoint, intelligent roads represent a substantially easier problem than creating an automatic driver that can cope with traffic situations as they currently exist. One major nontechnical barrier to creating an intelligent road system, however, is that it requires a large measure of cooperation between car manufacturers and the government agencies that manage our roads (and drivers). It is not an innovation that can be introduced to a small number of pioneering users; it needs to be implemented *all at once* in at least some roads and in all cars intending to access such roads. Presumably, cars not equipped with such automatic guidance equipment would not be allowed on intelligent roads. Again, cellular phone technology is similar: it was not feasible until both the portable phone equipment and the overall computerized communication systems were in place. Still, I would expect such a system to be introduced gradually. At first it would be featured on one or a few major highways on an experimental basis. If successful, it would spread from there.

The technology to accomplish this should be available in the first decade of the next century. But because political decision making is involved, it is difficult to predict when it will receive priority sufficient to warrant implementation. Though cellular-phone technology also involved the coordination of a complex system, it has grown rapidly because it created an entrepreneurial opportunity.

The more advanced scenario of the completely driverless car will take us well into the first half of the next century. Another approach might be to forget roads altogether and replace them with computer-controlled flying vehicles that can ascend and descend vertically. There is, after all, much more space in the three-dimensional open air than there is on our one-dimensional roads. There are already plans in place to install a satellite-based collision-avoidance system that will dramatically reduce airplane collisions by the end of the century. The flying vehicles envisioned here would be the size of today's cars and would be even easier to use.[17]

INVISIBLE CREDIT CARDS AND KEYS

During the 1990s we shall see highly reliable person-identification systems that use pattern-recognition techniques applied to fingerprint scanning and voice patterns. Such systems will typically combine the recognition of a personal attribute (a fingerprint or voice pattern) with a password that the user types in. The password prevents a misrecognition from causing a problem. The recognition system prevents unauthorized access by a thief who acquires someone's password.

Today we typically carry a number of keys and cards to provide us with access to, and use of, our homes, cars, offices, and a broad variety of financial services. All of these keys and cards could be eliminated by the widespread adoption of reliable person-identification technologies. The acceptance of this technology is likely to be hastened by the loss of confidence in hand-written signatures caused by the explosion of electronic publishing.

Unfortunately, this type of technology is also capable of helping Big Brother track and control individual transactions and movements.

INSTANT ASICS

One of the remarkable recent innovations in hardware technology is the advent of the application-specific integrated circuit (ASIC), in which an entire complex electronic system is placed on a single inexpensive chip. The advent of the ASIC has provided products that are complex, diverse, customized, and highly miniaturized. As Allen Newell points out, one might regard it as an almost magic technology: once an ASIC is developed, it provides enormous computational power at very low cost, takes up almost no space, and uses almost no electricity. The major barrier to greater deployment of this technology is the very long and expensive engineering cycles required to design such chips. The promise of *instant* ASICs is the ability to design an integrated circuit as easily as one writes a high-level computer program and, once designed, to have the actual chip available immediately.

The development of instant-ASIC technology is a major goal of the U.S. Department of Defense. Aside from its obvious military applications, it will also greatly accelerate the availability of innovative consumer products. Just as the difficulty of programming the early computers was quickly recognized as a major bottleneck, so is the difficult design process behind today's advanced chips. Indeed, the latter is receiving intense attention from all of the major players in the semiconductor industry. It is expected that in the early 1990s designers will be able to write chip programs (whose output is a piece of silicon) as easily and as quickly as computer programs.[18] The availability of instant-ASIC technology will eliminate for most purposes what little difference remains today between hardware and software engineering. It will accelerate the trend toward knowledge as the primary component of value in our products and services.

ARTIFICIAL PEOPLE AND THE MEDIA OF THE FUTURE

Rather than describe the vision I have in mind, I shall approach the idea of artificial people by starting with what is feasible today. Consider the artificial creatures that we interact with in computerized games and educational programs. An early and prime example is Pac Man, an artificial creature capable of responding and interacting with us in certain limited ways. One might not consider Pac Man to be much of a creature, let alone an artificial *person*. It is kind of a cartoon caricature of a fish with a limited repertoire of movements. Similarly, our range of emotional responses to it is very narrow, but it serves as a useful starting point.

Now consider what is feasible today, about a decade after the first introduction of Pac Man and other computer games. Reasonably lifelike video images of human faces can be completely synthesized and animated. Experiments at such advanced media laboratories as the MIT Media Laboratory have created completely synthetic yet realistic computer-generated images of human faces that can move and express a wide range of responses and emotions.

Let us imagine the next step: computer games and interactive educational programs that use synthetically generated human images. Rather than simply replaying a previously stored animated sequence, such programs would start with knowledge structures representing the concepts to be expressed and then translate each concept into spoken language and articulated facial and bodily movements. We would thus see and hear images—not prestored but created in real time—of people that are reasonably realistic. The motions and speech sounds would be computed as needed from an intent to express certain ideas and emotions. These artificial people would be responding to our actions within the context of the program.

Let us take several more steps. Add speech recognition and natural-language understanding. Add another several generations of improved image resolution and computing power for greatly enhanced visual realism. Add a more sophisticated problem-solving capability and more intelligence to provide greater subtlety of personality. Our artificial person is becoming more like a real person and less like Pac Man.

Applications would include very realistic games, movies that could include the viewer as a participant, and educational programs that would engage the student to learn from direct experience. Benjamin Franklin could take a child on a guided tour of colonial Philadelphia. Rather than a canned visual tour, this artificial Ben Franklin could answer questions, engage the child in dialogue, customize the tour on the basis of the child's own expressed interests, and otherwise provide an engaging experience. One could debate Abraham Lincoln or take Marilyn Monroe on a date. As with any creative medium, the possibilities are limited only by our imagination. As another example, the intelligent assistant could include a persona complete with appearance, accent, and personality. As time went on, such artificial persons would continue to grow in sophistication, realism, communicative and interactive ability, and of course intelligence. Ultimately, they would develop a sense of humor.

It should be noted that personality is not an attribute that can be stuck on an intelligent machine. A personality is almost certainly a necessary byproduct of any behavior complex enough to be considered intelligent. People already speak of the personalities of the software packages they use. Shaping the personality of intelligent machines will be as important as shaping their intelligence. After all, who wants an obnoxious machine?

Such artificial persons could eventually use three-dimensional projected holographic technology (a method for creating three-dimensional images that do not require the use of special glasses). Currently, most holograms are static three dimensional pictures, although some use multiple images to provide a sense of movement. The MIT Media Lab has succeeded in creating the world's first three-dimensional holographic image generated entirely by computer.[19] The ability to project a hologram entirely from computer data is an important step in imaging technology. If a computer can project one hologram, then it can be made to project any number. Ultimately, with sufficient computer power, the images could be generated fast enough to appear realistically to move. The movements would not be prestored but rather *computed* in real time to respond to each situation. Thus, our artificial people can ultimately be lifelike, life-size, three-dimensional images with sufficiently high resolution and subtlety of movement to be indistinguishable from real people. These future computer displays will also be able to project entire environments along with the people.

There is concern today regarding the power of television to shape our views and to engage our emotions and attention. Yet television is essentially noninteractive, of low resolution, and flat. A medium that provided nearly perfect resolution and three-dimensional images and interacted with us in an intelligent and natural way would be far more powerful in its emotional impact, possibly too powerful for many (real) people. Harnessing and regulating these media of the future will undoubtedly be an area of much debate and controversy.

Adoption of the advanced media technologies described here will begin in the late 1990s and mature over the first half of the next century. Applications include entertainment, education, conducting business transactions, even companionship.

ANOTHER APPROACH

An entirely different approach to the concept of artificial people lies in the area of robotics. Robots of the first generation, just like the first generation of computer-generated creature images (essentially pictures of robots), were neither intelligent nor realistic. We were as unlikely to mistake an early factory robot for a natural creature, let alone a person, as we were to mistake Pac Man for an image of a real animal. Here again, successive generations of technology have provided greater intelligence, subtlety, and naturalness. The primary drive for robotic technology lies in practical applications in the manufacturing and service industries. Admittedly, for most of these applications, resemblance to humans or to any other natural creature is of little relevance. Yet there will be applications for natural robots (sometimes called androids) as teachers, entertainers, and companions. Primitive robotic pets have already created a niche in the toy industry.

Creating a reasonably natural robotic imitation of a person is even more challenging than creating a convincing media image of a person. Any autonomous robot, humanoid or otherwise, has to be able to ambulate in a natural environment; this requires general-purpose vision and a high degree of fine motor coordination. Autonomous robots for exploring hostile environments, such as nuclear reactors and the surfaces of other planets, exist today. Routine use of autonomous robots in more conventional settings is likely to begin by the end of the century. Robots that are reasonably convincing artificial people will not appear until well into the next century.

Marvin Minsky has often said that a good source of insights into the realities of tomorrow's computer science can be found in today's science fiction. Isaac Asimov in his *Robots of Dawn* describes a society two centuries from now in which people live alongside a ubiquitous generation of robotic servants, companions, guards, and teachers. Two of the protagonists are a beautiful female scientist and her lover, a "male humaniform" robot.

PASSING THE TURING TEST

> Scientists from the University of Clear Valley reported today that a computer program they had created was successful in passing the famous Turing test. Computer scientists around the world are celebrating the achievement of this long-awaited milestone. Reached from his retirement home, Marvin Minsky, regarded as one of the fathers of artificial intelligence (AI), praised the accomplishment and said that the age of intelligent machines had now been reached. Hubert Dreyfus, a persistent critic of the AI field, hailed the result, admitting that he had finally been proven wrong.

The advent of computers passing the Turing test will almost certainly not produce the above sort of coverage. We will more likely read the following:

> Scientists from the University of Clear Valley reported today that a computer program they

had created was successful in passing the famous Turing test. Computer scientists reached at press time expressed considerable skepticism about the accomplishment. Reached from his retirement home, Marvin Minsky, regarded as one of the fathers of artificial intelligence (AI), criticized the experiment, citing a number of deficiencies in method, including the selection of a human "judge" unfamiliar with the state of the art in AI. He also said that not enough time had been allowed for the judge to interview the computer foil and the human. Hubert Dreyfus, a persistent critic of the AI field, dismissed the report as the usual hype we have come to expect from the AI world and challenged the researchers to use him as the human judge.

Alan Turing was very precisely imprecise in stating the rules of his widely accepted test for machine intelligence.[20] There is, of course, no reason why a test for artificial intelligence should be any less ambiguous than our definition of artificial intelligence. It is clear that the advent of the passing of the Turing test will not come on a single day. We can distinguish the following milestones:

Level 1. Computers arguably pass *narrow* versions of the Turing test of believability. A variety of computer programs are each successful in emulating human ability in some area: diagnosing illnesses, composing music, drawing original pictures, making financial judgements, playing chess, and so on.

Level 2. It is well established that computers can achieve human or higher levels of performance in a wide variety of intelligent tasks, and they are *relied upon* to diagnose illnesses, make financial judgements, etc.

Level 3. A single computer system arguably passes the *full* Turing test, although there is considerable controversy regarding test methodology.

Level 4. It is well established that computers are capable of passing the Turing test. No reasonable person familiar with the field questions the ability of computers to do this. Computers can engage in a relatively unrestricted range of intelligent discourse (and engage in many other intelligent activities) at human or greater levels of performance.

We are at level 1 today. A wide range of expert systems can meet or exceed human performance within narrowly defined (yet still intelligent) areas of expertise. The judgements of expert systems are beginning to be relied upon in a variety of technical and financial fields, although acceptance in the medical area is much slower. Also, computer success in a variety of artistic endeavors is beginning to be at least arguably comparable.

Level 2 is within sight and should be attained around the end of the century. As expert systems grow in sophistication and achieve more natural human interfaces, we will begin to rely on their expertise as much as (if not more than) human society relies on their idiot savant forebears today.

We will probably begin to see reports of level 3, and newspaper articles similar to the second one given above, during the first decade of the next century, with continued controversy for at least several decades thereafter. The first reports will almost certainly involve significant limitations to Turing's originally proposed challenge. We are close to having the underlying technology (if not the actual program) today if we use sufficiently naive judges and provide them with relatively little time to make their determinations.

Level 4 is what Turing had in mind when he predicted success by the year 2000. Achieving this level is *far* more difficult than any of the other three. It requires advanced natural-language understanding, vast knowledge bases of commonsense information, and decision-making algorithms capable of great subtlety and abstraction. Turing's prediction, made in 1950, will almost certainly not be fulfilled by the year 2000. I place the achievement of level 4 sometime between 2020 and 2070. If this turns out to be the case, then Turing will have been off by a factor of between 1.4 and 2.4 (70 to 120 years versus his prediction of 50 years), which actually is not bad for such a longterm prediction. Of course, there is no assurance that *my* prediction will be any more accurate than Turing's.

Hubert Dreyfus has indicated that he will concede that he was wrong (and has been wrong for his entire professional career) if he can be fooled as the human judge in a Turing test. Will this happen? If we assume that Dreyfus is in good health and further that continuing advances in bioengineering technology enable him (and the rest of us) to live longer than today's average life expectancy, then it is altogether possible. Personally, I would be willing to bet on it.

CONCLUSION

The above scenarios provide only a small sampling of the ways in which intelligent machines of the future can be expected to touch our lives. The computers of today, dumb as they are, have already infiltrated virtually every area of work and play. The bureaucracies of our society could hardly function without their extensive computer networks. If one adds just the sharply focused intelligence of the next phase of the age of intelligent machines to the already prodigious memory capacity and speed of today's computers, the combination will be a formidable one indeed. Our cars, watches, beds, chairs, walls, floors, desks, books, clothes, phones, homes, appliances, and virtually everything else we come into contact with will be intelligent, monitoring and servicing our needs and desires. The age of intelligent machines will not start on a particular day; it is a phenomenon that has started already, with the breadth and depth of machine intelligence growing each year. Turing predicted a time when people would talk naturally about machines thinking without expecting anyone to contradict them or be surprised. We are not there today, but the day will arrive so gradually that no one (except a few authors) will notice it when it does.

NOTES

1. Douglas R. Hofstadter, *Metamagical Themas: Questing for the Essence of Mind and Pattern* (New York: Basic Books, 1985), 128.
2. David Waltz, "The Prospects for Building Truly Intelligent Machines," *Daedelus* (Winter 1988): 204.
3. The preface in Tom Forester's *The Information Technology Revolution* examines similar issues. Cambridge, MA: MIT Press, 1985.
4. In the fall of 1987 an entire issue of *Scientific American* was devoted to this topic. In particular, see Abraham Peled, "The Next Computer Revolution," *Scientific American* (October 1987): 56–64.
5. See James D. Meindl, "Chips for Advanced Computing," *Scientific American* (October 1987): 79–81 and 86–88.
6. See Mark H. Kryder, "Data-Storage Technologies for Advanced Computing," *Scientific American* (October 1987): 117–125.
7. Such as massively parallel processors based possibly on superconductors. See Peter J. Denning, "Massive Parallelism in the Future of Science," *American Scientist* (January–February 1989): 16.
8. Marvin Minsky discusses this problem in "Easy Things are Hard," *Society of Mind* (New York: Simon & Schuster, 1985), 29.
9. Koji Kobayashi, *Computers and Communication: A Vision of C & C* (Cambridge, MA: MIT Press, 1986), 165–166.
10. See Marshall McLuhan, *Understanding Media: The Extension of Man* (New York: McGraw-Hill, 1964).
11. For a vision of an office system interfacing with a public communication network, see Koji Kobayashi, *Computers and Communication*, chapter 10. See also Roger Shank and Peter G. Childers, "The World of the Future," in *The Cognitive Computer* (Reading, MA: Addison-Wesley, 1984), 227–230.
12. David N. L. Levy, *All about Chess and Computers* (Rockland, MD: Computer Science Press, 1982). See also M. M. Botvinnik, *Computers in Chess: Solving Inexact Search Problems*. Trans. Arthur A. Brown (New York: Springer-Verlag, 1984).
13. An intriguing study of the relevance of comments made by master chess players during

play can be found in Jacques Pitrat, "Evaluating Moves Rather than Positions," in Barbara Pemici and Mareo Somalvico, eds., *III Convegno Internazionale L'Intelligenza Artificiale ed il Gioco Degli Scacchi* (Federazione Scacchistico Italiana, Regione Lombardia, Politecnico de Milano, 1981).

14. Evidence of this are early board-game programs modeled on master players' strategies. The 1959 checkers program of Arthur Samuels, for example, had 53,000 board positions in memory. See Peter W. Frey, "Algorithmic Strategies for Improving the Performance of Game-Playing Programs," in *Evolution, Games, and Learning: Models for Adaptation in Machines and Nature*, Proceedings of the Fifth Annual International Conference of the Center for Nonlinear Studies at Los Alamos, NM (May 20–24, 1985): 355.
15. Recursiveness and massive computational power allow for subtle (and hence enormously varied) solutions to algorithmic problems. See, for example, Gary Josin, "Neural Net Heuristics" *BYTE*, October 1987, pp. 183–192; and Douglas Lenat, "The Role of Heuristics in Learning by Discovery," in R. Z. Michalski, J. J. Carbonell, and T. M. Mitchell, eds., *Machine Learning: An Artificial Intelligence Approach* (Palo Alto, CA: Tioga Publishing, 1983). Also see Monroe Newborn, *Computer Chess* (New York, Academic Press, 1975): 8–15.
16. See John Hilusha's article, "Smart Roads Tested to Avoid Traffic Jams," *New York Times* (October 18, 1988).
17. Plans are already in place for the development and use of flying vehicles. See, for instance, "Simulation of an Air Cushion Vehicle Microform," final report for period January 1975–December 1976, Charles Stark Draper Laboratory, Cambridge, MA, 1977.
18. For recent advances in computer and chip design, see James D. Meindl, "Chips for Advanced Computing," *Scientific American* (October 1987): 78–88. An extensive but less current review of the technology may be found in Alan Burns, *The Microchip: Appropriate or Inappropriate Technology?* (New York: John Wiley & Sons, 1981).
19. See Stewart Brand, *The Media Lab: Inventing the Future at MIT* (New York: Viking Penguin), 83–91.
20. For a straightforward presentation of this famous test, see Isaac Malitz, "The Turing Machine," *BYTE* (November 1987): 348–358.

QUESTIONS

1. What effect do you think the advanced technologies described in this section will have on you as an individual? Which technology may have the most profound impact on you? Explain why and how.
2. How do you think these technologies might change the way factories of the future operate? Give examples.
3. How do you think these technologies might change office or service work? Give examples.

Index